DATE DUE

Atmospheric and Biological Environmental Monitoring

Young J. Kim · Ulrich Platt ·
Man Bock Gu · Hitoshi Iwahashi

Editors

Atmospheric and Biological Environmental Monitoring

Springer

Editors

Prof. Dr. Young J. Kim
Gwangju Institute of Science &
    Technology (GIST)
Department of Environmental Science
    and Engineering
261 Cheomdan-gwagiro
Bug-gu, Gwangju 500-712
Republic of Korea
yjkim@gist.ac.kr

Prof. Dr. Man Bock Gu
College of Life Sciences
    and Biotechnology,
Korea University
Seoul 136-701
Republic of Korea
mbgu@korea.ac.kr

Prof. Dr. Ulrich Platt
Institute of Environmental Physics
University of Heidelberg
Im Neuenheimer Feld 229
D-69120 Heidelberg
Germany
ulrich.platt@iup.uni-heidelberg.de

Prof. Dr. Hitoshi Iwahashi
Health Technology Research Center
National Institute of Advanced
Industrial Science and Technology
Midorigaoka, 1-8-31
Ikeda, Osaka 563-8577
Japan
hitoshi.iwahashi@aist.go.jp

ISBN 978-1-4020-9673-0   e-ISBN 978-1-4020-9674-7
DOI 10.1007/978-1-4020-9674-7
Springer Dordrecht Heidelberg London New York

Library of Congress Control Number: 2009921994

Cover images (from left to right): The automated observatory on PICO-NARE experimental site, photo
courtesy of Bruno Vieira; ADEMRC mobile atmospheric measurement laboratory, photo courtesy of
Young J. Kim; Photograph of genetically engineered bacteria emitting bioluminescence upon the dose of
toxic materials, courtesy of Man Bock Gu; SAPHIR Atmosphere Simulation Chamber, photo courtesy of
Andreas Wahner.

Cover design: deblik

Printed on acid-free paper

Springer is part of Springer Science+Business Media (www.springer.com)

Preface

The extent of harmful effects of pollution on atmospheric, terrestrial and aquatic environments can be translated into extreme temperature changes, dirty air, clean water shortages, and increased incidence of toxicity that harms every life on earth. Within a lifetime, our environment is changing drastically. Much of the information of environmental pollution impacts needs to be studied, from the mechanism of toxic nanoparticles on the molecular level to the detection of trace gases on the satellite perspective. It is therefore essential to develop advanced monitoring techniques, efficient process technologies and health impact assessment tools to fill the gaps in our scientific knowledge.

This edition of "Atmospheric and Biological Environmental Monitoring" is a handful of recent developments and techniques from environmental scientists in well-diversified fields. These collections of manuscripts are extracts from the recently concluded "7th International Symposium on Advanced Environmental Monitoring" organized by the ADvanced Environmental Monitoring and Research Center (ADEMRC), Gwangju Institute of Science and Technology (GIST), Korea and held on February 25–28, 2008 in Honolulu, Hawaii. The three parts highlight important aspects of emerging environmental monitoring technologies: Atmospheric Environment, Contaminants Control Process, and Environmental Toxicity Assessment. Observational tools presented in the first part ranges from in-situ measurements to satellite remote sensing for atmospheric monitoring. Highlighted in the second part is the recently developed water quality monitoring system for lake stratification and membrane technologies for detection and removal of contaminants. Lastly, toxicity monitoring of endocrine disruptors and nanoparticles are highlighted in the third part with new discoveries.

Our sincerest gratitude goes to the authors and researchers of these studies, for their participation and contribution to this book. We also like to thank all reviewers for providing scientific insights necessary to ensure the quality of this publication. We gratefully acknowledge Dr. Robert Doe, Publishing Editor and Nina Bennink of Springer, for their continued support and encouragement towards the fulfilment of this publication. Most of all, members of the symposium organizing committee deserves the most credit for the success of the symposium. The critical suggestions that all of you have shared were instrumental to the enhancement of this collection of manuscript. Finally, all these efforts would not be realized without the financial support of the Korea

Science and Engineering Foundation (KOSEF) through the Advanced Environmental Monitoring Research Center at Gwangju Institute of Science and Technology.

Gwangju, Republic of Korea   Young J. Kim
Heidelberg, Germany   Ulrich Platt
Seoul, Republic of Korea   Man Bock Gu
Osaka, Japan   Hitoshi Iwahashi

About the Editors

Young J. Kim
Editor
Director, Advanced
Environmental Monitoring
Research Center
Professor, Department of
Environmental Science
and Engineering
Gwangju Institute of
Science and Technology
261 Cheomdan-gwagiro
Bug-gu, Gwangju 500-712
Republic of Korea
E-mail: yjkim@gist.ac.kr

Ulrich Platt
Editor
Professor and Director
Institute of
Environmental Physics
University of Heidelberg
Im Neuenheimer Feld
229 D-69120 Heidelberg
Germany
E-mail: ulrich.platt@iup.uni-heidelberg.de

Man Bock Gu
Editor
Professor and Vice Dean,
College of Life Sciences
and Biotechnology,
Korea University
Seoul 136–701
Republic of Korea
E-mail: mbgu@korea.ac.kr

Hitoshi Iwahashi
Editor
Health Technology
Research Center
National Institute of
Advanced Industrial
Science and Technology
Midorigaoka, 1-8-31
Ikeda, Osaka 563-8577
Japan
E-mail: hitoshi.iwahashi@aist.go.jp

Contents

Part I Atmospheric Environment Monitoring   1

**Two- and Three Dimensional Observation of Trace Gas
and Aerosol Distributions by DOAS Techniques**   3
Ulrich Platt, Klaus-Peter Heue and Denis Pöhler

**Atmospheric Aerosol Monitoring from Satellite Observations:
A History of Three Decades** 13
Kwon H. Lee, Zhanqing Li, Young J. Kim and Alexander Kokhanovsky

**Digital Photographic Technique to Quantify Plume Opacity
During Daytime and Nighttime** 39
Ke Du, Mark J. Rood, Byung J. Kim, Michael R. Kemme, Bill
Franek and Kevin Mattison

**Scanning Infrared Remote Sensing System for Detection,
Identification and Visualization of Airborne Pollutants** 51
Ulrich Klenk, Eberhard Schmidt and Andreas Beil

**Remote Sensing of Tropospheric Trace Gases (NO_2 and SO_2)
from SCIAMACHY** 63
Chulkyu Lee, Randall V. Martin, Aaron van Donkelaar, Andreas
Richter, John P. Burrows and Young J. Kim

**An Advanced Test Method for Measuring Fugitive Dust
Emissions Using a Hybrid System of Optical Remote Sensing and
Point Monitor Techniques** 73
Ram A. Hashmonay, Robert H. Kagann, Mark J. Rood, Byung J.
Kim, Michael R. Kemme and Jack Gillies

**Aerosol Sampling Efficiency Evaluation Methods at the US Army
Edgewood Chemical Biological Center** 83
Jana Kesavan and Edward Stuebing

Smog Chamber Measurements 105
Seung-Bok Lee, Gwi-Nam Bae and Kil-Choo Moon

**Aerosol Concentrations and Remote Sources of Airborne
Elements Over Pico Mountain, Azores, Portugal** 137
Maria do Carmo Freitas, Adriano M.G. Pacheco, Isabel Dionísio and
Bruno J. Vieira

Part II Contaminants Control Process Monitoring   159

**Removal of Selected Organic Micropollutants from WWTP
Effluent with Powdered Activated Carbon and Retention
by Nanofiltration**   161
Kai Lehnberg, Lubomira Kovalova, Christian Kazner, Thomas
Wintgens, Thomas Schettgen, Thomas Melin, Juliane Hollender and
Wolfgang Dott

**Development of Vertically Moving Automatic Water Monitoring
System (VeMAS) for Lake Water Quality Management**   179
Dongil Seo and Eun Hyoung Lee

Part III Environmental Toxicity Monitoring and Assessment   191

Toxicity of Metallic Nanoparticles in Microorganisms- a Review   193
Javed H. Niazi and Man Bock Gu

**Environmental Monitoring by Use of Genomics
and Metabolomics Technologies**   207
Tetsuji Higashi, Yoshihide Tanaka, Randeep Rakwal, Junko Shibato,
Shin-ichi Wakida and Hitoshi Iwahashi

**A Gene Expression Profiling Approach to Study the Influence of
Ultrafine Particles on Rat Lungs**   219
Katsuhide Fujita, Yasuo Morimoto, Akira Ogami, Isamu Tanaka,
Shigehisa Endoh, Kunio Uchida, Hiroaki Tao, Mikio Akasaka,
Masaharu Inada, Kazuhiro Yamamoto, Hiroko Fukui, Mieko
Hayakawa, Masanori Horie, Yoshiro Saito, Yasukazu Yoshida,
Hitoshi Iwahashi, Etsuo Niki and Junko Nakanishi

**Effects of Endocrine Disruptors on Nervous System Related Gene
Expression: Comprehensive Analysis of Medaka Fish**   229
Emiko Kitagawa, Katsuyuki Kishi, Tomotaka Ippongi, Hiroshi
Kawauchi, Keisuke Nakazono, Katsunori Suzuki, Hiroyoshi Ohba,
Yasuyuki Hayashi, Hitoshi Iwahashi and Yoshinori Masuo

**Assessment of River Health by Combined Microscale Toxicity
Testing and Chemical Analysis**   241
Sagi Magrisso and Shimshon Belkin

Saccharomyces cerevisiae **as Biosensor for Cyto- and Genotoxic
Activity**   251
Jost Ludwig, Marcel Schmitt and Hella Lichtenberg-Fraté

**The Application of Cell Based Biosensor and Biochip
for Environmental Monitoring**   261
Junhong Min, Cheol-Heon Yea, Waleed Ahmed El-Said and Jeong-Woo Choi

**Fabrication of Electrophoretic PDMS/PDMS Lab-on-a-chip
Integrated with Au Thin-Film Based Amperometric Detection for
Phenolic Chemicals**   275
Hidenori Nagai, Masayuki Matsubara, Kenji Chayama, Joji Urakawa,
Yasuhiko Shibutani, Yoshihide Tanaka, Sahori Takeda and Shinichi Wakida

Swimming Behavioral Toxicity in Japanese Medaka (*Oryzias latipes*) Exposed to Various Chemicals for Biological Monitoring of Water Quality **285**
Ik Joon Kang, Junya Moroishi, Mitoshi Yamasuga, Sang Gyoon Kim and Yuji Oshima

The Effects of Earthworm Maturity on Arsenic Accumulation and Growth After Exposure to OECD Soils Containing Mine Tailings **295**
Byung-Tae Lee and Kyoung-Woong Kim

Abbreviations **303**

Index **307**

Contributors

Mikio Akasaka, Research Institute for Environmental Management Technology (EMTECH), National Institute of Advanced Industrial Science and Technology (AIST), Tsukuba, Ibaraki, 305-8569; Japan Industrial Technology Association (JITA), Tsukuba, Ibaraki, 305-0046, Japan.

Gwi-Nam Bae, Korea Institute of Science and Technology, 39-1 Hawolgok-dong, Seongbuk-gu, Seoul 136-791, Korea, gnbae@kist.re.kr

Andreas Beil, Bruker Daltonik GmbH, Permoserstr. 15, D-04318 Leipzig, Germany, www.bdal.com

Shimshon Belkin, Department of Plant and Environmental Sciences, Institute of Life Sciences, The Hebrew University of Jerusalem, Jerusalem 91904, Israel, shimshon@vms.huji.ac.il

John P. Burrows, Institute of Environmental Physics and Remote Sensing, University of Bremen, Bremen, Germany, john.burrows@iup.physik.uni-bremen.de

Kenji Chayama, Human Stress Signal Research Center (HSS), National Institute of Advanced Industrial Science and Technology (AIST), Midorigaoka 1-8-31, Ikeda, Osaka 563-8577, Japan; Facurity of Science and Technology, Konan University, 8-9-1 Okamoto, Higashinada-ku, Kobe 658-8501, Japan

Jeong-Woo Choi, College of Bionano technology, Kyungwon University, Seongnam, Gyunggi-Do, 461-701, Korea, jwchoi@sogang.ac.kr

Isabel Dionísio, Technological and Nuclear Institute; E.N. 10, 2686-953 Sacavém, Portugal, dionisio@itn.pt

Wolfgang Dott, RWTH Aachen University, Institute of Hygiene and Environmental Medicine, Pauwelsstr. 30, D-52074 Aachen, Germany, wolfgang.dott@post.rwth-aachen.de

Ke Du, Department of Civil and Environmental Engineering, University of Illinois, 205 N. Mathews Ave. Urbana, IL, USA 61801, kedu75@gmail.com

Waleed Ahmed El-Said, Interdisciplenary Program of Integrated Biotechnology, Sogang University, Seoul 121-742, Korea.

Shigehisa Endoh, Research Institute for Environmental Management Technology (EMTECH), National Institute of Advanced Industrial Science and Technology (AIST), Onogawa 16-1, Tsukuba, Ibaraki, 305-8569, Japan.

Bill Franek, Illinois Environmental Protection Agency, Bureau of Air, 9511 West Harrison Street, Des Plaines, IL 60016, USA

Maria do Carmo Freitas, Reactor-ITN, Technological and Nuclear Institute; E.N. 10, 2686-953 Sacavém, Portugal, cfreitas@itn.pt

Katsuhide Fujita, Health Technology Research Center (HTRC), National Institute of Advanced Industrial Science and Technology (AIST), Onogawa 16-1, Tsukuba, Ibaraki, 305-8569, Japan; Tel & Fax: +81-29-861-8260, ka-fujita@aist.go.jp

Hiroko Fukui, Health Technology Research Center (HTRC), National Institute of Advanced Industrial Science and Technology (AIST), Midorigaoka 1-8-31, Ikeda, Osaka, 563-8577, Japan

John Gillies, Division of Atmospheric Sciences, Desert Research Institute 2215 Raggio Parkway Reno NV 89512 USA. Tel: 775-764-7035 Fax:775-674-7016, jackg@dri.edu

Man Bock Gu, College of Life Sciences and Biotechnology, Korea University, Seoul 136-701 Republic of Korea, mbgu@korea.ac.kr

Ram A. Hashmonay, Advanced Air Monitoring Solutions, ARCADIS. 4915 Prospectus Drive, Suite F Durham, NC 27713, rhashmonay@arcadis-us.com

Mieko Hayakawa, Health Technology Research Center (HTRC), National Institute of Advanced Industrial Science and Technology (AIST), Midorigaoka 1-8-31, Ikeda, Osaka, 563-8577, Japan

Yasuyuki Hayashi, GeneFrontier, Corp., Todai-kashiwa-Plaza 306, 5-4-19, kashiwanoha, Kashiwa, Chiba 277-0882, Japan

Klaus-Peter Heue, Institute of Environmental Physics, INF 229, University of Heidelberg, Germany

Tetsuji Higashi, Human Stress Signal Research Center (HSS), National Institute of Advanced Industrial Science and Technology (AIST), Midorigaoka 1-8-31, Ikeda, Osaka 563-8577, Japan, s.wakida@aist.go.jp.

Juliane Hollender, Swiss Federal Institute of Aquatic Science and Technology (Eawag), Überlandstr. 133, CH 8600 Dübendorf, Switzerland

Masanori Horie, Health Technology Research Center (HTRC), National Institute of Advanced Industrial Science and Technology (AIST), Midorigaoka 1-8-31, Ikeda, Osaka, 563-8577, Japan

Masaharu Inada, Research Institute for Environmental Management Technology (EMTECH), National Institute of Advanced Industrial Science and Technology (AIST), Onogawa 16-1, Tsukuba, Ibaraki, 305-8569, Japan

Tomotaka Ippongi, GeneFrontier, Corp., Todai-kashiwa-Plaza 306, 5-4-19, kashiwanoha, Kashiwa, Chiba 277-0882, Japan

Hitoshi Iwahashi, Health Technology Research Center, National Institute of Advanced Industrial Science and Technology, Midorigaoka, 1-8-31 Ikeda, Osaka 563-8577 Japan, hitoshi.iwahashi@aist.go.jp

Robert H. Kagann, Advanced Air Monitoring Solutions, ARCADIS, North Carolina, USA

Ik Joon Kang, Aquatic Biomonitoring and Environmental Laboratory, Division of Bioresource and Bioenvironmental Sciences. Kyushu University Graduate School, Hakozaki 6-10-1, Higashi-ku, Fukuoka 812-8581, Japan

Hiroshi Kawauchi, GeneFrontier, Corp., Todai-kashiwa-Plaza 306, 5-4-19, kashiwanoha, Kashiwa, Chiba 277-0882, Japan

Christian Kazner, RWTH Aachen University, Department of Chemical Engineering, Turmstr. 46, D-52056 Aachen, Germany

Michael R. Kemme, U.S. Army Engineer Research and Development Center - Construction Engineering Research Laboratory (ERDC-CERL), Farber Drive Champaign, IL 61826-9005, USA, Michael.R.Kemme@usace.army.mil.

Jana Kesavan, US ARMY Edgewood Chemical Biological Center, AMSRD-ECB-RT-TA E5951, 5183 Blackhawk Road, Aberdeen Proving Ground, Maryland 21010, USA, Jana.Kesavan@US.ARMY.MIL

Byung J. Kim, U.S. Army Engineer Research and Development Center-Construction Engineering Research Laboratory, Champaign, IL 61826, USA

Kyoung-Woong Kim, Department of Environmental Science & Engineering, Gwangju Institute of Science and Technology (GIST), Gwangju 500-712, South Korea, kwkim@gist.ac.kr

Sang Gyoon Kim, Bio monitoring Group, SEIKO Electric Co., Ltd., Tenjin 3-20-1, Koga, Fukuoka, 811-3197, Japan.

Young J. Kim, Advanced Environmental Monitoring Research Center. Professor, Department of Environmental Science and Engineering, Gwangju Institute of Science and Technology, 261 Cheomdan-gwagiro Bug-gu, Gwangju 500-712 Republic of Korea, yjkim@gist.ac.kr

Katsuyuki Kishi, Japan Pulp & Paper Research Institute, Inc., 5-13-11, Kannondai, Tsukuba, Ibaraki 300-2635, Japan.

Emiko Kitagawa, Human Stress Signal Research Center, National Institute of Advanced Industrial Science and Technology (AIST), Tsukuba West, 16-1 Onogawa, Tsukuba 305-8569, Japan

Ulrich Klenk, University of Wuppertal, Department of Safety Engineering/Environmental Protection, D-42097 Wuppertal – Germany, www.uws.uni-wuppertal.de, klenk@uni-wuppertal.de

Alexander Kokhanovsky, Institute of Environmental Physics, University of Bremen, Bremen, Germany

Lubomira Kovalova, RWTH Aachen University, Institute of Hygiene and Environmental Medicine, Pauwelsstr. 30, D-52074 Aachen, Germany; and, Swiss Federal Institute of Aquatic Science and Technology (Eawag), Überlandstr. 133, CH 8600 Dübendorf, Switzerland

Byung-Tae Lee, Department of Chemistry & Geochemistry, Colorado School of Mines, Golden, CO 80401, USA, btlee@mines.edu

Chulkyu Lee, Department of Physics and Atmospheric Sciences, Dalhousie University, Halifax, Nova Scotia, Canada, chulkyu.lee@dal.ca

Eun Hyoung Lee, M-Cubic Inc., Migun Technoworld, 533 Yongsan-dong, Yuseong-gu, Daejeon, 305-500, Korea, lehmmm@empal.com

Kwon H. Lee, Earth System Science Interdisciplinary Center, Department of Atmospheric and Ocean Science, University of Maryland, College Park, MD 20740, USA, Kwonlee@umd.edu

Seung-Bok Lee, Korea Institute of Science and Technology, 39-1 Hawolgok-dong, Seongbuk-gu, Seoul 136-791, Korea, sblee2@kist.re.kr

Kai Lehnberg, RWTH Aachen University, Institute of Hygiene and Environmental Medicine, Pauwelsstr. 30, D-52074 Aachen, Germany

Zhanqing Li, Earth System Science Interdisciplinary Center, Department of Atmospheric and Ocean Science, University of Maryland, College Park, MD 20740, USA

Hella Lichtenberg-Fraté, University of Bonn, IZMB, Molekular Bioenergetics, Kirschallee 1, 53115 Bonn, Germany, h.lichtenberg@uni-bonn.de

Jost Ludwig, University of Bonn, IZMB, Molekular Bioenergetics, Kirschallee 1, 53115 Bonn, Germany

Sagi Magrisso, Department of Plant and Environmental Sciences, Institute of Life Sciences, The Hebrew University of Jerusalem, Jerusalem 91904, Israel

Randall V. Martin, Department of Physics and Atmospheric Sciences, Dalhousie University, Halifax, Nova Scotia, Canada;and, Harvard-Smithsonian Center for Astrophysics, Cambridge, MA, USA, randall.martin@dal.ca

Yoshinori Masuo, Human Stress Signal Research Center, National Institute of Advanced Industrial Science and Technology (AIST), Tsukuba West, 16-1 Onogawa, Tsukuba 305-8569, Japan, y-masuo@aist.go.jp

Masayuki Matsubara, Human Stress Signal Research Center (HSS), National Institute of Advanced Industrial Science and Technology (AIST), Midorigaoka 1-8-31, Ikeda, Osaka 563-8577, Japan; and, Facurity of Science and Technology, Konan University, 8-9-1 Okamoto, Higashinada-ku, Kobe 658-8501, Japan

Kevin Mattison, Illinois Environmental Protection Agency, Bureau of Air, 9511 West Harrison Street, Des Plaines, IL 60016, USA

Thomas Melin, RWTH Aachen University, Department of Chemical Engineering, Turmstr. 46, D-52056 Aachen, Germany

Junhong Min, College of Bionano technology, Kyungwon University, Seongnam, Gyunggi-Do, 461-701, Korea

Kil-Choo Moon, Korea Institute of Science and Technology, 39-1 Hawolgok-dong, Seongbuk-gu, Seoul 136-791, Korea, kcmoon@kist.re.kr

Yasuo Morimoto, Institute of Industrial Ecological Sciences, University of Occupational and Environmental Health, 1-1, Iseigaoka, Yahata nishi, Kitakyushu, Fukuoka, 807-8555, Japan

Junya Moroishi, Aquatic Biomonitoring and Environmental Laboratory, Division of Bioresource and Bioenvironmental Sciences. Kyushu University Graduate School, Hakozaki 6-10-1, Higashi-ku, Fukuoka 812-8581, Japan

Hidenori Nagai, Health Technology Research Center (HTRC), National Institute of Advanced Industrial Science and Technology (AIST), Midorigaoka 1-8-31, Ikeda, Osaka 563-8577, Japan

Junko Nakanishi, Research Institute of Science for Safety and Sustainability (RISS), National Institute of Advanced Industrial Science and Technology (AIST), Onogawa 16-1, Tsukuba, Ibaraki, 305-8569, Japan

Keisuke Nakazono, GeneFrontier, Corp., Todai-kashiwa-Plaza 306, 5-4-19, kashiwanoha, Kashiwa, Chiba 277-0882, Japan

Javed H. Niazi, College of Life Sciences and Biotechnology, Korea University, Anam-dong, Seongbuk-Gu, Seoul 136-701, South Korea, javedkolkar@gmail.com

Etsuo Niki, Health Technology Research Center (HTRC), National Institute of Advanced Industrial Science and Technology (AIST), Midorigaoka 1-8-31, Ikeda, Osaka, 563-8577, Japan

Akira Ogami, Institute of Industrial Ecological Sciences, University of Occupational and Environmental Health, 1-1, Iseigaoka, Yahata nishi, Kitakyushu, Fukuoka, 807-8555, Japan

Byung-Keun Oh, Departen of Chemical and Bomolecular Engineering, Sogang University; Interdisciplenary Program of Integrated Biotechnology, Sogang University, Seoul 121-742, Korea

Hiroyoshi Ohba, GeneFrontier, Corp., Todai-kashiwa-Plaza 306, 5-4-19, kashiwanoha, Kashiwa, Chiba 277-0882, Japan

Yuji Oshima, Laboratory of Marine Environmental Science, Division of Bioresource and Bioenvironmental Sciences, Kyushu University Graduate School, Hakozaki 6-10-1, Higashi-ku, Fukuoka 812-8581, Japan

Adriano M.G. Pacheco, CERENA-IST, Technical University of Lisbon; Av. Rovisco Pais 1, 1049-001 Lisboa, Portugal, apacheco@ist.utl.pt

Ulrich Platt, Institute of Environmental Physics, University of Heidelberg, INF 229, D-69120 Heidelberg Germany, ulrich.platt@iup.uni-heidelberg.de

Denis Pöhler, Institute of Environmental Physics, University of Heidelberg, INF 229, Heidelberg, Germany

Randeep Rakwal, Health Technology Research Center (HTRC), National Institute of Advanced In-dustrial Science and Technology (AIST), 1-8-31 Midorigaoka, Ikeda, Osaka 563-8577; 16-1 Onogawa, Tsukuba, Ibaraki 305-8569, Japan

Andreas Richter, Institute of Environmental Physics and Remote Sensing, University of Bremen, Bremen, Germany, andreas.richter@iup.phusik.uni-bremen.de

Mark J. Rood, Ivan Racheff Professor of Environmental Engineering, Env. Eng. & Sci. Program, Department of Civil and Environmental Engineering, University of Illinois, 205 N. Mathews Ave. Urbana, IL 61801, USA, mrood@illinois.edu website: http://aqes.cee.uiuc.edu/

Yoshiro Saito, Health Technology Research Center (HTRC), National Institute of Advanced Industrial Science and Technology (AIST), Midorigaoka 1-8-31, Ikeda, Osaka, 563-8577, Japan

Thomas Schettgen, RWTH Aachen University, Department and Outpatient Clinic of Occupational and Social Medicine, Pauwelsstr. 30, D-52074 Aachen, Germany

Eberhard Schmidt, University of Wuppertal, Department of Safety Engineering/Environmental Protection, D-42097 Wuppertal, Germany, www.uws.uni-wuppertal.de

Marcel Schmitt, University of Bonn, IZMB, Molekular Bioenergetics, Kirschallee 1, 53115 Bonn, Germany

Dongil Seo, Department of Environmental Engineering, Chungnam National University, Daejeon, 305-764, Korea, seodi@cnu.ac.kr

Junko Shibato, Health Technology Research Center (HTRC), National Institute of Advanced In-dustrial Science and Technology (AIST), 1-8-31 Midorigaoka, Ikeda, Osaka 563-8577; 16-1 Onogawa, Tsukuba, Ibaraki 305-8569, Japan

Yasuhiko Shibutani, Osaka Institute of Technology, 5-16-1 Omiya, Asahi, Osaka 535-8585, Japan

Edward Stuebing, US ARMY Edgewood Chemical Biological Center, AMSRD-ECB-RT-TA E5951, 5183 Blackhawk Road, Aberdeen Proving Ground, MD 21010, USA

Katsunori Suzuki, GeneFrontier, Corp., Todai-kashiwa-Plaza 306, 5-4-19, kashiwanoha, Kashiwa, Chiba 277-0882, Japan

Sahori Takeda, Reseach Institute for Innovation in Sustainable Chemistry, National Institute of Advanced Industrial Science and Technology (AIST), Midorigaoka 1-8-31, Ike-da, Osaka 563-8577, Japan

Isamu Tanaka, Institute of Industrial Ecological Sciences, University of Occupational and Environmental Health, 1-1, Iseigaoka, Yahata nishi, Kitakyushu, Fukuoka, 807-8555, Japan

Yoshihide Tanaka, Human Stress Signal Research Center (HSS), National Institute of Advanced Industrial Science and Technology (AIST), Midorigaoka 1-8-31, Ikeda, Osaka 563-8577, Japan

Hiroaki Tao, Research Institute for Environmental Management Technology (EMTECH), National Institute of Advanced Industrial Science and Technology (AIST), Onogawa 16-1, Tsukuba, Ibaraki, 305-8569, Japan

Kunio Uchida, Research Institute for Environmental Management Technology (EMTECH), National Institute of Advanced Industrial Science and Technology (AIST), Onogawa 16-1, Tsukuba, Ibaraki, 305-8569, Japan

Joji Urakawa, Human Stress Signal Research Center (HSS), National Institute of Advanced Industrial Science and Technology (AIST), Midorigaoka 1-8-31, Ikeda, Osaka 563-8577, Japan; and, Osaka Institute of Technology, 5-16-1 Omiya, Asahi, Osaka 535-8585, Japan

Aaron van Donkelaar, Department of Physics and Atmospheric Sciences, Dalhousie University, Halifax, Nova Scotia, Canada, aaron.van.donkelaar@dal.ca

Bruno J. Vieira, Reactor-ITN, Technological and Nuclear Institute; E.N. 10, 2686-953 Sacavém, Portugal, bvieira@itn.pt

Shin-ichi Wakida, Health Technology Research Center (HTRC), National Institute of Advanced In-dustrial Science and Technology (AIST),1-8-31 Midorigaoka, Ikeda, Osaka 563-8577; 16-1 Onogawa, Tsukuba, Ibaraki 305-8569, Japan

Thomas Wintgens, RWTH Aachen University, Department of Chemical Engineering, Turmstr. 46, D-52056 Aachen, Germany

Kazuhiro Yamamoto, Research Institute of Instrumentation Frontier (RIIF), National Institute of Advanced Industrial Science and Technology (AIST), Higashi 1-1-1, Tsukuba, Ibaraki, 305-8565, Japan

Mitoshi Yamasuga, Bio monitoring Group, SEIKO Electric Co., Ltd., Tenjin 3-20-1, Koga, Fukuoka, 811-3197, Japan

Cheol-Heon Yea, Department of Chemical and Bomolecular Engineering, Sogang University, Seoul 121-742, Korea

Yasukazu Yoshida, Health Technology Research Center (HTRC), National Institute of Advanced Industrial Science and Technology (AIST), Midorigaoka 1-8-31, Ikeda, Osaka, 563-8577, Japan

Part I
Atmospheric Environment Monitoring

Two- and Three Dimensional Observation of Trace Gas and Aerosol Distributions by DOAS Techniques

Ulrich Platt, Klaus-Peter Heue and Denis Pöhler

Abstract Spatially resolved measurements of trace gas abundances by satellite have revolutionised the field of large-scale tropospheric chemistry observation and modelling during recent years. Now a similar revolution is imminent on local and regional scales. A key role in these advances is played by spatially resolving spectroscopic techniques like active and passive – DOAS tomographic measurements of two-dimensional trace gas distributions, as well as ground based and airborne Imaging DOAS (I-DOAS) observation of 2D- and 3D- trace gas patterns. A particularly promising approach is the combination of tomographic techniques with imaging – DOAS on airborne platforms, which can provide three-dimensional trace gas distributions. While satellite-based 2D – mapping of trace gas distributions is now in widespread use for global and regional investigations aircraft based instruments allow complementary studies at much higher spatial resolution (tens of meters instead of tens of km). Since state of the art instruments can be employed rather than technology from the last decade (which is dictated by reliability requirements and long lead times of satellite experiments) novel approaches like tomographic techniques or Short-Wave Infra-Red (SW-IR) observations can be applied. Technological approaches and sample results are discussed.

Keywords Trace gas · Tropospheric chemistry · Imaging DOAS · Airborne

1 Introduction

Spatially resolved measurements of trace gas abundances by satellite have revolutionised the field of large-scale tropospheric chemistry observation and modelling during recent years. Now time is ripe for a similar revolution of local and regional atmospheric measurements. Similar progress of modelling and associated improvement in our understanding of atmospheric processes on local and regional scales can be expected. The trace gas measurements discussed here rely on the well-known Differential Optical Absorption Spectroscopy (DOAS), see e.g. Platt et al. (1979), Platt and Stutz (2008). This technique can be applied in active mode, i.e. using an artificial light source or passive mode relying on natural light sources i.e. solar radiation.

2 Spatially Resolved Doas – Measurements

Spatially resolving spectroscopic techniques can be divided in several categories:

(1) Range resolved LIDAR techniques.
(2) Tomographic techniques, which derive the spatial information from path averaged measurements over a multitude of (intersecting) paths.
(3) Imaging DOAS (I-DOAS) observation of trace gas patterns.

A particularly promising approach is the combination of tomographic techniques with I-DOAS on airborne platforms, to provide three-dimensional trace gas distributions.

U. Platt (✉)
Institute for Environmental Physics, INF 229,
University of Heidelberg, Heidelberg, Germany
e-mail: Ulrich.Platt@iup.uni-heidelberg.de

Y.J. Kim et al. (eds.), *Atmospheric and Biological Environmental Monitoring*,
DOI 10.1007/978-1-4020-9674-7_1, © Springer Science+Business Media B.V. 2009

2.1 Active DOAS Tomography

The active Long-Path DOAS (LP-DOAS) technique is a combination of multiple LP-DOAS measurements with tomographic inversion techniques. Typically several dozen individual light paths are used to probe the concentration field from different directions. Ideally the measurements are performed simultaneously; however, frequently hardware limitations dictate sequential measurements at the individual paths. For example in a recent experiment encompassing the city centre of Heidelberg (Pöhler et al. 2007) three "multi-beam" DOAS instruments (Pundt and Mettendorf 2005) at different locations were combined with several reflectors each for a total of 20 light paths (see Fig. 1), about half of which could be operated simultaneously. Nevertheless, one complete measurement cycle required less than 10 min. The area covered is about 3.5 by 4 km^2. For each light path the average concentration of several trace gases (see Fig. 2) are evaluated according to DOAS principles. Using special tomographic inversion techniques (e.g. Laepple et al. 2004, Hartl et al. 2006) which consider the irregular arrangement of light paths within the probed area two-dimensional trace gas distributions are derived using least-squares minimum norm solutions. The reconstruction grid employed is indicated by the yellow lines in Fig. 1 (bottom). Concentration fields of five trace gases (NO_2, SO_2, O_3, HCHO, and HONO, see Fig. 2) could be simultaneously retrieved at a time resolution of 15 min.

Highest NO_2 levels were found in the late evening and morning hours especially in the regions near high-traffic roads (Fig. 3). Due to the wind direction from south west, the emissions are dispersed to north-east. We can conclude that the local high NO_2 concentrations occur due to emissions from traffic during rush hours. The highest SO_2 levels (peaking in the south) arose in the early morning hours, probably due to most small home heating systems resuming operation. Areas in the north and north-west heated by a district heating network or by gas furnaces show no increased SO_2 concentration. We conclude that the high SO_2 levels are mainly due to emissions from small oil-fired home heating systems.

Overall, the results show that active LP-DOAS tomography is a suitable technology for monitoring spatial variations of trace gas distributions at good temporal resolution. Nevertheless, the effort required in technology and manpower was found to be considerable, therefore simpler approaches are desired. A potentially much simpler technique is the Topographic Target Light Scattering – DOAS described in the following Section 2.2, however it has the disadvantage of requiring sun-light, thus nighttime measurements are not possible. Efforts in making active DOAS instruments more reliable and easier to use resulted in the replacement of thermal light sources (i.e. Xe-arc lamps) by light emitting diodes (Kern et al. 2006). Due to their long lifetime and low power consumption operation of active LP-DOAS instruments become much simple and more cost effective. Within their (rather narrow) spectral emission interval (10–20 nm) comparable brightness can be achieved, potentially detrimental Fabry-Perot etalon effects can be overcome by techniques described by Sihler (2007). Another novel approach to simplify active LP-DOAS instruments is based on the replacement of the traditional dual Newton-type transmitting/receiving telescope by a fibre coupled telescope, where different quartz fibres in a bundle act as transmitters and receivers. While greatly simplifying the optical setup and its mechanical stability this approach at the same time can enhance the light throughput by a factor of three (Tschritter 2007, Merten 2008, Merten et al. 2009).

2.2 Topographic Target Light Scattering – DOAS (ToTaL-DOAS)

An approach potentially providing path averaged trace gas measurements suitable for tomographic inversion with little logistic effort is the topographic target-DOAS technique (Frins et al. 2006). In brief, this technique derives trace gas column densities on horizontal paths defined by the instrument at the close end (S_{close}) and a topographic target (e.g. a building) at the far end (S_{far}) as sketched in Fig. 4. By ratioing scattered sun-light spectra obtained by pointing the telescope at a diffuser plate close to the instrument and the radiation returned from a distant topographic target the solar Fraunhofer structure and the sections of the light paths between the (near and far) targets and the sun cancel. In effect one obtains the difference $S = S_{far} - S_{close}$, from which the average concentration along the nearly horizontal section of the light path (path 2 in Fig. 4) can be derived by dividing S by the known distance L to the

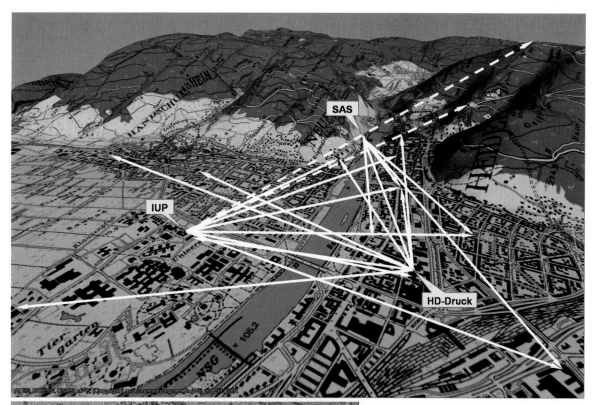

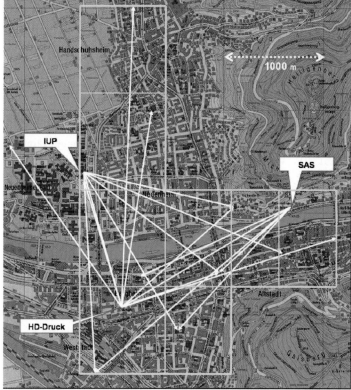

Fig. 1 Measurement geometry in Heidelberg with three Multi-beam DOAS-telescopes located at the top of the buildings labelled "IUP", "SAS" and "HD-Druck". The light beams (*white lines*) are directed to 20 retro reflector arrays. *Top panel*: Sketch of the 3-D arrangement of telescopes and retro reflector arrays. *Bottom panel*: Map of Heidelberg with light beams (*white lines*). The *yellow lines* indicate the grid used for the tomographic 2D reconstruction of the trace gases shown in Figs. 2 and 3

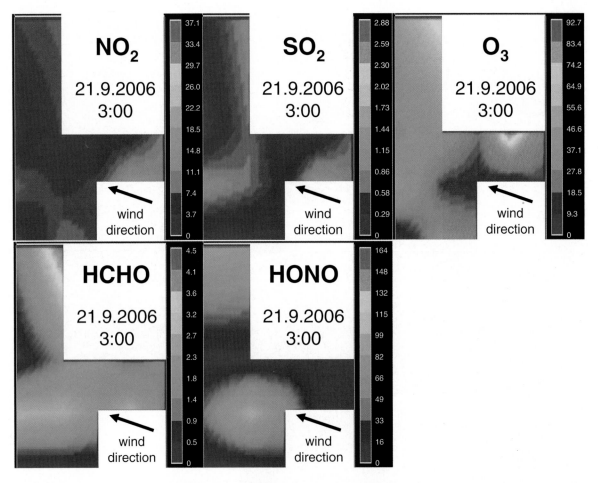

Fig. 2 Tomographic reconstructed distributions (in ppb) of NO$_2$, SO$_2$, O$_3$, HCHO, and HONO in the centre of Heidelberg as measured on Sept. 21, 2006 between 3:00 am and 6:00 am local time. Colour scales give mixing ratios in ppb, note different scales for the different species

far target. This simple approach assumes that the trace gas concentration in the vertical sections (paths 1, 3) are equal at above close and far target. This may actually be the case if a plume exists between instrument and far target. Also, even if there are differences between paths 1, 3 their influence may not be large since the horizontal section (several km) may be much longer than the inversion height determining the column densities across paths 1, 3. Finally, possible larger differences will show up in the tomographic inversion of the 2D-trace gas distribution and can be taken into account in a further iteration.

In other words a small number of, passive (and thus simple) instruments can determine dozens or even hundreds (depending on geometry and desired temporal resolution) of path-averaged trace gas concentrations. Disadvantages is the requirement of daylight for the measurement, also the useable spectral range is limited to wavelengths where radiation is provided by the sun and can penetrate the Earth's atmosphere (in particular only radiation with wavelengths longer than 300 nm will be available). While already measurements using this technique were reported (Frins et al. 2006, 2008, Louban et al. 2008), its large scale application in arrangements allowing tomographic inversions still have to be explored.

2.3 Airborne Imaging – DOAS (I-DOAS)

Another approach to obtain 2D trace gas concentration fields is the Airborne Imaging DOAS technique. This technique was originally developed

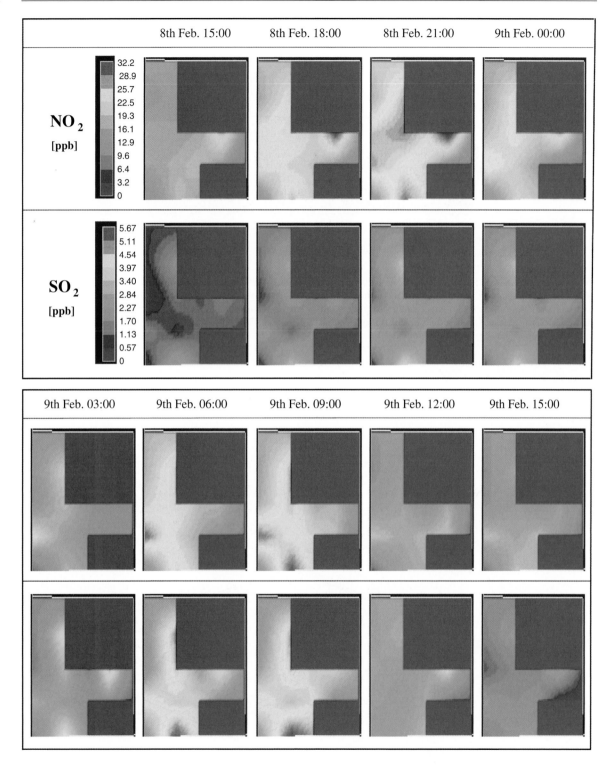

Fig. 3 Tomographic reconstructed distributions (in ppb, see colour scales) of NO₂ (rows 1, 3) and SO₂ (rows 2, 4) on February 8/9, 2006. Data are 3 h averages with beginning at the given local time. The area and the reconstruction grid are indicated by *yellow lines* in the map of Fig. 1. The main wind direction during this period was from south-west

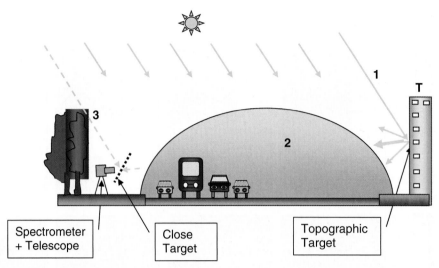

Fig. 4 The principle of ToTaL-DOAS (Frins et al. 2008). Trace gas column measurements are made alternately directing the telescope to the far (topographic) target and to the close target (*dashed line*). In both cases scattered sunlight is the light source, with the far target trace gas concentrations are averaged over paths 1, 2 resulting in S_{far}; using the close target only light path 3 is employed giving By taking the difference $S_{far} - S_{close}$ the average concentration along the nearly horizontal section of the light path (2) can be derived

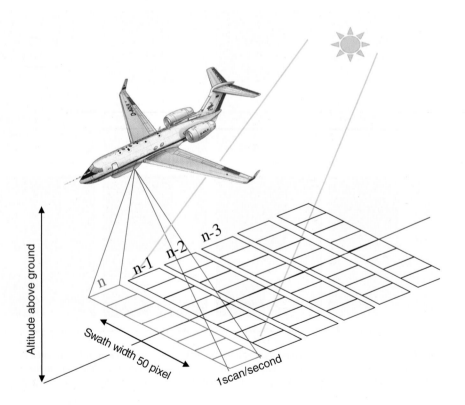

Fig. 5 The principle of airborne imaging DOAS mapping of 2D trace gas distributions

for ground based mapping of two-dimensional (horizontal and vertical) mapping of trace gas plumes (Lohberger et al. 2004, Louban et al. 2008). This technique makes use of a one dimensional imaging spectrograph, which produces a separate spectrum for each point along the entrance slit. This is accomplished by replacing the usual one dimensional (diode array or CCD) by a two-dimensional version with one dimension (e.g. line direction) yielding spectral information and the other (e.g. column) yielding spatial information. DOAS evaluation of each spectrum (line) yields a "column" consisting of up to several hundred trace gas optical densities corresponding to different positions of light entrance along the spectrograph entrance slit. In airborne I-DOAS applications (see Fig. 5) the motion of the plane results in longitudinal information while the imaging spectrometer (equipped with a two-dimensional detector) generates a swath perpendicular to the flight direction (lateral).

The longitudinal resolution is usually determined by the exposure time of the spectrograph and the aircraft ground speed. The lateral resolution is given by magnification and resolution of the optical system and the flight altitude, as described by Heue et al. (2008). Obviously the number of lateral pixles can not exceed the numbers of lines on the detector.

A series of successful airborne measurements of NO_2 and SO_2 distributions in the Highveld area (South Africa) were conducted. Examples illustrating the capability of the technique to map emission plumes are shown in Figs. 6 and 7. More results and evaluations e.g. to determine trace gas emissions by the plants are reported by Heue et al. (2008). Figure 6 shows a section of the NO_2 slant-column density distribution in the plume originating from the Secunda fuel refinery (South Africa) measured on Oct. 5, 2006 overlaid on a Google earth photograph. The aircraft was flying at an altitude of 4500 m above ground. The flight trajec-

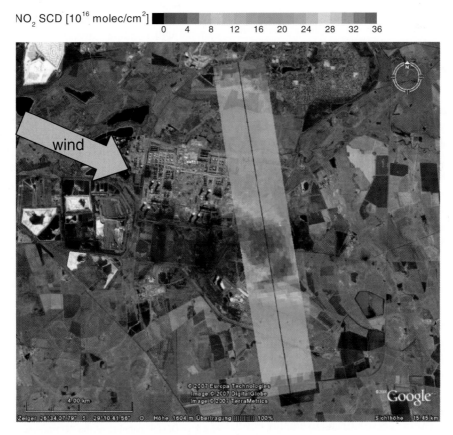

Fig. 6 Airborne I-DOAS mapping of the 2D NO_2 slant-column distribution in the plume originating from the Secunda fuel refinery (South Africa) as measured on Oct. 5, 2006, at 4500 m flight altitude above ground. The measurements (*coloured stripe*) cover a total distance of 11.9 km, at a swath Width of 1.9 km, the swath is 27 pixles wide (73 m by 148 m each)

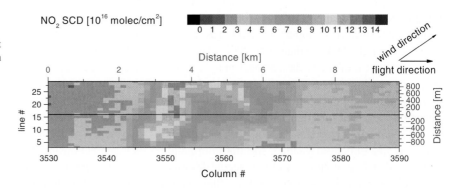

Fig. 7 Airborne I-DOAS mapping of the 2D NO_2 distribution at the Duvha plant (South Africa) as measured on Oct. 5, 2006, at 4500 m flight altitude above ground. The measurements cover a total distance of 3.6 km, at a swath width of 1.7 km, the swath is 27 pixles wide (63 m by 163 m each)

tory intersected the plume at nearly right angle (wind direction was approx. west-north-west, see arrow in Fig. 6) showing peak NO_2 SCD's in the centre of the plume approaching $4 \cdot 10^{17}$ molec.cm^{-2}. Assuming a plume of circular cross section with a radius of 500 m this SCD would correspond to an average NO_2 concentration of $4 \cdot 10^{12}$ molec.cm^{-3} (or a mixing ratio around 170 ppb).

3 Summary and Outlook

In recent years several techniques relying on active as well as on passive DOAS principles have been developed, which allow two-dimensional mapping of atmospheric trace gases relevant for atmospheric chemistry including NO_2, SO_2, O_3, CH_2O, CHOCHO, HONO, BrO, and aromatic compounds at high sensitivity. Spatial resolution ranges from several 10 m to several 100 m. It is anticipated that the applications of these techniques will lead to a similar advance in our understanding of local and lower regional scale physico-chemical processes in the atmosphere as satellite measurements have provided for larger regional and global scales.

Active tomographic techniques have shown their strength in a first, promising application (Pöhler et al. 2007), however further technological development will be required to make instruments simpler and more reliable. Promising technologies include developments to replace the thermal light sources by Light emitting diodes (Kern et al. 2006) and to employ novel fibre optics (Merten et al. 2009).

As showed in Section 2.3 the two-dimensional distribution of trace gases can be determined in a horizontal plane (along the flight track and perpendicular to the flight direction), however no information about the third dimension, i.e. the altitude distribution of the observed trace gases can be gained directly. Although indirect information on the altitude can be deduced from independent data e.g. stack height or boundary layer thickness. To address the aspect directly future systems will combine several imaging spectrometers observing the same air mass under different viewing directions as illustrated in Fig. 8 showing an example with three independent instruments.

Tomographic reconstructions, as shown in Heue 2005, will allow to derive 2D trace gas concentrations in a (nearly) vertical plain defined by the direction of flight and the vertical (i.e. the line from the aircraft to the centre of a particular ground pixel) using the information from the three instruments recorded at a series of successive aircraft positions. Combining this information with the I-DOAS approach true three-dimensional reconstructions of trace gas distributions below the aircraft will be derived.

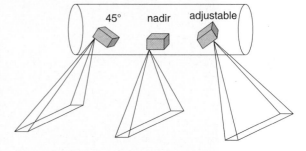

Fig. 8 I-DOAS mapping of the 3D trace gas distribution combining imaging DOAS and tomographic principles

References

Frins, E., Bobrowski, N., Platt, U. and Wagner, T. 2006. Tomographic MAX-DOAS observations of sun illuminated targets: a new technique providing well defined absorption paths in the boundary layer. *Appl. Opt.*, 45, 24, 6227–6240.

Frins, E., Platt, U. and Wagner, T. 2008. High spatial resolution measurements of NO₂ applying topographic target light scattering-differential optical absorption spectroscopy (ToTaL-DOAS). *Atmos. Chem. Phys.*, 8, 7595–7601.

Hartl, A., Song, B.C. and Pundt, I. 2006. 2D reconstruction of atmospheric concentration peaks from horizontal long path DOAS tomographic measurements: parameterisation and geometry within a discrete approach. *Atmos. Chem. Phys.*, 6, 847–861.

Heue, K.-P. 2005. Airborne multi AXis DOAS instrument and measurements of two-dimensional tropospheric trace gas distributions, dissertation, institut für umweltphysik, Universität Heidelberg.

Heue, K.-P., Wagner, T., Broccardo, S.P., Piketh, S.J., Ross, K.E. and Platt, U. 2008. Direct observation of two-dimensional trace gas distributions with an airborne imaging DOAS instrument. *Atmos. Chem. Phys.*, 8, 6707–6717.

Kern, C., Trick, S., Rippel, B. and Platt, U. 2006. Applicability of light-emitting diodes as light sources for active DOAS measurements. *Appl. Opt.*, 45, 2077–2088.

Laepple, T., Knab, V., Mettendorf, K.-U. and Pundt, I. 2004. Longpath DOAS tomography on a motorway exhaust gas plume: numerical studies and application to data from the BAB II campaign. *Atmos. Chem. Phys.*, 4, 1323–1342.

Lohberger, F., Hönninger, G. and Platt, U. 2004, Ground-based imaging differential optical absorption spectroscopy of atmospheric gases. *Appl. Opt.*, 43, 24, 4711–4717.

Louban, I., Píriz, G., Platt, U. and Frins, E. 2008. Measurement of SO₂ and NO₂ applying ToTaL-DOAS from a remote site. *J. Opt. A: Pure Appl. Opt.*, 10, 104017 (6pp), doi: 10.1088/1464-4258/10/10/104017.

Merten, A. 2008. Neues design von langpfad-DOAS-instrumenten basierend auf faseroptiken und anwendungen der untersuchung der urbanen atmosphäre. *Doctoral Thesis Ruprecht-Karls University, Heidelberg.*

Merten, A., Tschritter, J. and Platt, U. 2009. New design of DOAS-long-path telescopes based on fiber optics. *submitted to Appl. Opt.*

Platt, U., Perner, D. and Pätz, H. 1979. Simultaneous measurements of atmospheric CH₂O, O₃ and NO₂ by differential optical absorptions. *J. Geophys. Res.*, 84, 6329–6335.

Platt, U. and Stutz, J. 2008. Differential optical absorption spectroscopy, principles and applications, *Springer, XV, 597 p. 272 illus., 29 in color.* (Physics of Earth and Space Environments), ISBN 978-3-540-21193-8.

Pöhler, D., Hartl, A. and Platt, U. 2007. Tomographic LP-DOAS measurements of 2D trace gas distributions above the city of Heidelberg, Germany, Proc. 6th Internatl. Conf. on Urban Air Quality, Limasso, Cyprus, March 27–29.

Pundt, I. and Mettendorf, K.U. 2005. Multibeam long-path differential optical absorption spectroscopy instrument: a device for simultaneous measurements along multiple light paths. *Appl. Opt.*, 44, 23, 4985–4994.

Sihler, H. 2007, Light-emitting diodes as light sources in spectroscopic measurements of atmospheric trace gases. *Diploma Thesis, Friedrich-Schiller-University, Jena.*

Tschritter, J. 2007. Entwicklung einer DOAS-Optik der 3. Generation und ein vergleich mit herkömmlichen systemen. *Diploma Thesis, Ruprecht-Karls University, Heidelberg.*

Atmospheric Aerosol Monitoring from Satellite Observations: A History of Three Decades

Kwon H. Lee, Zhanqing Li, Young J. Kim and Alexander Kokhanovsky

Abstract More than three decades have passed since the launch of the first satellite instrument used for atmospheric aerosol detection. Since then, various powerful satellite remote sensing technologies have been developed for monitoring atmospheric aerosols. The application of these new technologies to different satellite data have led to the generation of multiple aerosol products, such as aerosol spatial distribution, temporal variation, fraction of fine and coarse modes, vertical distribution, light absorption, and some spectral characteristics. These can be used to infer sources of major aerosol emissions, the transportation of aerosols, interactions between aerosols and energy and water cycles, and the involvement of aerosols with the dynamic system. The synergetic use of data from different satellite sensors provides more comprehensive information to better quantify the direct and indirect effects of aerosols on the Earth's climate. This paper reviews how satellite remote sensing has been used in aerosol monitoring from its earliest beginnings and highlights future satellite missions.

Keywords Satellite · Instrument · Remote sensing · Aerosol · Monitoring

1 Introduction

Atmospheric aerosols are defined as suspended particles (solid or liquid) in a gas medium. The particles that compose aerosols range in size from nanometers to tens of micrometers, depending on whether they originate from natural sources (e.g., pollens, sea-salt, wind-blown dust, volcanic ash) or from man-made sources (e.g., smoke, soot, biomass burning). Aerosols can contribute to a reduction in visibility (Trijonis et al. 1991) and a decline in human health (Davidson et al. 2005) as well as affecting climate change (IPCC 2007). To fully understand aerosol effects, their characteristics (quantity, composition, size distribution, and optical properties) must be known on local to global scales (Kaufman et al. 2002).

Aerosol properties have been typically acquired using ground-based point measurements. Details concerning aerosol properties have been obtained from in-situ measurements, such as from aircraft or balloons, but these were limited to a few aerosol intensive measurement campaigns. Examples of such campaigns include the International Global Atmospheric Chemistry (IGAC) programs (IGAC 1996), the Tropospheric Aerosol Radiation Forcing Observation Experiment (TARFOX) (Russell et al. 1999) and three Aerosol Characterization Experiments such as ACE-1 (Bates et al. 1998), ACE-2 (Raes et al. 2000), and ACE-Asia (Huebert et al. 2003). The use of satellites to monitor aerosols has the advantage of providing routine measurements on a global scale and is an important tool for use in improving our understanding of aerosol properties.

The first visual observations of atmospheric aerosol effects were made from the manned spacecrafts. Cosmonaut Yuri Gagarin observed clouds and their shadows, as well as optical phenomena due to the presence of aerosols, during the first manned space flight on the spacecraft Vostok on April 12, 1961. These first observations were visual in nature but in

K.H. Lee (✉)
Department of Atmospheric and Ocean Science, Earth System Science Interdisciplinary Center, University of Maryland, College Park, MD 20740, USA
e-mail: Kwonlee@umd.edu

Y.J. Kim et al. (eds.), *Atmospheric and Biological Environmental Monitoring*, DOI 10.1007/978-1-4020-9674-7_2, © Springer Science+Business Media B.V. 2009

subsequent space flights, photography was used by cosmonaut G. S. Titov (Vostok-2, August 6, 1961), cosmonaut V. V. Tereshkova (Vostok-6, June 16, 1963), K. P. Feoktistov (Voskhod, October 12, 1964), A. A. Leonov (Voskhod-2, March 18, 1965), and others. They took photos of the horizon in order to estimate the vertical distribution of aerosols. A. G. Nikolaev and V. I. Sevastyanov (Soyuz-9, June 1, 1970) used hand-held spectrophotometers to measure the spectrometry of the twilight and daylight horizons, as well as that of clouds and snow. This instrument was also used in several follow-up missions. Stratospheric aerosol measurements using a hand-held sun photometer were made on the Apollo-Soyuz in 1975 (Pepin and McCormick 1976). Further information on the first instrumental observations of the planet from manned aircrafts is given by Lazarev et al. (1987).

The first detection of aerosols from an un-manned spacecraft was achieved by the Multi Spectral Scanner (MSS) onboard the Earth Resources Technology Satellite (ERTS-1) (Griggs 1975; Fraser 1976; Mekler et al. 1977) and the first operational aerosol products were generated from the TIROS-N satellite launched on 19 October 1978. The Advanced Very High Resolution Radiometer (AVHRR) onboard TIROS-N was originally intended for weather observations but its capability was expanded to the detection of aerosols. The Nimbus-7 was launched on 25 October 1978, carrying the Stratospheric Aerosol Measurement instrument (SAM) (McCormick et al. 1979) and the Total Ozone Mapping Spectrometer (TOMS). While the TOMS was not originally designed for aerosol monitoring, it has since provided the longest measurement record of global aerosols from space (Herman et al. 1997; Torres et al. 2002). These launches thus marked the beginning of an era of satellite-based remote sensing of aerosols that has lasted over three decades to date.

Advances in satellite monitoring capabilities have resulted in the generation of many valuable scientific datasets from local to global scales, which are useful to researchers, policy makers, and the general public. Satellite instruments give us the ability to make more accurate measurements on a nearly daily basis across a broader geographic area and across a longer time frame. This paper reviews various spaceborne sensors used in the remote sensing of aerosols and the associated data products retrieved from satellite measurements. Section 2 presents an overview of

satellite remote sensing data and instruments. Various aerosol retrieval techniques applied to satellite data is introduced in Section 3. In Section 4, the acquisition of satellite data and applications, including intercomparisons, climatologies, and synergy studies, are discussed. The prospects for future missions are highlighted as well.

2 Satellite Observations for Aerosol Monitoring

Space agencies, such as the National Aeronautics and Space Administration (NASA), the National Ocean and Atmosphere Administration (NOAA), the European Space Agency (ESA), le Centre National d'Etudes Spatiales (CNES) in France, the Japanese Aerospace Exploration Agency (JAXA), the China Meteorological Administration, the Royal Netherlands Meteorological Institute (KNMI), and the German Aerospace Centre (DLR), have launched many satellite instruments. Table 1 shows a timeline of satellite missions from 1972 to 2006 and a summary of the features for each sensor. Aerosol monitoring from space has, in the past, been accomplished using satellite data not explicitly designed with this application in mind. Historical satellite observations still in operation are the TOMS and AVHRR series. The AVHRR has been primarily used for the surveillance of weather systems and the monitoring of sea surface temperatures (SST) and land vegetation indices (VI). The TOMS was originally designed for deriving the total ozone content in the atmosphere. As a by-product, aerosol information has been successfully extracted from both sensors, such as aerosol optical depth/thickness (AOD/AOT, τ) from the AVHRR (Stowe et al. 1997) and the UV-absorbing aerosol index (AI) from the TOMS (Herman et al. 1997; Hsu et al. 1999).

Information concerning aerosols was also inferred from other later sensors, such as the Sea-viewing Wide Field-of-view Sensor (SeaWiFS), and the Moderate Resolution Imaging Spectro-radiometer (MODIS); the near-future Visual/Infrared Imager Radiometer Suite (VIIRS) will continue in this vein. The SeaWiFS, developed for studying marine biogeochemical processes, has been employed to produce aerosol

Table 1 The history of platforms and sensors used to derive aerosol properties from space

Launch	End	Platform	Instrument	# of bands (wavelengths (μm))	Accuracy	Reference[a]
1972	1978	Landsat(ERTS-1)	MSS	4(0.5–1.1)	$\tau(10\%)$	Griggs (1975)
1974	1981	SMS-1, 2	VISSR	5(0.65–12.5)		
1975	Present	GOES-1~12	VISSR	5(0.65–12.5)	$\tau(18\sim34\%)^b$	Knapp et al. (2002)
1975	1975	Apollo-Soyuz	SAM	0.83	–	McCormick et al. (1979)
1977	2005	GMS-1~5	VISSR	4(0.45–12.5)	–	–
1978	1980	TIROS-N	AVHRR	4(0.58–11.5)	–	–
1978	1993	Nimbus-7	SAM-2,	1	$\sigma_{ext}(10\%)$	McCormick et al. (1979)
			CZCS,	6(0.443–11.5)	–	–
			TOMS	6(0.312–0.380)	–	–
1979	1981	AEM-B	SAGE	4(0.385,0.45,0.6,1.0)	$\sigma_{ext}(10\%)$	Chu and McCormick (1979)
1979	Present	NOAA-6~16	AVHRR	5(0.58–12)	$\tau(10\%)^c$, $\tau(3.6\%)^d$	Stowe et al. (1997)
						Mishchenko et al. (1999)
1984	2005	ERBS	SAGE-2	4(0.386–1.02)	$\sigma_{ext}(10\%)$	Chu et al. (1989)
1997	Present	TRMM	VIRS	5(0.63–12)	$\tau(35\%)$, $\alpha(\pm0.5)$	Ignatov and Stowe (2000)
1991	1996	SPOT-3	POAM-2	9(0.353–1.060)	$\sigma_{ext}\ (\sim20\%)$	Randall et al. (1996)
1991	1999	ERS-1	ATSR,	4(1.6, 3.7, 11, 12)	–	–
			GOME	4(0.24–0.79)	–	Torricella et al. (1999)
1992	2005	UARS-	HALOE	8(2.45–10.01)	$r_{eff}(\pm15\%)$, $\sigma_{ext}(\pm5\%)$	Hervig et al. (1998)
1994	1994	SSD	LITE	3(0.355, 0.532, 1.064)	$\beta(\lambda_1)/\beta(\lambda_2)(<5\%)$	Gu et al. (1997)
1995	Present	ERS-2	ATSR-2,	7(0.55–12)	$\tau(<0.03)$, $\tau(30\%)$	Veefkind et al. (1999)
			GOME	0.24–0.79		
1996	Present	Earth Probe	TOMS	6(0.309–0.360)	$\tau(20\sim30\%)^e$	Torres et al. (2002)
1996	1997	ADEOS	POLDER,	9(0.443–0.910)	$\tau(20\sim30\%)^f$,	Herman et al. (1997)
			ILAS,	2(0.75–0.78, 6.21–11.77)	–	–
			OCTS	7(0.412–0.865)	–	–
1997	Present	OrbView-2	SeaWiFS	8(0.412–0.865)	$T(5\sim10\%)$	Gordon and Wang (1994)
1998	Present	SPOT-4	POAM-3	9(0.354–1.018)	$\sigma_{ext}(\pm30\%)$	Randall et al. (2001)
1999	Present	TERRA	MODIS,	36 (0.4–14.4)	$\tau(5\sim15\%)^g$	Remer et al. (2005)
			MISR	4 (0.45~0.87)	$\tau(10\sim20\%)$	Kahn et al. (2005)
2001	2005	METEOR-3M	SAGE-3	9(0.385–1.545)	$\sigma_{ext}(5\%)$, $\tau(5\%)$	Thomason et al. (2007)
2001	Present	PROBA	CHRIS	62(0.4–1.05)	–	Barnsley et al. (2004)
2001	Present	Odin	OSIRIS	0.274–0.810	$\sigma_{ext}(15\%)$	Bourassa et al. (2007)
2002	Present	AQUA	MODIS	–	–	–
2002	Present	ENVISAT	AATSR,	7(0.55–12.0)	$\tau(0.16)$,	Grey et al. (2006)
			MERIS,	15(0.4–1.05)	$\tau(\sim0.2)$,	Vidot et al. (2008)
			SCIAMACHY	0.24–2.4	$AI(\sim0.4)$	Graaf and Stammes (2005)
2002	2003	ADEOS-2	POLDER-2,	9(0.443–0.910)	–	–
			ILAS-2,	4(0.75–12.85)	–,	Zasetsky and Sloan (2005)
			GLI	36(0.38–12)	$\tau(\sim0.1)$	Murakami et al. (2006)
2002	Present	MSG-1	SEVIRI	12(0.6–13.4)	$\tau(0.08)$	Popp et al. (2007)
2003	2003-	ICEsat	GLAS	2(0.532, 1.064)	$\sigma_{ext}(10\%)$, $\tau(20\%)$	Palm et al. (2002)
2004	Present	AURA	OMI,	3(0.27–0.5)	$\tau(30\%)$,	Torres et al. (2007),
			HIRDLS	21(6–18)	$\sigma_{ext}(5\sim25\%)$	Froidevaux and Douglass (2001)
2004	Present	PARASOL	POLER-3	8(0.44–0.91)	–	–
2006	Present	CALIPSO	CALIOP	2(0.532, 1.064)	–	–

[a] References of the validation study for accuracy listed here.
[b] Accuracy for operational GOES aerosol retrieval may apply for other GOES series.
[c] Accuracy for single channel AVHRR aerosol retrieval algorithm may apply for other AVHRR series.
[d] Accuracy for two channel AVHRR aerosol retrieval algorithm may apply for other AVHRR series.
[e] Accuracy for TOMS AOT retrieval from Nimbus-7 to Earth Probe.
[f] may apply for the POLDER-2 and -3.
[g] same to the MODIS/Aqua.

data required for atmospheric correction (Gordon and Wang 1994). With the launch of Terra (EOS AM-1), more advanced instruments like MODIS and the Multiangle Imaging SpectroRadiometer (MISR) provide substantially improved aerosol retrievals (Remer et al. 2005; Diner et al. 1998). The same applies to the Medium Resolution Imaging Spectrometer (MERIS) and Advanced Along-Track Scanning Radiometer (AATSR) onboard the ESA EnviSAT. The launch of POLDER on ADEOS II added more capabilities by virtue of its polarization measurements of backscattered solar light (Leroy et al. 1997). Space-borne light detection and ranging (LIDAR) observations from the Lidar In-space Technology Experiment (LITE) (Winker et al. 1996), the Geoscience Laser Altimeter System (GLAS) (Spinhirne et al. 2005a, b), and the most recently launched Cloud-Aerosol Lidar and Infrared Pathfinder Satellite Observations (CALIPSO) (Vaughan et al. 2004) allow for global-scale assessments of the vertical distribution of aerosols, backscatter, extinction, and depolarization ratios.

In terms of information content, satellite data may be classified into three general categories. The first is aimed at portraying the spatial and temporal dynamics of aerosol loading. The second is concerned with columnar aerosol properties retrievals (e.g., aerosol columnar mass retrievals) through use of spectral, polarization, and angular characteristics of backscattered solar light. The third provides information on the vertical profile of aerosols from the surface into the stratosphere. Contingent upon the need of a particular aerosol attribute, a single sensor or combination of sensors may be used.

There are two basic types of satellite instruments depending on the observation geometry, namely vertical and horizontal measurements (Fig. 1). By vertical (or nadir viewing) observation, the instrument faces to nadir or near-nadir and senses the radiation coming from the Earth. Most instruments employ this concept to provide column integrated products. Observation in horizontal direction including Limb-viewing and occultation sounding, probes the Earth's limb at various depths in the atmosphere. This observation is characterized by the altitude and the geolocation of the tangent point. Especially, solar occultation instruments can retrieve aerosol extinction profile from measurement of sunlight extinction through the atmospheric limb during sunrise and sunset. All these methods require accurate calibration of instruments and sound

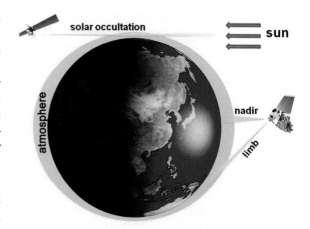

Fig. 1 Vertical (nadir) and horizontal (limb and solar occultation) satellite observation concept. Nadir viewing is looking straight down to measure columnar observation. Limb viewing provides a much longer path through the atmosphere, and also makes it easier to determine the altitudes of the observed substances

treatment of unknown optical properties of aerosols, surface reflectivity, and gaseous absorption.

Inference of aerosol properties from satellite relies on the interaction of electromagnetic radiation scattered and/or absorbed by the atmospheric constituents and the surface target as illustrated in Fig. 2. Radiation is received by two basic types of sensors: passive and active. Passive sensors record radiation emitted by the Sun and reflected back to the sensor while active sensors receive energy emitted by the sensor itself (laser beam). Aerosol remote sensing is an ill-posed problem because the number of variables to be determined is larger than the number of parameters, which can be in principle found and constrained from the satellite measurements themselves. The essence of aerosol remote sensing is to decompose mixed signals emanating from atmospheric gases, aerosols, and the surface, after clouds are filtered out. Reflectance, the ratio of radiances received by a sensor over that reaching the top of atmosphere (TOA) in a particular direction, can be expressed by the following equation:

$$\rho_{TOA}(\theta_0, \theta_S, \phi) = \rho_{atm}(\theta_0, \theta_S, \phi) \\ + \frac{T_0(\theta_0) \cdot T_S(\theta_S) \cdot A_g}{1 - s \cdot A_g} \quad (1)$$

where $\rho_{atm}(\theta_0, \theta_S, \phi)$ is the reflectance by the atmosphere, and $T_0(\theta_0)$ and $T_S(\theta_S)$ are downward and

Fig. 2 Basic scheme of radiative transfer processes for passive and active satellite remote sensing

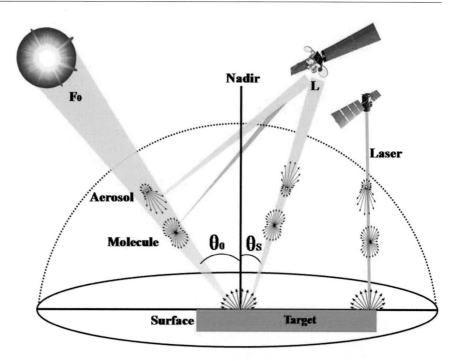

upward total transmission (diffuse plus direct); θ_S is the satellite zenith angle, θ_0 is the solar viewing angle, and ϕ is the relative azimuth angle. The spherical albedo is given by s and A_g is the surface reflectance. It follows from Eq. (1) that the signal received by a satellite sensor is dictated by atmospheric variables (gases, aerosols, cloud hydrometeors, etc.) and surface variables. When a cloud is present, reflection by the cloud is often overwhelming. As such, the first step is to identify the presence of clouds. Aerosol remote sensing is only valid under clear-sky conditions. Any cloud contamination can easily confuse the faint signal of aerosols, whereas excessive cloud screening may remove pixels containing heavy aerosol loading. Due to the delicacy of cloud screening, it remains the largest uncertainty in aerosol retrievals (Jeong and Li 2005). The second step is to account for molecular scattering due to atmospheric molecules and gas absorption. The Rayleigh path radiance can be determined using the spectral dependence of the well-known Rayleigh optical depth (ROD) and the Rayleigh phase function. The third step is to remove surface reflection from satellite-received signal. Early attempts at aerosol retrievals (Griggs 1975; Mekler et al. 1977; Durkee et al. 1986; Stowe 1991; Higurashi and Nakajima 1999; Mishchenko et al. 1999; Deuzé et al. 1999) were limited to dark surfaces with low and uniform

reflectivities, such as oceans. However, aerosol remote sensing over brighter land surfaces is very important for environmental and climate studies because most aerosols originate from continental sources such as bare soil, deserts, urban, industrial, and agricultural areas. Aerosol retrieval over land requires accurate knowledge of surface reflectance, and its spectral and angular dependence. The first attempt of aerosol retrieval over land is found in Kaufman and Joseph (1982). Thanks to the advent of new remote sensing techniques developed by taking advantage of multi-angle and multi-spectral measurements (Kaufman et al. 1997; Martonchik et al. 1998, 2002; Hsu et al. 2004; Remer et al. 2005; Levy et al. 2007b), such a limitation has been eliminated or lessened considerably.

3 Satellite Aerosol Remote Sensing Techniques

Many algorithms have been developed for aerosol detection using satellite measurements made at single- or multiple-wavelengths, nadir view and multi-angle views, with or without polarization, and low earth or geostationary orbits. Some of the algorithms are used for routine applications, while others are used for research and development. Numerous attempts were also made to compare and assess different satellite aerosol products, including those from the MODIS, MISR, AVHRR, TOMS, SeaWiFS, MERIS, AATSR, and other instruments (Myhre et al. 2004; Jeong et al. 2005; Jeong and Li 2005; Kokhanovsky et al. 2007; Kokhanovsky and de Leeuw 2009). The accuracies of various aerosol retrievals are summarized in Table 1. It must be remarked, however, that the estimations of the errors of the aerosol retrieval algorithms are given usually after analyzing large statistical ensembles of coincident ground and satellite spectral AOT measurements. Therefore, errors for any given measurement can be much larger as compared to the average error for the ensemble. The principles and limitations of these algorithms for tropospheric aerosols were previously reviewed by King et al. (1999). This paper reviews a number of aerosol retrieval techniques and categorizes them according to location in the atmosphere (troposphere and stratosphere) and from single- and multi-sensor data. An overview of the different techniques is outlined below.

3.1 Geostationary Satellite Algorithm

Although the AOT inferred from polar (or low orbit) satellite measurements provide global coverage map with fine spatial resolution, these observations are limited in space and time. The geostationary satellite measurements provide a unique tool for quantifying aerosol properties with high temporal resolution. Aerosol retrieval from geostationary observations has advantage in the obtaining of surface reflectance information from 'background image' acquired from composited minimum reflectance values among numerous views of the same location for a period. The AOT can be then retrieved by comparing imagery to this 'background image' (Knapp and Stowe 2002; Knapp et al. 2005). The uncertainty of the operational GOES AOT retrieval was reported as ±18–34% (Knapp and Stowe 2002; Knapp et al. 2002). Other geostationary satellites such as METEOSAT and GMS have been frequently used to derive aerosol properties (Dulac et al. 1992; Moulin et al. 1997; Wang et al. 2003; Costa et al. 2006).

3.2 Single-Channel AVHRR Algorithm

The most used single channel for aerosol retrievals is channel 1 of the AVHRR (the wavelength $\lambda = 0.63\,\mu m$) (Rao et al. 1989; Stowe 1991; Stowe et al. 1997; Ignatov et al. 1995a). The AVHRR algorithms are generally developed based on the look-up table (LUT) calculated using radiative transfer codes such as Dave (1973) and by assuming certain types of aerosol models. In an earlier algorithm (Stowe 1991), non-absorbing ($n = 1.5$–$0.0i$) aerosols with a size distribution following a modified Junge size distribution were assumed.

$$
\begin{aligned}
\frac{dN}{dr} &= 0 \quad (r < r_{\min}, r > r_{\max}) \\
&= A(r_{\min} \leq r \leq r_m) \\
&= A\left(\frac{r}{r_m}\right)^{-(\upsilon+1)} \quad (r_m \leq r \leq r_{\max})
\end{aligned} \tag{2}
$$

where, $r_{\min}$, r_m, $r_{\max}$ are particle radii equal to 0.03, 0.1, and 10 μm, respectively; size parameter $\upsilon = 3.5$, the normalized constant A. The retrieval results were validated against ship-borne sun-photometer measurements made within ±2 hours of the satellite overpass (Ignatov et al. 1995b). The comparison shows a negative bias, i.e. $\tau_{sat} = 0.64 \cdot \tau_{sp} - 0.02$ (Stowe 1997).

The algorithm currently used for generating operational AVHRR aerosol products, known as AVHRR Pathfinder Atmosphere (PATMOS) (Stowe et al. 2002; Jacobowitz et al. 2003) uses a lognormal aerosol size distribution

$$
\frac{dN}{dr} = \frac{A}{\sqrt{2\pi}r\ln\sigma}\exp\left[-\frac{1}{2}\cdot\left(\frac{\ln r - \ln r_m}{\ln\sigma}\right)^2\right] \tag{3}
$$

where $r_m = 0.1\,\mu m$, $\sigma = 2.03$ with a refractive index $n = 1.4$–$0.0i$, and the Fresnel model to account for the bidirectional reflectance of a calm ocean surface (Viollier et al. 1980; Gordon and Morel 1983). These adjustments bring satellite AOT retrievals into agreement with surface observations to better than 10%. The liner regression between the two is $\tau_{sat} = 0.91\tau_{sp} + 0.01$ (Stowe et al. 1997).

3.3 Dual-Channel AVHRR Algorithm

The Ångström exponent (α), a parameter used to denote aerosol particle size, can be derived using both AVHRR shortwave channels ($\lambda = 0.65$, $0.85 \, \mu m$) (Stowe et al. 1997; Mishchenko et al. 1999; Geogdzhayev et al. 2002). The two-channel algorithm has been applied to the International Satellite Cloud Climatology Project (ISCCP) cloud-free product (Rossow et al. 1996) to generate the Global Aerosol Climatology Product (GACP). Aerosols are assumed to be spherical with the power-law size distribution and a refractive index of $1.5-0.003i$. In principle, two-channel algorithms are expected to provide more accurate retrievals than one-channel algorithms. However, because there is no onboard calibration of the instrument, the accuracy of the algorithms is more susceptible to calibration errors in both channels. Both single-channel and dual-channel algorithms are most sensitive to cloud screening errors, which is by far the largest source of errors in retrieving aerosol parameters.

It is worth noting that the retrieval of AOT is very sensitive to the choice of aerosol size distribution and complex refractive indices. For the same TOA reflectance, use of two distinct distribution functions (power law and bi-modal log-normal distributions), as adopted by the GACP and MODIS algorithms, can account for a large portion of the discrepancies in AOT retrievals (Jeong et al. 2005). Geogdzhayev et al. (2002) and Knapp et al. (2002) also showed that the imaginary part of the refractive index can also affect AOT retrieval. Unfortunately, there is no consensus as to which size distribution is more representative on a global scale. Many factors can change the aerosol size distribution, such as aerosol type, humidity, season and location, etc.

3.4 TOMS Algorithm

The TOMS instrument has flown on Nimbus-7, ADEOS and EP-TOMS since 1978, providing the longest record of data for monitoring ozone depletion. Hsu et al. (1996) found that the ratio of its two channels (331 and 360 nm) is sensitive to absorbing aerosols and

an aerosol index (AI) was defined as (Herman et al. 1997) and is given by:

$$AI = -100 \log_{10} \left[\left(\frac{I_{340}}{I_{380}} \right)_{meas} - \left(\frac{I_{340}}{I_{380}} \right)_{calc} \right] \quad (4)$$

where I_{meas} and I_{calc} are the measured and calculated backscattered radiances at the two wavelengths. Under the existence of absorbing aerosols, I_{meas} is smaller than I_{calc} predicted by the Dave's Lambert Equivalent Reflectivity (LER) model (McPeters et al. 1996) so produces positive residues, and vice-versa for non-absorbing aerosols. One of the unique strengths of this technique is that since clouds produce nearly zero residues, the presence of subpixel clouds does not affect the detection of aerosols (Herman et al. 1997). Daily global TOMS AI products have been generated and are widely employed to detect and monitor the spatial and temporal variations of elevated smoke and dust and other types of absorbing aerosols.

Attempts were also made to extract additional quantitative aerosol parameters, such as AOT and single scattering albedo (SSA) (Herman et al. 1997; Torres et al. 1998). Unlike the AI, which is mainly sensitive to UV-absorbing aerosols, the TOMS near-UV AOT retrieval algorithm is sensitive to all aerosol types. However, this retrieval is affected by the aerosol layer altitude, the single-scattering albedo, and subpixel cloud contamination due to its large footprint (about $40 \, km^2$ at nadir) (Herman et al. 1997; Torres et al. 1998, 2002). Torres et al. (2002) presented the first long-term (1979 to present) nearly-global climatology of AOT over both land and ocean with a retrieval uncertainty of $\sim 30\%$ relative to Aerosol Robotic Network (AERONET) observations, while the AOT of non-absorbing aerosols agreed to within 20%. The SSA derived from TOMS generally agrees within 0.03 of AERONET retrievals (Torres et al. 2005). The main constraint on the capability of the technique lies in the lack of information on aerosol type, vertical distribution and surface reflectance. The retrieval algorithm, called the 'near-UV algorithm', uses two backscattered radiances at near-UV bands. Three major aerosol types are assumed for the construction of the LUT and the examination of the variability of the relationship between the spectral contrast and the radiance at the longer wavelength. These LUTs are used to determine AOT and SSA.

3.5 Ocean Color Algorithms (CZCS, SeaWiFS, OCTS, MODIS)

The aerosol retrieval from ocean color sensors begins with the following equation (Gordon and Wang 1994):

$$\rho_{TOA}(\lambda) = \rho_r(\lambda) + \rho_a(\lambda) + \rho_{ra}(\lambda)$$
$$+ \rho_g(\lambda) + t \cdot \rho_w(\lambda) \tag{5}$$

where $\rho_r(\lambda)$, $\rho_a(\lambda)$, $\rho_{ra}(\lambda)$, $\rho_g(\lambda)$, and $\rho_w(\lambda)$ represent reflectances due to multiple scattering by air molecules (Rayleigh scattering), aerosols, the interaction between molecular and aerosol scattering which is negligible in the single-scattering case, the rough ocean surface which is also negligible because of low reflection over the ocean and the tilting sensor, and the water-leaving reflectance, respectively. The atmospheric transmission is represented by t. By using a set of aerosol models, aerosol effects at near-infrared (NIR) bands can be evaluated from Eq. (5) because $\rho_w(\lambda)$ at these bands are usually negligible for the open ocean waters due to strong water absorption (Hale and Querry 1973; Smith and Baker 1981).

Aerosol products are by-products from the atmospheric correction for the ocean color algorithm (Gordon and Wang 1994). Using Eq. (5), the $\rho_a(\lambda)$ values are derived from ocean color observations, then used to select the two most appropriate aerosol models from a set of LUTs. The current SeaWiFS and MODIS ocean color data processing algorithms use 12 aerosol models for generating the LUTs (Wang et al. 2005). They are the Oceanic model with 99% RH, the Maritime model and the Coastal model with an RH of 50, 70, 90, and 99%, and the Tropospheric model with an RH of 50, 90, and 99%, respectively. A weight that is best-fit to the measured NIR radiances from the radiances computed using the two selected aerosol models. Using the two aerosol models with the weight and the satellite measured radiance, the AOT and Ångström exponent can then be retrieved (Gordon and Wang 1994).

3.6 Polarization (POLDER, POLDER-2, POLDER-3)

The POLDER measures the polarization, directional, and spectral characteristics of solar light reflected by aerosols. A scientific goal of the POLDER experiment was to determine the physical and optical properties of aerosols so as to classify them and study their variability and cycles (Herman et al. 1997; Deuzé et al. 1999). The POLDER instrument is a push-broom-type, wide field-of-view, multi-band imaging radiometer and polarimeter with eight narrow spectral bands in the visible and near infrared (0.443, 0.490, 0.565, 0.665, 0.763, 0.765, 0.865, and 0.910 μm). The spectral variation allows the derivation of the aerosol size and thus their scattering phase function, as well as the AOT. The polarization provides some information on the aerosol refractive index and shape (spherical or non-spherical), which improves the determination of the scattering phase function. The algorithm is based on LUTs from POLDER directional, spectral and polarized measurements for several aerosol models. Using this unique information from POLDER measurements, Breon et al. (2002) found that the effect of aerosols on cloud microphysics is significant and occurs on a global scale. The accuracy in AOT retrieval was reported as 30% (Herman et al. 1997). The Ångström exponent derived from POLDER data correlated well with AERONET data, although it is also systematically underestimated by 30% (Goloub et al. 1999).

3.7 Multi-Channel Algorithm (SeaWiFS, MODIS, MERIS)

The MODIS instrument is deployed on both the Terra (EOS-AM) and Aqua (EOS-PM) satellites and measures upwelling radiances in 36 bands for wavelengths ranging from 0.4 to 14.5 μm. With a spatial resolution of 250, 500 m, or 1 km at nadir, MODIS data have been employed to generate the most comprehensive aerosol products including AOT, fine mode fraction (FMF), effective radius of aerosol particles (equal to the ratio of the third to the second moment of the aerosol size distribution), and mass concentration (Kaufman et al. 1997; Tanré et al. 1997, 1999; Remer et al. 2005). The retrieval uncertainty of the MODIS AOT products falls within the expected range of $\pm 0.03 \pm 0.05 \tau_{sat}$ over ocean and $\pm 0.05 \pm 0.15 \tau_{sat}$ over land (Remer et al. 2005; Chu et al. 2002). While the expected accuracy is met in general, significantly larger errors are found in certain regions (Levy et al. 2005; Li et al. 2007), especially where no or few ground measurements were available to train the algorithm. To remedy some of the problems, modifications were introduced by Levy et al. (2007a) to better account for the effects of surface spectral and bidirectional reflectance, as well as aerosol absorption. The modified algorithm is now used to generate the Collection 5 (C005) product (Remer et al. 2006). Over land, the C005 product has a significantly improved accuracy when compared to the earlier version of the product, as was shown in some validation studies using ground-based AERONET data (Levy et al. 2007b; Mi et al. 2007) and hand-held sunphotometer data in China (Li et al. 2007).

Retrieving aerosol properties from satellite remote sensing over a bright surface is a challenging problem. The Bremen Aerosol Retrieval (BAER) is capable of retrieving AOT over land surfaces and was first developed by von Hoyningen-Huene et al. (2003). It is based on the assumption that the surface reflectance is comprised of the mixed spectra from vegetation and bare soil. The fraction of vegetation in the pixel is estimated in an iterative way tuned by the NDVI. This method is very flexible to use for aerosol retrieval with visible satellite observation data. Applications of the BAER algorithm and validation has been reported for the SeaWiFS (von Hoyningen-Huene et al. 2003; Lee et al. 2004), MERIS (von Hoyningen-Huene et al. 2006), SCIAMACHY (von Hoyningen-Huene et al. 2005), and MODIS (Lee et al. 2005, 2006a, b, 2007a).

Hsu et al. (2004) proposed a new approach, called 'Deep Blue', to retrieve aerosol properties over bright land surfaces such as arid, semiarid, and urban areas. Those areas are typically very bright in the red to the NIR spectral region, but are relatively darker in the blue-band region. Using the global surface reflectance database of 0.1×0.1-degree resolution from the minimum reflectivity technique (e.g., finding the clearest scene during each season for a given location), the contribution of the surface-reflected radiance can be separated from the satellite-receiving radiance. Aerosol properties including AOT and aerosol type can then be determined simultaneously in the algorithm using LUTs. Comparisons of the satellite AOT and the AERONET AOT indicate good agreement (i.e., within 30%) over sites in Nigeria and Saudi Arabia (Hsu et al. 2004) and over East Asia (Hsu et al. 2006).

3.8 Multi-Angle, Multi-Channel (MISR)

The MISR instrument shares the Terra platform with the MODIS and uses nine individual CCD-based push-broom cameras to view Earth at nine different view angles: one at nadir and eight symmetrical views at 26.1, 45.6, 60.0, and 70.5 degrees forward and aft of nadir. Each camera obtains images at four spectral bands (443, 558, 672, and 866 nm) with a horizontal resolution of 1.1 km in non-red bands and 275 m in the red band (Diner et al. 1998). To retrieve spectral AOT and additional properties such as the Ångström exponent, SSA, number fraction, and volume fraction, the MISR offers a unique combination of multiple bands and multi- angles that convey richer information about aerosols (Martonchik and Diner 1992; Martonchik et al. 1998, 2002; Diner et al. 2008). The retrieval algorithm differs over water, dense dark vegetation (DDV), and heterogeneous land (Martonchik et al. 1998, 2002). For dark water, zero water-leaving radiances at red and near-infrared wavelengths are considered, which is similar to the ocean color algorithm. The algorithm for DDV uses an angular shape for the surface bidirectional reflectance factor (BRF) with angular measurements. For heterogeneous land, empirical orthogonal functions derived from the spectral contrast by multi-angle observations are used to determine AOT and the aerosol model.

Validation of MISR AOTs using AERONET AOTs has been reported in many studies. The comparisons show a positive bias of 0.02 with an overestimation of 10% over southern Africa (Diner et al. 2001), an overestimation of about 0.05 over China (Christopher and Wang 2004), a linear relationship of $\tau_{\text{sat}} = 0.92\tau_{\text{sp}} + 0.02$ ($R^2 = 0.90$) and a retrieval error of $0.04\pm0.18\tau_{\text{sp}}$ over the United States (Liu et al. 2004), an uncertainty of 0.08 in desert areas (Martonchik et al. 2004), and linear relationships in the red and blue bands of $\tau_{\text{sat}} = 0.74\tau_{\text{sp}} + 0.11$ ($R = 0.87$) and $\tau_{\text{sat}} = 0.83\tau_{\text{sp}} + 0.03$ ($R = 0.86$), respectively, over various AERONET sites (Abdou et al. 2005). Kahn et al. (2005) reported that from a two-year comparison, about two-thirds of the MISR-retrieved AOT values fall within 0.05 or 20% of AERONET AOTs and more than a third are within 0.03 or 10% of AERONET AOTs.

3.9 Active Sensing (LITE, GLAS, CALIPSO)

Passive instruments have great difficulty with vertically resolving information about aerosols. However, space-borne lidars can provide a global view of the vertical structure of aerosol extinction from the Earth's surface through to the middle stratosphere, depending upon the presence of cloud and the aerosol density. Aerosol extinction from lidar measurements can be interpreted using the lidar equation. The single-scatter lidar equation is often written as:

$$P(R) = J \frac{c}{2} \frac{A}{R^2} \beta(R) T_{opt} T^2(R) \qquad (6)$$

where $P(R)$ is the instantaneous optical power returned from a sample volume at range R, J is the laser pulse energy, c is the speed of light, A is the receiver area, β is the volume backscatter cross section ($km^{-1}sr^{-1}$), and T_{opt} is the transmission of the lidar optics. The term $T^2(R)$ is the two-way transmission between the lidar and the sample volume and is given by:

$$T^2(R) = \exp \left[-2 \int_0^R \sigma(z)\,dz \right] \qquad (7)$$

where σ is the volume extinction coefficient, which includes the effects of both scattering and absorption. Then Eq. (7) is applicable to find the vertically distributed aerosol extinction.

The first spaceborne instrument, LITE, was a three-wavelength (1064, 532, and 256 nm) backscatter lidar developed by NASA and flown on the space shuttle Discovery for 10 days in September 1994 (McCormick et al. 1993; Winker et al. 1996). The LITE mission demonstrated that spaceborne lidar offers an effective means for detecting the spatial features of significant regional aerosol concentrations resulting, for example, from Saharan dust (Powell et al. 1997; Berthier et al. 2006), and African and South American biomass burning and anthropogenic sources (Grant et al. 1997; Hoff and Strawbridge 1997; Kent et al. 1998). The LITE mission stimulated the development of new space lidars such as the GLAS onboard ICESat (Spinhirne et al. 2005a, b) and the CALIPSO satellite which are currently generating aerosol products. The CALIPSO is flying in formation in a constellation of satellites called the A-Train. In addition to data from the A-train, CALIPSO data provides a more complete and understandable aerosol data set, which is used for various modeling studies.

3.10 Limb Sounding (SAGE, SAGE-2. POAM-2. POAM-3, HALOE, ILAS, SCIAMACHY)

The the Stratospheric Aerosol and Gas Experiment (SAGE) III instrument contains 12 spectral channels over the wavelength region of 0.28–1.54 µm and is essentially an improved version of its predecessors, SAGE I and II. The Naval Research Laboratory (NRL)'s the Polar Ozone and Aerosol Measurement (POAM) II onboard the French satellite SPOT-3 is a solar occultation instrument. It is designed as a simpler version of the SAGE II instrument to measure the vertical profiles of aerosols, O_3, NO_2, and H_2O in nine channels between 0.35 and 1.06 µm, with a 1 km vertical resolution (Glaccum et al. 1996). Aerosol products include the vertical profiles of polar region aerosols (Randall et al. 1996). Following a sensor improvement, the POAM III onboard the SPOT 4 satellite has been operational. NASA's HALOE instrument is able to measure vertical profiles of aerosol extinction. However, because it uses broadband and gas-filter radiometry methods (Russell et al. 1993) in the spectral range between 2.45 and 10.04 µm, it can provide stratospheric microphysical aerosol information when the aerosol loading is high, such as during volcanic eruptions (Hervig et al. 1998). Another occultation instrument called the Improved Limb Atmospheric Spectrometer (ILAS) can measure the vertical profiles of aerosol extinction in the infrared band between 6.21 and 11.77 µm and in a visible band centered near 0.78 µm. The SCanning Imaging Absorption spectroMeter for Atmospheric CHartographY (SCIAMCHY) instrument onboard EnviSAT is a high-resolution spectrometer designed to measure sunlight transmitted, reflected, and scattered by the Earth 's atmosphere or surface in the UV, visible, and NIR wavelength regions (0.24–2.38 µm, $\Delta\lambda = 0.24 \sim 1.48$ nm). It performs measurement not only in the limb but also in the nadir mode. Due to its wide wavelength range and spectral resolution, SCIAMACHY measurements turn out to be well-suited for the retrieval of atmospheric aerosol and have an uncertainty of 13~20% (Nicolantonio et al. 2006). Additionally, SCIMAMCHY can provide UV-absorbing AI, which is similar to the TOMS AI (Graaf and Stammes 2005). The main problem of this instrument as related to the aerosol remote sensing is the large footprint (typically, $30 \times 60 \, \text{km}^2$). Therefore, the number of clear pixels is very limited.

3.11 Neural Network (AVHRR, OCTS, MODIS)

The neural network (NN) approach has been proven to be a useful tool to solve such nonlinear problems as the retrieval of aerosol from satellite radiance measurements. Li et al. (2001) used both a commercial NN package and multi-threshold techniques to identify smoke from biomass burning using AVHRR imagery data. NN can learn complex linear and nonlinear relationships in the radiometric data between smoke, clouds, and land. This method was applied to process daily AVHRR images acquired across Canada and the results showed reasonable correspondence with TOMS AI. Another NN technique is a LUT-based method. Okada et al. (2001) used a LUT-based NN technique for aerosol retrieval from OCTS/ADEOS data. In this method, LUTs storing TOA reflectance values with various atmosphere-ocean conditions are used for tuning the NN. Geometric conditions and reflectances at two ADEOS/OCTS bands (0.67 and 0.865 μm) are input into the NN, and aerosol properties are extracted. The NN technique has proven to reduce processing times, and is promising for effective aerosol retrievals on a global scale.

3.12 Multi-Sensor (GOME and ATSR2, SCIAMACHY and AATSR, MODIS, CALIPSO-CloudSat)

Given the unique information content of individual sensors, use of data from multiple sensors sheds a new light for aerosol remote sensing. Holzer-Popp et al. (2002a) proposed a synergetic aerosol retrieval method, called the Synergetic Aerosol Retrieval (SYNAER), which was applied to the combinations of GOME-ATSR2 onboard the ERS-2 satellite (Holzer-Popp et al. 2002a, b) and SCIAMACHY-AATSR onboard the EnviSAT (Holzer-Popp et al. 2008). The advantage of these combinations is that they provide complementary information from a radiometer and a spectrometer aboard one satellite platform to extract AOT and the most plausible aerosol type. The retrieval accuracy is 0.1 at three visible wavelengths (Holzer-Popp et al. 2002b).

Use of data from the same type of instrument but onboard different satellites is another approach. Tang et al. (2005) tested an aerosol retrieval method by exploiting the synergy between MODIS/Terra and MODIS/Aqua (SYNTAM) for various surface conditions including bright surfaces. Most recently, new techniques were proposed using active spaceborne LIDAR and passive radar measurements. Josset et al. (2008) showed a substantially improved accuracy of 1% bias and 0.07 standard deviation in aerosol retrievals over ocean by combining CALIPSO and CloudSat data.

4 Data Products and Applications

4.1 Operational Data Products

The successive satellite missions provide operational aerosol products based on state-of-art retrieval algorithms and users can easily acquire the data. Figures 3 and 4 show the global distributions of representative monthly-averaged aerosol products. Each product is used for aerosol monitoring separately or in combination. In the early days of aerosol monitoring, retrievals of aerosol properties from satellite data was only possible over oceans as shown in Fig. 3(a) and (b). However, the AOT is much larger near the continents than over oceans as shown in Fig. 4. This suggested that aerosol detection over land could be quite important so retrievals over land began in the 1980s. These efforts are well documented in several studies (e.g. Kaufman and Joseph 1982; Kaufman and Fraser 1983; Fraser et al. 1984; Herman et al. 1997; Kaufman et al. 1997; Torres et al. 1998; Veefkind and Leeuw 1998; Knapp 2002; Knapp and Stowe 2002; Knapp et al. 2002; von Hoyningen et al. 2003; Hsu et al. 2004; Remer et al. 2005; Levy et al. 2007a). The development of aerosol retrieval algorithms over

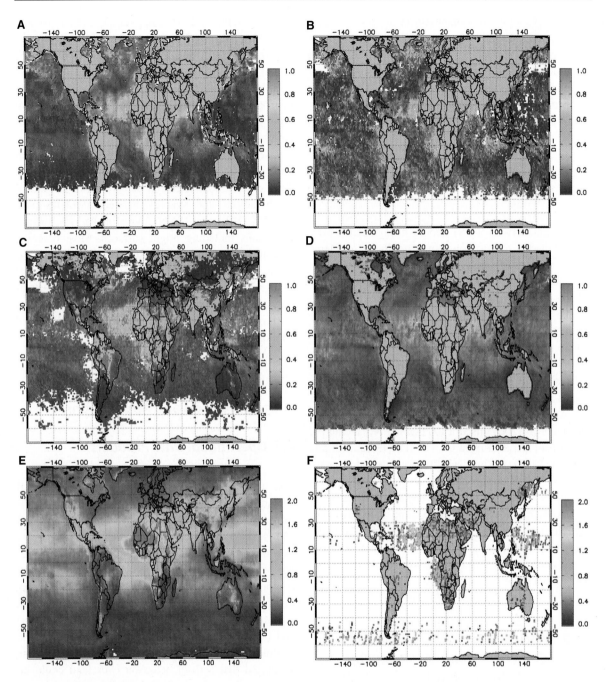

Fig. 3 Monthly-averaged aerosol products for August 1998 before Terra

land and the interest in aerosol monitoring research has accelerated since the launch of Terra in 1999. The first and the second generations of satellite aerosol monitoring can be distinguished before and after the launch of Terra.

In Figs. 3 and 4, large AOTs are generally shown to the west of the middle and northern parts of Africa and the Asian continent, and are mainly due to desert dust, biomass burning, and man-made pollution. The AOTs over the eastern and western coasts of North America are also mainly due to man-made pollution. Biomass burning in South America and southern Africa are the main sources of carbonaceous aerosols. The general spatial distribution pattern looks similar in all retrievals but significant differences among the sensors still exist due to various reasons: sensor calibration, aerosol model assumptions, cloud screening, treatment of surface reflectance, among others.

4.2 Inter-Comparisons

Algorithms for aerosol retrievals are sensor-specific because of the different characteristics of the satellite instruments. Moreover, different types and versions of the aerosol retrieval algorithms have been developed for the same instrument. When applied to the same data

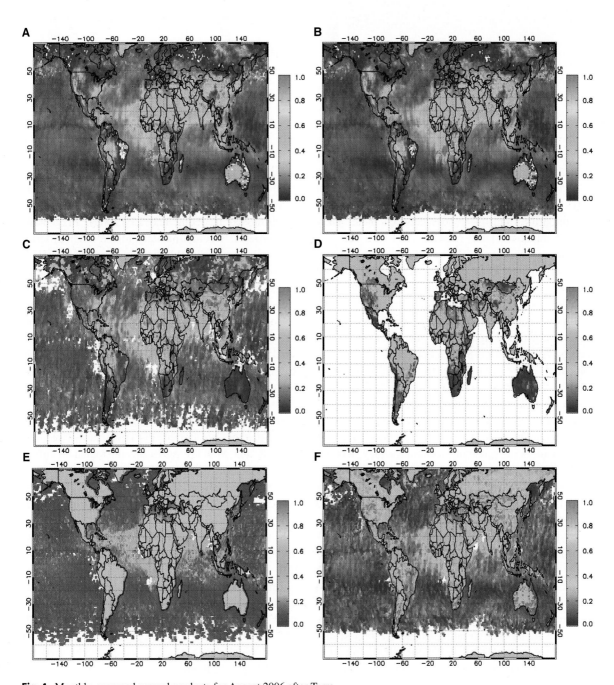

Fig. 4 Monthly-averaged aerosol products for August 2006 after Terra

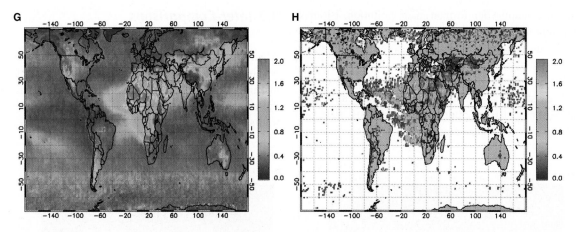

Fig. 4 (continued)

set, they yield rather different results. This can be confusing to users of the products and policy makers, with damaging consequences.

Abdou et al. (2005) showed comparisons between AOTs retrieved by the MODIS and by MISR to explore the similarities and differences between them. They showed that over land, MODIS AOTs at 470 and 660 nm are larger than MISR AOTs by about 35 and 10%, on average. Over oceans, MISR AODs at 470 and 660 nm are generally higher than MODIS AOTs by about 0.1 and 0.05, respectively. In Myhre et al. (2004), an 8-month period of AOTs derived from five different retrieval algorithms applied to 4 satellite instruments, such as AVHRR, OCTS, POLDER, and TOMS, are compared. There is at least a factor of 2 differences between the AOTs from these retrievals. In Fig. 5, the largest uncertainties are found in the Southern Hemisphere and the smallest differences are mostly located near the continents in the Northern Hemisphere. Differences in cloud screening techniques may account for the large discrepancy.

The monthly mean AOTs over oceans from a total of 9 aerosol retrievals during a period of 40 months (September 1997–December 2000) was made by Myhre et al. (2005). In most ocean regions, significant differences in AOT were identified. Figure 6 shows the zonal mean AOT for the entire 40-month period (a), the 8-month period from January to August 1998 (b), and the 10-month period from March to December 2000 (c). In this figure, the largest differences are found at high latitudes and the largest differences between MODIS and MISR are also found at high latitudes. For the entire period, the differences are largest in the southern hemisphere; during the period of March–December 2000, large differences are also found in the northern hemisphere.

Inter-comparisons between the long-term (1983–2000) aerosol products from AVHRR and TOMS are discussed in Jeong and Li (2005). In general, each product is complimentary to the other in terms of global aerosol distribution. However, the AVHRR cannot represent continental sources except for large aerosol plumes. The TOMS AI and AOT have differences in where higher aerosol loading over land are located. These differences may arise from inherent problems found in aerosol retrievals, such as cloud, ocean color contamination, and induced oceanic aerosols. Figure 7 shows a common problem originated from cloud screening in AVHRR aerosol retrieval. The choice of aerosol size distribution is another problematic issue. Jeong et al. (2005) demonstrated that the considerable discrepancies between the AVHRR and MODIS AOTs are attributed to differences in aerosol size distribution functions used in the retrievals, namely, the power law (AVHRR) and bimodal log normal (MODIS) size distribution functions (Fig. 8).

Kokhanovsky et al. (2007) provided the first brief discussion concerning an inter-comparison study of AOT at 0.55 μm retrieved using six different satellite instruments and ten different algorithms for a single scene over central Europe on October 13, 2005. The spatially averaged AOT was equal to 0.14 for MISR, NASA MODIS and POLDER products and is smaller by 0.01 for the ESA MERIS and larger by 0.04 for the MERIS BAER product. The AOT from AATSR

Fig. 5 Averaged statistics for the five aerosol retrievals from November 1996 to June 1997: (**a**) mean and (**b**) standard (after Myhre et al. 2004)

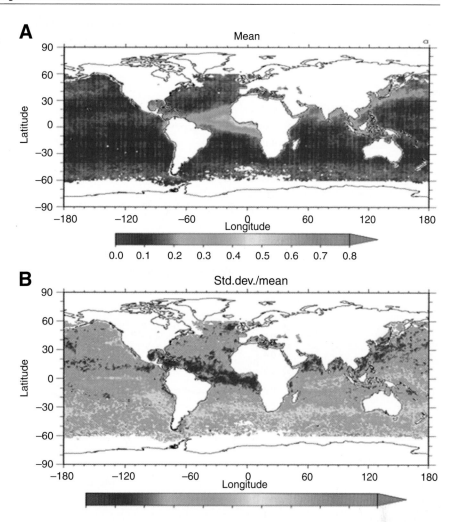

gives on average larger values than those from all other instruments, while SCIAMACHY retrievals underestimate aerosol loading. In Fig. 9, validation against AERONET shows that MERIS provides the most accurate AOT retrievals for this scene.

4.3 Aerosol Climatology

The longest climatology of AOT is generated from AVHRR using the two-channel retrieval algorithm for the period extending from August 1981 to June 2005 (Mishchenko et al. 2007a, b). A slightly decreasing trend in AOT is seen over this time period (Fig. 10). Analyses of regional trends reveal decreases over Europe, part of the Atlantic Ocean, and increases along a portion of the western coast of Africa, along the southern and south-east coasts of Asia, and over the

45–60°S latitudinal belt (Mishchenko et al. 2007b). An unsurprising result is that the northern hemispheric mean AOT systematically exceeds that averaged over the southern hemisphere. The effects of two major volcanic eruptions (El Chichon and Mt. Pinatubo) are clearly visible, consistent with the SAGE stratospheric AOT. Bauman et al. (2003a, b) used data from SAGE II and CLAES during December 1984–August 1999 to develop stratospheric aerosol climatology. They found that it took 5 years after the Mt. Pinatubo eruption for the stratospheric aerosol loading to return to its pre-eruption level.

4.4 Synergy

A number of satellite-derived aerosol products are available, and the synergy among these satellite

Fig. 6 Zonal mean AOT as a function of latitude for the entire period of investigation, as well as for two selected periods (after Myhre et al. 2004)

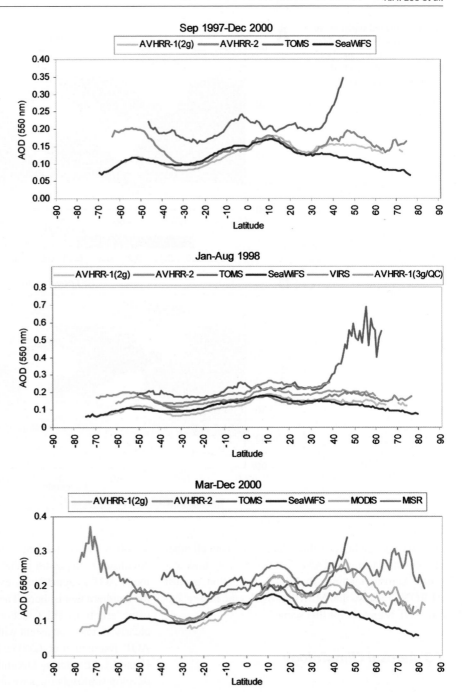

products may lead to a new and improved data set. Global aerosol products from satellites such as AVHRR, TOMS, and MODIS are particularly useful in this regard. The spatial distributions of the aerosol products from these instruments are complimentary in revealing different aspects of aerosol characteristics. For example, aerosol SSAs can be derived from the

combination of AOT and AI (Hu et al. 2007) and a combination of satellite and ground-based measurements (Lee et al. 2007b). Aerosol type classification by absorption and Ångström exponent was first suggested by Higurashi and Nakajima (2002). Improved aerosol type classification methods by merging essential information about aerosol mass loading from AOT,

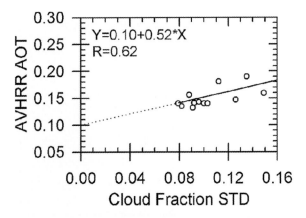

Fig. 7 AOT as a function of standard deviation (STD) of the ISCCP cloud fraction over the coastal areas of Peru and Chile (10–30° S, 70–90° W) where a region of enhanced AVHRR AOT is shown (after Jeong and Li 2005). The cloud fraction STD was found to be positively correlated with the AVHRR AOT with a correlation coefficient equal to 0.62

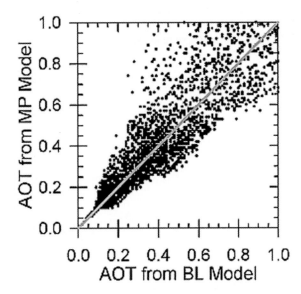

Fig. 8 Scatterplot of AOT from the modified power law models versus that from bimodal lognormal models (after Jeong et al. 2005). Note that they exhibit very large discrepancies by up to a factor of two. This suggests that the selection of a particular aerosol model is an important factor influencing the retrieval of the AOT

size from Ångström or fine mode fraction, and absorption by AI were developed (Barnaba and Gobbi 2004; Jeong and Li 2005; Lee et al. 2006c; Kim et al. 2007). Finally, the synergy between different products from the same instrument can overcome some of the limitations of surface monitoring networks and enhance daily air quality forecasts associated with particle pollution as shown in Fig. 11 (Al-Saadi et al. 2005).

5 Future Instruments

5.1 APS/GLORY

The Aerosol Polarimetry Sensor (APS) onboard the Glory satellite (http://glory.gsfc.nasa.gov/index.html) is scheduled for launch into a low Earth orbit (LEO) in December 2008 (Mishchenko et al. 2007c). The APS is designed to measure the properties of aerosols for the long-term effects on the Earth climate record and it will enable a greater understanding of the seasonal variability of aerosol properties. The APS has the ability to collect multi-angle, multi-spectral photopolarimetric measurements of the atmosphere and the underlying surface along the satellite ground track. APS observations will provide accurate retrievals of aerosol microphysical parameters and are expected to improve global aerosol assessments.

5.2 NPP

The National Polar-orbiting Operational Environmental Satellite System Preparatory Project (NPP) is a joint mission with the NPOESS Integrated Program Office (IPO) (http://jointmission.gsfc.nasa.gov/). The NPP will provide continuity in global Earth Science observations of the atmosphere, land, and oceans after the EOS Terra and Aqua missions are over. NPP will take measurements of atmospheric and SST, humidity soundings, land and ocean biological productivity, and cloud and aerosol properties using three different sensors: the Visible Infrared Imaging Spectroradiometer Suite (VIIRS), the Crosstrack Infrared Sounder (CrIS), and the Advanced Technology Microwave Sounder (ATMS). The launch is planned for September 2009, with a mission duration of 5 years.

5.3 EarthCARE

The Earth Clouds, Aerosols, and Radiation Explorer (EarthCARE), due for launch in 2012 (ESA 2004), is a joint European-Japanese mission addressing the need for a better understanding of the interactions between cloud, radiative and aerosol processes that play a role in climate regulation. Four distinct instruments are planned for deployment on EarthCARE:

Fig. 9 Comparison of satellite and ground measurements of AOT at 0.55 μm (after Kokhanovsky et al. 2007)

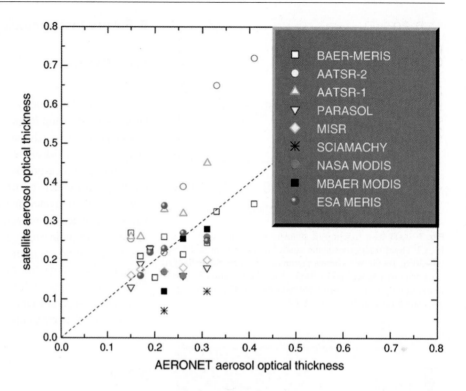

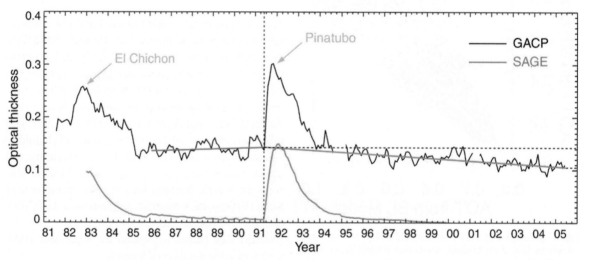

Fig. 10 Time series of the globally-averaged column AOT over the oceans and the SAGE record of globally-averaged stratospheric AOT (after Mishchenko et al. 2007a)

the Backscatter Lidar (ATLID), the Cloud Profiling Radar (CPR), the Multi-Spectral Imager (MSI), and the Broadband Radiometer (BBR). The goal of this mission is to improve the representation and understanding of the Earth's radiative balance in climate and numerical weather forecast models by acquiring vertical profiles of clouds and aerosols, as well as the radiances at the top of the atmosphere.

5.4 MSI/Sentinel-2

Sentinel-2 is the medium spatial resolution optical mission of the Global Monitoring for Environment and Security (GMES) program (http://www.esa.int/esapub/bulletin/bulletin131/bul131b_martimort.pdf). The multi-spectral instrument (MSI) will gen-

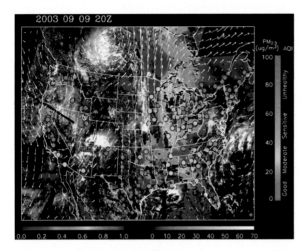

Fig. 11 Example of an air quality forecast (after Al-Saddi et al. 2005). The colored and *white–black* contour areas represent MODIS AOT and cloud. Continuous PM2.5 monitoring stations are shown as a circle color-coded by the hourly concentration [scale on the right, with associated AQI (EPA 1999)]. Fire locations are shown as diamonds and color-coded according to the fire probability: *bright pink* for higher probability fires and *violet* for lower probability fires. The 850 mb wind direction and speed from the NCEP Eta Model are shown as *white* vectors

erate optical images in 13 spectral channels in the visible and short-wave infrared range (443–2190 nm) down to 10, 20, and 60 m spatial resolution with an image width of 290 km. The MSI features a three-mirror anastigmat (TMA) telescope with a pupil diameter of about 150 mm; it is the key to the high image quality across the wide field of view (290 km). The telescope structure and mirrors are made of silicon carbide to minimize thermal deformation. The launch of the first Sentinel-2 satellite is planned for 2012.

5.5 Sentinel-3

Another GMES satellite, the Sentinel-3, will continue EnviSAT's mission and includes enhancements to meet the operational revisit requirements and to facilitate new products and evolution of services (http://www.esa.int/esapub/bulletin/bulletin131/bul131c_aguirre.pdf). The two optical instruments, Ocean and Land Colour Instrument (OLCI) based on MERIS/EnviSAT and The Sea and Land Surface Temperature Radiometer (SLSTR) based on AATSR/EnviSAT, will provide a common quasi-simultaneous view of the Earth to help develop syn-

ergetic products. The first launch is expected in 2011/2012.

5.6 TRAQ

The Tropospheric composition and Air Quality (TRAQ) is a mission focused on understanding the tropospheric system for air quality, sources and sinks, and climate change (Levelt et al. 2006). New synergistic sensors include a UV/VIS/NIR/SWIR instrument (TROPOMI), which is a follow-on instrument to OMI, an FTIR instrument (SIFTI), a cloud detector (CLIM) and an instrument resembling POLDER (OCAPI), which will extend the spectral range up to 2.2 μm for pollution aerosol detection. The TRAQ mission will be the first mission fully dedicated to air quality and the science issues concerning tropospheric composition and global change.

6 Summary

Since its beginning three decades ago, satellite remote sensing of atmospheric aerosols offers more complete spatial coverage and includes vertical profile and spectral/optical information when compared to the point-measurement data typically used to evaluate local-scale aerosols. Operational satellite aerosol products such as AOT, Ångström exponent, mode fraction, single scattering albedo, and the vertical profile are currently available from space agencies such as NOAA, NASA, ESA, CNES, KNMI, JAXA, etc. Several aerosol retrieval algorithms for specific sensor characteristics are reviewed. The independent satellite measurements can be use to study the spatial distribution, transport, origin, and climatology of aerosols. A combination of satellite measurements can be used to examine their compatibility and synergy for producing improved products. Satellite data can also be used for air quality forecasting. These efforts will be base of future satellite missions.

Acknowledgments The authors are grateful to the many agencies that have providing the satellite data used here. They include PATMOS data from NESDIS/NOAA, POLDER/PARASOL data from the ICARE thematic center, AVHRR data from GISS/NASA, MODIS data from NASA's LAADS Web, MISR from ASDC/NASA, GOME and SCIAMACHY data from KNMI's TEMIS, SeaWiFS data from GSFC/NASA's ocean color web and TOMS data from GSFC/NASA.

Abbreviations

AATSR	Advance Along Track Scanning Radiometer
ACE	Aerosol Characterization Experiments
AERONET	Aerosol Robotic Network
AI	Aerosol Index
AOD	Aerosol Optical Depth
AOT	Aerosol Optical Thickness
APS	Aerosol Polarimetry Sensor
AQI	Air Quality Index
ATBD	Algorithm Theoretical Basis Document
ATMS	Advanced Technology Microwave Sounder
AVHRR	Advanced Very High Resolution Radiometer
BAER	Bremen Aerosol Retrieval
BBR	Broadband Radiometer
BRF	Bidirectional Reflectance Factor
CALIPSO	Cloud-Aerosol Lidar and Infrared Pathfinder Satellite Observations
CHRIS	Compact High Resolution Imaging Spectrometer
CNES	le Centre National d'Etudes Spatiales
CPR	Cloud Profiling Radar
CrIS	Crosstrack Infrared Sounder
DDV	Dense Dark Vegetation
DLR	German Aerospace Centre
EarthCARE	Earth Clouds, Aerosols, and Radiation Explorer
ENVISAT	Environment Satellite (http://envisat.esa.int)
EOS	Earth Observation System
ERS	European Remote Sensing satellite
ERTS	Earth Resoureces Technology Satellite
ESA	European Space Agency (http://www.esa.it/export/esaCP/index.html)
FMF	Fine Mode Fraction
GACP	Global Aerosol Climatology Product
GLAS	Geoscience Laser Altimeter System
GMES	Global Monitoring for Environment and Security
GMS	Geostationary Meteorology Satellite
IGAC	International Global Atmospheric Chemistry Observation
ILAS	Improved Limb Atmospheric Spectrometer
ISCCP	International Satellite Cloud Climatology Project
JAXA	Japan Aerospace Exploration Agency
JPL	Jet Propulsion Laboratory
KNMI	Royal Netherlands Meteorological Institute
L1, L2	Level 1, Level 2
LEO	Low Earth Orbit
LER	Lambert Equivalent Reflectivity
LIDAR	Light Detection and Ranging
LITE	Lidar In-space Technology Experiment
LUT	Look Up Table
MERIS	Medium Resolution Imaging Spectrometer Instrument
MISR	Multiangle Imaging SpectroRadiometer
MODIS	Moderate Resolution Imaging Spectroradiometer
MSI	Multi-Spectral Imager
NASA	National Aeronautics and Space Administration
NIR	Near InfraRed
NN	Neural Network
NOAA	the National Ocean and Atmosphere Administration
NPP	National Polar-orbiting Operational Environmental Satellite System Preparatory Project
NRL	Naval Research Laboratory
PATMOS	Pathfinder Atmosphere
POAM	Polar Ozone and Aerosol Measurement
POLDER	POLarization and Directionality of the Earth's Reflectances
PROBA	Project for On-Board Autonomy
ROD	Rayleigh optical depth
SAM	Stratospheric Aerosol Measurement
SCIAMACHY	SCanning Imaging Absorption 5 spectroMeter for Atmospheric CHartographY

SeaWiFS	Sea-viewing Wide Field-of-view Sensor
SLSTR	Sea and Land Surface Temperature Radiometer
SPOT	Satellite Pour l'Observation de la Terre
SSA	Single Scattering Albedo
SSD	Space Shuttle Discovery
SST	Sea Surface Temperature
SWIR	Short Wave Infra Red
SYNAER	Synergetic Aerosol Retrieval
SYNTAM	Synergy a combination of MODIS/Terra and MODIS/Aqua
TARFOX	the Tropospheric Aerosol Radiation Forcing Experiment
TIR	Thermal InfraRed
TMA	three-mirror anastigmat
TOA	Top of Atmosphere
TOMS	Total Ozone Mapping Spectrometer
TRAQ	Tropospheric composition and Air Quality
UV	Ultra Violet
VI	Vegetation Index
VIIRS	Visual/Infrared Imager Radiometer Suite
VIRS	Visualisation and analysis tool
VNIR	Visible Near Infrared

References

Abdou WA, DJ Diner, JV Martonchik, CJ Bruegge, RA Kahn, BJ Gaitley, KA Crean, LA Remer, and B Holben (2005) Comparison of coincident MISR and MODIS aerosol optical depths over land and ocean scenes containing AERONET sites. *J. Geophys. Res.* doi:10.1029/2004JD004693.

Al-Saadi J, J Szykman, B Pierce, C Kittaka, D Neil, DA Chu, L Remer, L Gumley, E Prins, L Weinstock, C McDonald, R Wayland, and F Dimmick (2005) Improving national air quality forecasts with satellite aerosol observations. *Bull. Am. Met. Soc.* 86(9): 1249–1261.

Barnaba F and GP Gobbi (2004) Aerosol seasonal variability over the Mediterranean region and relative impact of maritime, continental and Saharan dust particles over the basin from MODIS data in the year 2001. *Atmos. Chem. Phys.* 4: 2367–2391.

Barnsley M, J Settle, M Cutter, D Lobb, and F Teston (2004) The PROBA/CHRIS mission: A low-cost smallsat for hyperspectral, multi-angle, observations of the earth surface and atmosphere. *IEEE Trans. Geosci. Remote Sens.* 42: 1512–1520.

Bates TS, BJ Huebert, JL Gras, FB Griffiths, and PA Durkee (1998) International Global Atmospheric Chemistry (IGAC) Project's First Aerosol Characterization Experiment (ACE 1): Overview. *J. Geophys. Res. 103*(D13): 16297–16318.

Bauman JJ, PB Russell, MA Geller, and P Hamill (2003a) A stratospheric aerosol climatology from SAGE II and CLAES measurements: 1. Methodology. *J. Geophys. Res.* 108(D13), doi:10.1029/2002JD002992.

Bauman JJ, PB Russell, MA Geller, and P Hamill (2003b) A stratospheric aerosol climatology from SAGE II and CLAES measurements: 2. Results and comparisons, 1984–1999. *J. Geophys. Res.* 108(D13): 4383, doi:10.1029/2002JD002993.

Berthier S, P Chazette, P Couvert, J Pelon, F Dulac, F Thieuleux, C Moulin, and T Pain (2006) Desert dust aerosol columnar properties over ocean and continental Africa from Lidar in-Space Technology Experiment (LITE) and Meteosat synergy. *J. Geophys. Res.* 111(D21202), doi:10.1029/2005JD006999.

Bourassa AE, DA Degenstein, RL Gattinger, and EJ Llewellyn (2007) Stratospheric aerosol retrieval with optical spectrograph and infrared imaging system limb scatter measurements. *J. Geophys. Res.* 112(D10217), doi:10.1029/2006JD008079.

Breon FM, D Tanre, and S Generoso (2002) Aerosol effect on cloud droplet size monitored from satellite. *Science.* 295: 834–838.

Christopher SA and J Wang (2004) Intercomparison between multi-angle imaging spectroradiometer (MISR) and sunphotometer aerosol optical thickness in dust source regions over China: Implications for satellite aerosol retrievals and radiative forcing calculations. *Tellus.* B56(5): 451–456, doi:10.1111/j.1600-0889.2004.00120.x.

Chu DA, YJ Kaufman, C Ichoku, LA Remer, D. Tanré, and BN Holben (2002) Validation of MODIS aerosol optical depth retrieval overland. *Geophys. Res. Lett.* 29, doi:10.1029/2001GL013205.

Chu WP and MP McCormick (1979) Inversion of stratospheric aerosol and gaseous constituents from spacecraft solar extinction data in the 0.38–1.0-Mum wavelength region. *Appl. Opt.* 18: 1404–1413.

Chu WP, MP McCormick, J Lenoble, C Brognoiz, and P Pruvost (1989) SAGE II inversion algorithm. *J. Geophys. Res.* 94: 8339–8351.

Costa MJ, BJ Sohn, V Levizzani, AM Silva (2006) Radiative forcing of Asian dust determined from the synergized GOME and GMS satellite data – A case study. *J. Meteorol. Soc. Jpn.* 84: 85–95.

Dave J (1973) Development of the programs for computing characteristics of ultraviolet radiation: Scalar case. NAS5-21680, NASA Goddard Space Flight Cent., Greenbelt, Md.

Davidson CI, RF Phalen, PA Solomon (2005) Airborne particulate matter and human health: A review. *Aerosol Sci. Technol.* 39(8): 737–749.

Deuzé JL, M Herman, P Goloub, D Tanré, and A Marchand (1999) Characterization of aerosols over ocean from POLDER/ADEOS-1. *Geophys. Res. Lett.* 26(10): 1421–1424.

Diner DJ, WA Abdou, TP Ackerman, K Crean, HR Gordon, RA Kahn, JV Martonchik, S McMuldroch, SR Paradise, B Pinty, MM Verstraete, M Wang, and RA West (2008) MISR level 2 aerosol retrieval algorithm theoretical basis. Jet Propul. Lab. Pasadena, CA, JPL-D 11400, Rev. G.

Diner DJ, WA Abdou, CJ Bruegge, JE Conel, KA Crean, BJ Gaitley, MC Helmlinger, RA Kahn, JV Martonchik, SH Pilorz, and BN Holben (2001) MISR aerosol optical depth retrievals over southern Africa during the SAFARI-2000 dry season campaign. *Geophys. Res. Lett.* 28: 3127–3130.

Diner DJ, JC Beckert, TH Reilly, CJ Bruegge, JE Conel, RA Kahn, JV Martonchik, TP Ackerman, R Davies, SAW Gerstl, HR Gordon, JP Muller, RB Myneni, PJ Sellers, B Pinty, and MM Vertraete (1998) Multi-angle Imaging Spectro Radiometer (MISR) instrument description and experiment overview. *IEEE Trans. Geosci. Remote Sens.* 36: 1072–1087.

Dulac F, S Tanré, G Bergametti, P Buat-Ménard, M Desbois, and D Sutton (1992) Assessment of the African airborne dust mass over the western Mediterranean Sea using Meteosat data. *J. Geophys. Res.* 97: 2489–2506.

Durkee PA, DR Jensen, EE Hindman, and TH Vonder Haar (1986) The relationship between marine aerosols and satellite detected radiance. *J. Geophys. Res.* 91: 4063–4072.

EPA (1999) Guideline for reporting of daily air quality – air quality index (AQI). EPA-454/R-99-010, 25 pp. [Available online at http://www.epa.gov/ttn/ oarpg/t1/memoranda/rg701.pdf.]

European Space Agency (ESA) (2004) Reports for mission selection, The six candidate earth explorer missions: EarthCARE. ESA SP-1279(1).

Fraser RS (1976) Satellite measurement of mass of Sahara dust in the atmosphere. *Appl. Opt.* 15: 2471–2479.

Fraser RS, YJ Kaufman, and RL Mahoney (1984) Satellite measurements of aerosol mass and transport. *Atmos. Environ.* 18: 2577–2584.

Froidevaux L and A Douglass (2001) Earth Observing System (EOS) aura science data validation plan (available from http://eos-aura.gsfc.nasa.gov/mission/validation.html).

Geogdzhayev IV, MI Mishchenko, WB Rossow, B Cairns, and AA Lacis (2002) Global two-channel AVHRR retrievals of aerosol properties over the ocean for the period of NOAA-9 observations and preliminary retrievals using NOAA-7 and NOAA-11 data. *J. Atmos Sci.* 59: 262–278.

Glaccum W, R Lucke, RM Bevilacqua, EP Shettle, JS Hornstein, DT Chen, JD Lumpe, SS Krigman, DJ Debrestian, MD Fromm, F Dalaudier, E Chassefiere, C Deniel, CE Randall, DW Rusch, JJ Olivero, C Brogniez, J Lenoble, and R Kremer (1996) The polar ozone and aerosol measurement instrument. *J. Geophys. Res.* 101: 14479–14487.

Goloub P, D Tanré, JL Deuzé, M Herman, A Marchand, and FM Bréon (1999) Validation of the first algorithm for deriving the aerosol properties over the ocean using the POLDER/ADEOS measurements. *IEEE Trans. Geosci. Remote Sens.* 37(3): 1586–1596.

Gordon HR and AY Morel (1983) Remote Assessment of Ocean Color for Interpretation of Satellite Visible Imagery: A Review. Springer-Verlag, New York.

Gordon HR and M Wang (1994) Retrieval of water-leaving radiance and aerosol optical thickness over the oceans with SeaWiFS: A preliminary algorithm. *Appl. Opt.* 33: 443–452.

Graaf M de and P Stammes (2005) SCIAMACHY absorbing aerosol index – calibration issues and global results from 2002–2004. *Atmos. Chem. Phys.* 5: 2385–2394, SRef-ID:1680-7324/acp/2005-5-2385.

Grant WB, EV Browell, CF Butler, and GD Nowicki (1997) LITE measurements of biomass burning aerosols and comparisons with correlative airborne lidar measurements of

multiple scattering in the planetary boundary layer. In: A Ansmann, R Neuber, P Rairoux, and U Wandinger (eds) *Advances in Atmospheric Remote Sensing with Lidar.* Springer-Verlag, Berlin, 153–156.

Grey WMF, PRJ North, and SO Los (2006) Computationally efficient method for retrieving aerosol optical depth from ATSR-2 and AATSR data. *Appl. Opt.* 45: 2786–2795.

Griggs M (1975) Measurements of atmospheric aerosol optical thickness over water using ERTS-1 data. *J. Air Pollut. Control. Assoc.* 25: 622–626.

Gu YY, CS Gardner, PA Castelberg, GC Papen, and MC Kelley (1997) Validation of the Lidar In-space Technology Experiment: Stratospheric temperature and aerosol measurements. *Appl. Opt.* 36: 5148–5157.

Hale GM, and MR Querry (1973) Optical constants of water in the 200 nm to 200 μm wavelength region. *Appl. Opt.* 12: 555–563.

Herman JR, PK Bhartia, O Torres, C Hsu, C Seftor, and E Celarier (1997) Global distribution of UV-absorbing aerosols from Nimbus 7/TOMS data. *J. Geophys. Res.* 102: 16911–16922.

Herman M, JL Deuzé, C Devaux, Ph Goloub, FM Bréon, and D Tanré (1997) Remote sensing of aerosols over land surfaces, including polarization measurements; application to some airborne POLDER measurements. *J. Geophys. Res.* 102: 17039–17049.

Hervig ME, T Deshler, and JM Russell III (1998) Aerosol size distributions obtained from HALOE spectral extinction measurements. *J. Geophys. Res.* 103: 1573–1583.

Higurashi A and T Nakajima (1999) Development of a two channel aerosol retrieval algorithm on global scale using NOAA AVHRR. *J. Atmos. Sci.* 56: 924–941.

Higurashi A and T Nakajima (2002) Detection of aerosol types over the East China Sea near Japan from four-channel satellite data. *Geophys. Res. Lett.* 29(17): 1836, doi:10.1029/2002GL015357.

Hoff RM and KB Strawbridge (1997) LITE observations of anthropogenically produced aerosols. In: A Ansmann, R Neuber, P Rairoux, and U Wandinger (eds) *Advances in Atmospheric Remote Sensing with Lidar,* Springer-Verlag, Berlin, 145–148.

Holzer-Popp T, M Schroedter-Homscheidt, H Breitkreuz, L Klüser, D Martynenko (2008) Synergistic aerosol retrieval from SCIAMACHY and AATSR onboard ENVISAT. *Atmos. Chem. Phys. Discuss.* 8: 1–49.

Holzer-Popp T, M Schroedter, and G Gesell (2002a) Retrieving aerosol optical depth and type in the boundary layer over land and ocean from simultaneous GOME spectrometer and ATSR-2 radiometer measurements, 1, Method description. *J. Geophys. Res.* 107(D21): 4578, doi:10.1029/2001JD002013.

Holzer-Popp T, M Schroedter, and G Gesell (2002b) Retrieving aerosol optical depth and type in the boundary layer over land and ocean from simultaneous GOME spectrometer and ATSR-2 radiometer measurements, 2, Case study application and validation. *J. Geophys. Res.* 107(D24): 4770, doi:10.1029/2002JD002777.

Hsu NC, JR Herman, PK Bhartia, CJ Seftor, O Torres, AM Thompson, JF Gleason, TF Eck, and BN Holben (1996) Detection of Biomass Burning Smoke from TOMS Measurements. *Geophys. Res. Lett.* 23(7): 745–748.

Hsu NC, JR Herman, O Torres, BN Holben, D Tanré, TF Eck, A Smirnov, B Chatenet, and F Lavenu (1999) Comparison of the TOMS aerosol Index with Sun-photometer aerosol optical thickness: Results and application. *J. Geophy. Res.* 23: 745–748.

Hsu NC, SC Tsay, MD King, and JR Herman (2004) Aerosol properties over bright-reflecting source regions. *IEEE Trans. Geosci. Remote Sens.* 42(3): 557–569.

Hsu NC, SC Tsay, MD King, and JR Herman (2006) Deep blue retrievals of Asian aerosol properties during ACE-Asia. *IEEE Trans. Geosci. Remote Sens.* 44: 3180–3195.

Hu RM, RV Martin, and TD Fairlie (2007) Global retrieval of columnar aerosol single scattering albedo from space-based observations. *J. Geophys. Res.* 112(D02204), doi:10.1029/2005JD006832.

Huebert BJ, T Bates, PB Russell, G Shi, YJ Kim, K Kawamura, G Carmichael, and T Nakajima (2003) An overview of ACE-Asia: Strategies for quantifying the relationships between Asian aerosols and their climatic impacts. *J. Geophys. Res.* 108(D23): 8633, doi:10.1029/2003JD003550.

Ignatov A and L Stowe (2000) Physical basis, premises, and self-consistency checks of aerosol retrievals from TRMM VIRS. *J. Appl. Meteor.* 39(12): 2259–2277.

Ignatov A, L Stowe, SM Sakerin, and GK Korotaev (1995a) Validation of the NOAA/NESDIS satellite aerosol product over the North Atlantic in 1989. *J. Geophys. Res.* 100(D3): 5123–5132.

Ignatov A, L Stowe, R Singh, D Kabanov, and I Dergileva (1995b) Validation of NOAA AVHRR aerosol retrievals using sun-photometer measurements from RV Akademik Vernadsky in 1991. *Adv. Space Res.* 16(10): 95–98.

International Global Atmospheric Chemistry Project (IGAC) (1996) Atmospheric aerosols: A new focus of the International Global Atmospheric Chemistry Project. In: PV Hobbs, and BJ Huebert (eds), *IGAC Core Project Office*, Mass. Inst. Technol., Cambridge.

Intergovernmental Panel on Climate Change (IPCC) (2007) Climate change 2007: The physical science basis. Contribution of working group I to the fourth assessment report of the intergovernmental panel on climate change. In: S Solomon, D Qin, M Manning, Z Chen, M Marquis, KB Averyt, M Tignor, and HL Miller (eds), Cambridge University Press, Cambridge, UK and New York, USA, 996.

Jacobowitz H, L Stowe, G Ohring, A Heidinger, K Knapp, and NR Nalli (2003) The Advanced Very High Resolution Radiometer Pathfinder Atmosphere (PATMOS) climate 25 dataset: A resource for climate research. *Bull. Am. Meteor. Soc.* 84: 785–793.

Jeong MJ, and Z Li (2005) Quality, compatibility, and synergy analyses of global aerosol products derived from the advanced very high resolution radiometer and Total Ozone Mapping Spectrometer. *J. Geophys. Res.* 110, D10S08, doi:10.1029/2004JD004647.

Jeong MJ, Z Li, DA Chu, and SC Tsay (2005). Quality and compatibility analyses of global aerosol products derived from the advanced very high resolution radiometer and Moderate Resolution Imaging Spectroradiometer, *J. Geophys. Res.* 110, D10S09, doi:10.1029/2004JD004648.

Josset D, J Pelon, A Protat, and C Flamant (2008) New approach to determine aerosol optical depth from combined CALIPSO and CloudSat ocean surface echoes. *Geophys. Res. Lett.* 35, L10805, doi:10.1029/2008GL033442.

Kahn RA, BJ Gaitley, JV Martonchik, DJ Diner, KA Crean, and B Holben (2005) Multiangle Imaging Spectrora-

diometer (MISR) global aerosol optical depth validation based on 2 years of coincident Aerosol Robotic Network (AERONET) observations. *J. Geophys. Res.* 110, D10S04, doi:10.1029/2004JD004706.

Kaufman YJ and RS Fraser (1983) Light extinction by aerosols during summer air pollution, *J. Appl. Meteorol.* 22: 1694–1706.

Kaufman YJ and JH Joseph (1982) Determination of Surface Albedos and Aerosol Extinction Characteristics from Satellite Imagery. *J. Geophys. Res.* 87(C2): 1287–1299.

Kaufman YJ, D Tanré, and O Boucher (2002) A satellite view of aerosols in the climate system. *Nature.* 419: 215–223.

Kaufman YJ, D Tanré, LA Remer, EF Vermote, A Chu, and BN Holben (1997) Operational remote sensing of tropospheric aerosol over land from EOS moderate resolution imaging spectroradiometer. *J. Geophys. Res.* 102(D14): 17051–17067.

Kent GS, CR Trepte, KM Skeens, and DM Winker (1998) LITE and SAGE II measurements of aerosols in the southern hemisphere upper troposphere. *J. Geophys. Res.* 103(D15): 19111–19127.

Kim J, J Lee, HC Lee, A Higurashi, T Takemura, and CH Song (2007) Consistency of the aerosol type classification from satellite remote sensing during the Atmospheric Brown Cloud–East Asia Regional Experiment campaign. *J. Geophys. Res.* 112, D22S33, doi:10.1029/2006JD008201.

King MD, YJ Kaufman, D Tanré, and T Nakajima (1999) Remote sensing of tropospheric aerosols from space: Past, present, and future. *Bull. Am. Meteorol. Soc.* 80: 2229–2259.

Knapp KR, R Froulin, S Kondragunta, and AI Prados (2005) Towards aerosol optical depth retrievals over land from GOES visible radiances: Determining surface reflectance. *Int. J. Remote Sens.* 26(18): 4097–4116.

Knapp KR and LL Stowe (2002) Evaluating the potential for retrieving aerosol optical depth over land from AVHRR pathfinder atmosphere data. *J. Atmos. Sci.* 59(3): 279–293.

Knapp KR, TH Vonder Harr, and YJ Kaufman (2002) Aerosol optical depth retrieval from GOES-8: Uncertainty study and retrieval validation over South America. *J. Geophys. Res.* 107(D7): 4055, doi:10.1029/2001JD000505.

Knapp, KR (2002) Quantification of aerosol signal in GOES 8 visible imagery over the United States. *J. Geophys. Res.* 107(D20): 4426, doi:10.1029/2001JD002001.

Kokhanovsky AA, FM Breon, A Cacciari, E Carboni, D Diner, WD Nicolantonio, RG Grainger, WMF Grey, R Höller, KH Lee, Z Li, PRJ North, A Sayer, G Thomas, W von Hoyningen-Huene (2007) Aerosol remote sensing over land: Satellite retrievals using different algorithms and instruments. *Atmos. Res.* 85: 372–394.

Kokhanovsky AA and G de Leeuw (2009) Satellite aerosol remote sensing over land, Berlin, Springer-Praxis.

Lazarev AI, VV Kovalenok, and SV Avakyan (1987) Investigation of Earth from manned spacecraft, Leningrad, Gidrometeoizdat.

Lee KH, YJ Kim, and W. von Hoyningen-Huene (2004) Estimation of regional aerosol optical thickness from satellite observations during the 2001 ACE-Asia IOP. *J. Geophys. Res.* 109 D19S16, doi:10.1029/2003JD004126

Lee KH, JE Kim, YJ Kim, J Kim, and W von Hoyningen-Huene (2005) Impact of the smoke aerosol from Russian forest fires on the atmospheric environment over

Korea during may 2003. *Atmos. Environ.* 39(2): 85–99, doi:10.1016/j.atmosenv.2004.09.032.

Lee KH, YJ Kim, and MJ Kim (2006a) Characteristics of aerosol observed during two severe haze events over Korea in June and October 2004. *Atmos. Environ.* 40: 5146–5155. doi:10.1016/j.atmosenv.2006.03.050.

Lee KH, YJ Kim, W von Hoyningen-Huene, and JP Burrows (2006b) Influence of land surface effects on MODIS aerosol retrieval using the BAER method over Korea. *Int. J. Remot. Sens.* 27(14): 2813–2830.

Lee KH, YJ Kim, W von Hoyningen-Huene, and JP Burrows (2007a) Spatio-Temporal Variability of Atmospheric Aerosol from MODIS data over Northeast Asia in 2004. *Atmos. Environ.* 41(19): 3959–3973. doi:10.1016/j.atmosenv. 2007.01.048.

Lee DH, KH Lee, and YJ Kim (2006c) Application of MODIS aerosol data for aerosol type classification. *Korean J. Remot Sens.* 22(6): 495–505 (In Korean).

Lee KH, Z Li, MS Wong, J Xin, Y Wang, WM Hao, and F Zhao (2007b) Aerosol single scattering albedo estimated across China from a combination of ground and satellite measurements, *J. Geophys. Res.* 112(D22S15), doi:10.1029/ 2007JD009077.

Leroy M, JL Deuzé, FM Bréon, O Hautecoeur, M Herman, JC Buriez, D Tanré, S Bouffies, P Chazette, and JL Roujean (1997) Retrieval of atmospheric properties and surface bidirectional reflectances over land from POLDER/ADEOS. *J. Geophys. Res.* 102: 17023–17038.

Levelt P, C Camy-Perret, H Eskes, M van Weele, D Hauglustaine, I Aben, D Tanré, L Lavanant, C Clerbaux, M de Maziere, R Jogma, M Dobber, P Veefkind, J Leon, and P Coheur (2006) A mission for TRoposheric composition and Air Quality (TRAQ), American Geophysical Union, Fall Meeting 2006, abstract #A31B-0888, 2006AGUFM.A31B0888L.

Levy RC, LA Remer, and O Dubovik (2007b) Global aerosol optical properties and application to Moderate Resolution Imaging Spectroradiometer aerosol retrieval over land, *J. Geophys. Res.* 112(D13210), doi:10.1029/2006JD007815.

Levy RC, LA Remer, JV Martins, YJ Kaufman, A Plana-Fattori, J Redemann, PB Russell, and B Wenny (2005) Evaluation of the MODIS aerosol retrievals over ocean and land during CLAMS. *J. Atmos. Sci.* 62: 974–992.

Levy RC, LA Remer, S Mattoo, EF Vermote, and YJ Kaufman (2007a) A second-generation algorithm for retrieving aerosol properties over land from MODIS spectral reflectance. *J. Geophys. Res.* 112(D13211).

Li Z, A Khananian, R Fraser, and J Cihlar (2001) Automatic detection of fire smoke using artificial neural networks and threshold approaches applied to AVHRR imagery. *IEEE Trans. Geosci. Remote Sens.* 39: 1859–1870.

Li Z, F Niu, KH Lee, J Xin, WM Hao, B Nordgren, Y Wang, and P Wang (2007) Validation and understanding of MODIS aerosol products using ground-based measurements from the handheld sunphotometer network in China. *J. Geophy. Res.* 112(D22S07), doi:10.1029/2007JD008479.

Liu Y, JA Sarnat, BA Coull, P Koutrakis, and DJ Jacob (2004) Validation of Multiangle Imaging Spectroradiometer (MISR) aerosol optical thickness measurements using Aerosol Robotic Network (AERONET) observations over the contiguous United States. *J. Geophys. Res.* 109(D06205), doi:10.1029/2003JD003981.

Martonchik JV and DJ Diner (1992) Retrieval of aerosol and land surface optical properties from multi-angle satellite imagery. *IEEE Trans. Geosci. Remote Sens.* 30: 223–230.

Martonchik JV, DJ Diner, KA Crean, and MA Bull (2002) Regional aerosol retrieval results from MISR. *IEEE Trans. Geosci. Remote Sens.* 40: 1520–1531.

Martonchik JV, DJ Diner, RA Kahn, TP Ackerman, MM Verstraete, B Pinty, and HR Gordon (1998) Techniques for the retrieval of aerosol properties over land and ocean using multiangle imaging. *IEEE Trans. Geosci. Remote Sens.* 36: 1212–1227.

Martonchik JV, DJ Diner, R Kahn, B Gaitley, and BN Holben (2004) Comparison of MISR and AERONET aerosol optical depths over desert sites. *Geophys. Res. Lett.* 31(L16102), doi:10.1029/2004GL019807.

McCormick MP, P Hamill, TJ Pepin, WP Chu, TJ Swissler, and LR McMaster (1979) Satellite studies of the stratospheric aerosol. *Bull. Am. Meteorol. Soc.* 60(9):1038–1046.

McCormick MP, DM Winker, EV Browell, JA Coakley, CS Gardner, RM Hoff, GS Kent, SH Melfi, RT Menzies, CMR Menzies, DA Randall, and JA Reagan (1993) Scientific investigations planned for the Lidar In-space Technology Experiment (LITE). *Bull. Am. Meteorol. Soc.* 74(2): 205–214.

McPeters, RD, PK Bhartia, AJ Krueger, JR Herman, and O Torres (1996) Nimbus 7 Total Ozone Mapping Spectrometer (TOMS) Data Products User's Guide, NASA Reference Publication.

Mekler Y, H Quenzel, G Ohring, and I Marcus (1977) Relative atmospheric aerosol content from ERTS observations. *J. Geophys. Res.* 82: 967–972.

Mi W, Z Li, X Xia, B Holben, R Levy, F Zhao, H Chen, and M Cribb (2007) Evaluation of the Moderate Resolution Imaging Spectroradiometer aerosol products at two Aerosol Robotic Network stations in China. *J. Geophys. Res.* 112(D22S08), doi:10.1029/2007JD008474.

Mishchenko MI, B Cairns, G Kopp, CF Schueler, BA Fafaul, JE Hansen, RJ Hooker, T Itchkawich, HB Maring, and LD Travis (2007c). Precise and accurate monitoring of terrestrial aerosols and total solar irradiance: Introducing the Glory mission. *Bull. Am. Meteorol. Soc.* 88: 677–691, doi:10.1175/BAMS-88-5-677.

Mishchenko MI, IV Geogdzhayev, B Cairns, WB Rossow, and AA Lacis (1999) Aerosol retrievals over the ocean by use of channels1 and 2 AVHRR data: Sensitivity analysis and Preliminary results. *Appl. Opt.* 38: 7325–7341.

Mishchenko MI, IV Geogdzhayev, WB Rossow, B Cairns, BE Carlson, AA Lacis, L Liu, and LD Travis (2007a) Long-term satellite record reveals likely recent aerosol trend. *Science.* 315(1543), doi:10.1126/science.1136709.

Mishchenko MI and IV Geogdzhayev (2007b) Satellite remote sensing reveals regional tropospheric aerosol trends. *Opt. Express.* 15: 7423–7438.

Moulin C, F Guillard, F. Dulac, and CE Lambert (1997) Long-term daily monitoring of Saharan dust load over marine areas using Meteosat ISCCP-B2 data, Part I: Methodology and preliminary results for the Western Mediterranean. *J. Geophys. Res.* 102: 16947–16958.

Murakami H, K Sasaoka, K Hosoda, H Fukushima, M Toarani, R Frouin, BG Mitchell, M Kahru, PY Deschamps, D Clark, S Flora, M Kishino, S Saitoh, I Asanuma, A Tanaka, H Sasaki, K Yokouchi, Y Kiyomoto, H Saito, C Dupouy,

A Siripong, S Matsumura, and J Ishizaka (2006) Validation of ADEOS-II GLI ocean color products using in-situ observations. *J. Oceanogr.* 62: 373–393.

Myhre G, F Stordal, M Johnsrud, DJ Diner, IV Geogdzhayev, JM Haywood, BN Holben, T Holzer-Popp, A Ignatov, R Kahn, YJ Kaufman, N Loeb, J Martonchik, MI Mishchenko, NR Nalli, LA Remer, M Schroedter-Homscheidt, D Tanré, O Torres, M Wang (2005) Intercomparison of satellite retrieved aerosol optical depth over ocean during the period September 1997 to December 2000, *Atmos. Chem. Phys.* 5: 1697–1719.

Myhre G, F Stordal, M Johnsrud, A Ignatov, MI Mishchenko, IV Geogdzhayev, D Tanré, JL Deuzé, P Goloub, T Nakajima, A Higurashi, O Torres, and BN Holben (2004) Intercomparison of satellite retrieved aerosol optical depth over ocean. *J. Atmos. Sci.* 61: 499–513.

Nicolantonio WD, A Cacciari, S Scarpanti, G Ballista, E Morisi, and R Guzzi (2006) SCIAMACHY TOA reflectance correction effects on aerosol optical depth retrieval, Proc. of the First 'Atmospheric Science Conference', ESRIN, Frascati, Italy 8–12 May 2006 (ESA SP-628, July 2006)

Okada Y, S Mukai, and I Sano (2001) Neural network approach for aerosol retrieval. *IGARSS.* 4: 1716–1718.

Palm S, W Hart, D Hlavka, EJ Welton, A Mahesh, and J Spinhirne (2002) GLAS atmospheric data products. GLAS algorithm theoretical basis document (ATBD). Version 4.2. Lanham, MD: Science Systems and Applications, Inc.

Pepin TJ and MP McCormick (1976) Stratospheric Aerosol Measurement – Experiment MA-007, Apollo-Soyuz Test Project, Preliminary Science Report, NASA-JSC, TM-X-58173: 9.1–9.8.

Popp C, A Hauser, N Foppa, and S Wunderle (2007) Remote sensing of aerosol optical depth over central Europe from MSG-SEVIRI data and accuracy assessment with ground-based AERONET measurements. *J. Geophys. Res.* 112(D24S11), doi:10.1029/2007JD008423.

Powell KA, CR Trepte, and GS Kent (1997) Observation of Saharan dust by LITE. In: A Ansmann, R Neuber, P Rairoux, and U Wandinger (eds). *Advances in Atmospheric Remote Sensing with Lidar*, Springer-Verlag, Berlin, 149–152.

Raes F, T Bates, F McGovern, and M VanLiedekerke (2000) The 2nd Aerosol Characterization Experiment (ACE-2): General overview and main results. *Tellus.* 52:111–125.

Randall CE, DW Rusch, JJ Olivero, RM Bevilacqua, LR Poole, JD Lumpe, MD Fromm, KW Hoppel, JS Hornstein, and EPShettle (1996) An overview of POAM II aerosol measurements at 1.06 m. *Geophys. Res. Lett.* 23: 3195–3198.

Rao CRN, EP McClain, and LL Stowe (1989) Remote sensing of aerosols over the oceans using AVHRR data theory, practice, and applications. *Int. J. Remote Sens.* 10(4–5): 743–749.

Remer LA, YJ Kaufman, D Tanré, S Mattoo, DA Chu, JV Martins, RR Li, C Ichoku, RC Levy, RG Kleidman, TF Eck, E Vermote, BN Holben (2005) The MODIS aerosol algorithm, products and validation, *J. Atmos. Sci.* 62: 947–973.

Remer L, D Tanré, and YJ Kaufman (2006) Algorithm for Remote Sensing of Tropospheric Aerosols from MODIS: Collection 5, Algorithm Theoretical Basis Document, http://modis.gsfc.nasa. 5 gov/data/atbd/atmos atbd.php.

Rossow WB, AW Walker, DE Beuschel, and MD Roiter (1996) International Satellite Cloud Climatology Project ~ISCCP documentation of new cloud data sets, World Climate

Research Programme Rep. WMOyTD 737 ~World Meteorological Organization, Geneva, Switzerland.

Russell III JM, LL Gordley, JH Park, SR Drayson, WD Hesketh, RJ Cicerone, AF Tuck, JE Frederick, JE Harries, and PJ Crutzen (1993) The halogen occulation experiment. *J. Geophys. Res.* 98(D6): 10,777–10,797.

Russell PB, PV Hobbs, and LL Stowe (1999) Aerosol properties and radiative effects in the United States East Coast haze plume: An overview of the Tropospheric Aerosol Radiative Forcing Observational Experiment (TARFOX). *J. Geophys. Res.* 104(D2): 2213–2222.

Smith RC and KS Baker (1981) Optical properties of the clearest natural waters. *Appl. Opt.* 20: 177–184.

Spinhirne JD, SP Palm, WD Hart, DL Hlavka, and EJ Welton (2005b) Cloud and aerosol measurements from GLAS: Overview and initial results. *Geophys. Res. Lett.* 32(L22S03), doi:10.1029/2005GL023507.

Spinhirne JD, SP Palm, DL Hlavka, WD Hart, EJ Welton (2005a) Global aerosol distribution from the GLAS polar orbiting lidar instrument. *Remote Sensing of Atmospheric Aerosols, 2005. IEEE Workshop on aerosols.* 2–8, DOI: 10.1109/AEROSOL.2005.1494140.

Stowe LL (1991) Cloud and aerosol products at NOAA/NESDIS. *Paleogeogr. Paleoclimatol. Paleoecol.* 90: 25–32.

Stowe LL, AM Ignatov, and RR Singh (1997) Development, validation, and potential enhancements to the second-generation operational aerosol product at the National Environmental Satellite, Data, and Information Service of the National Oceanic and Atmospheric Administration. *J. Geophys. Res.* 102: 16923–16934.

Stowe LL, H Jacobowitz, G Ohring, K Knapp, and N Nalli (2002) The Advanced Very High Resolution Pathfinder Atmosphere (PATMOS) climate dataset: Initial analysis and evaluations. *J. Climate.* 15: 1243–1260.

Tang J, Y Xue, T Yuc, and YN Guan (2005) Aerosol optical thickness determination by exploiting the synergy of TERRA and AQUA MODIS. *Remot Sens. Environ.* 94: 327–34.

Tanré D, YJ Kaufman, M Herman, and S Mattoo (1997) Remote sensing of aerosol properties over oceans using the MODIS/EOS spectral radiances. *J. Geophys. Res.* 102(D14): 16971–16988.

Tanré D, LA Remer, YJ Kaufman, S Mattoo, PV Hobbs, JM Livingston, PB Russell, and A Smirnov (1999) Retrieval of aerosol optical thickness and size distribution over ocean from the MODIS airborne simulator during TARFOX. *J. Geophys. Res.* 104(D2): 2261–2278.

Thomason LW, LR Poole, and CE Randall (2007) SAGE III aerosol extinction validation in the Arctic winter: Comparisons with SAGE II and POAM III. *Atmos. Chem. Phys.* 7: 1423–1433.

Torres O, PK Bhartia, JR Herman, Z Ahmad, and J Gleason (1998) Derivation of aerosol properties from satellite measurements of backscattered ultraviolet radiation: Theoretical basis. *J. Geophys. Res.* 103: 17099.

Torres O, PK Bhartia, JR Herman, A Sinyuk, and B Holben (2002) A long term record of aerosol optical thickness from TOMS observations and comparison to AERONET measurements. *J. Atm. Sci.* 59:398–413.

Torres O, PK Bhartia, A Sinyuk, EJ Welton, and B Holben (2005) Total Ozone Mapping Spectrometer measurements of aerosol absorption from space: Comparison to SAFARI 2000 ground-based observations. *J. Geophys. Res.* 110(D10S18), doi:10.1029/2004JD004611.

Trijonis J, R Charlson, R Husar, WC Malm, M Pitchford, W White (1991) Visibility: Existing and historical conditions – causes and effects. In: *Acid Deposition: State of Science and Technology*. Report 24. National Acid Precipitation Assessment Program.

Vaughan M, S Young, D Winker, K Powell, A Omar, Z Liu, Y Hu, and C Hostetler (2004) Fully automated analysis of space-based lidar data: An overview of the CALIPSO retrieval algorithms and data products. *Proc. SPIE.* 5575: 16–30.

Veefkind JP, G de Leeuw, PA Durkee, PB Russell, PV Hobbs, and JM Livingston (1999) Aerosol optical depth retrieval using ATSR-2 and AVHRR data during TARFOX. *J. Geophys. Res.* 104(D2): 2253–2260.

Veefkind, JP and G de Leeuw (1998) A new aerosol retrieval algorithm to determine the spectral aerosol optical depth from satellite radiometer measurements. *J. Aerosol Sci.* 29: 1237–1248.

Vidot J, R Santer, and O Aznay (2008) Evaluation of the MERIS aerosol product over land with AERONET. *Atmos. Chem. Phys. Discuss.* 8: 3721–3759

Viollier M, D Tanré, and P Deschamps (1980) An algorithm for remote sensing of water color from space. *Boundary Layer Meteorol.* 18: 247–267.

von Hoyningen-Huene W, M Freitag, and JP Burrows (2003) Retrieval of aerosol optical thickness over land surfaces from top-of-atmospher radiance. *J. Geophys. Res.* 108: 4260, doi:10.1029/2001JD002018.

von Hoyningen-Huene W, AA Kokhanovsky, JP Burrows, V Bruniquel-Pinel, P Regner, F Baret (2006) Simultaneous determination of aerosol and surface chracteristics from MERIS top-of-atmosphere reflectance. *Adv. Space Res.* 37: 2172–2177.

von Hoyningen-Huene W, AA Kokhanovsky, M Wuttke, M Buchwitz, S Noel, K Gerilowski, JP Burrows, B Latter, R Siddans, and BJ Kerridge (2005) Validation of SCIA-MACHY top-of-atmosphere reflectance for aerosol remote sensing using MERIS L1 data, *Atmos. Chem. Phys. Discuss.* 6: 673–699.

Wang J, SA Christopher, F Brechtel, J Kim, B Schmid, J Redemann, PB Russell, P Quinn, and BN Holben (2003) Geostationary satellite retrievals of aerosol optical thickness during ACE-Asia. *J. Geophys. Res.* 108(D23): 8657, doi:10.1029/2003JD003580.

Wang M, KD Knobelspiesse, and CR McClain (2005) Study of the Sea-Viewing Wide Field-of-View Sensor (SeaWiFS) aerosol optical property data over ocean in combination with the ocean color products. *J. Geophys. Res.* 110(D10S06), doi:10.1029/2004JD004950.

Winker DM, RH Couch, and MP McCormick (1996) An overview of LITE: NASA's Lidar In-space Technology Experiment. *Proc. IEEE.* 84: 164–180.

Zasetsky AY and JJ Sloan (2005) Monte Carlo approach to identification of the composition of stratospheric aerosols from infrared solar occultation measurements. *Appl. Opt.* 44: 4785–4790.

Digital Photographic Technique to Quantify Plume Opacity During Daytime and Nighttime

Ke Du, Mark J. Rood, Byung J. Kim, Michael R. Kemme, Bill Franek and Kevin Mattison

Abstract United States Environmental Protection Agency (USEPA) developed opacity standards for sources of visible emissions to protect the visual quality of ambient air. Method 9 is USEPA's standard method to quantify plume opacity by visual observations from qualified human observers during daytime. These observers are required to be certified twice a year at a "smoke school". "Smoke school" is more formally referred to as "Visible Emissions Training", which generally consists of a lecture session and a certification event where observers are field tested for their ability to determine the opacity of plumes. However, the use of observations by humans to quantify plume opacity introduces subjectivity, and is expensive due to semi-annual certification requirements of the observers. In addition, sources may emit plumes during nighttime that also need to be monitored to determine compliance for those sources. Digital Optical Method (DOM) was developed to quantify plume opacity from digital photographs for both daytime and nighttime conditions. Daytime field campaigns were completed during smoke schools with the Illinois Environmental Protection Agency (IEPA) and with industrial stacks during daytime. Field campaigns were also completed during nighttime with a smoke generator operated by IEPA. The field tests demonstrated that DOM has advantages when compared to Method 9 by its lower cost, improved objectivity, and availability of photographs of the visible emissions and their environments. Errors in results from DOM when compared to a reference in-stack transmissometer or a Method 9 observer are within USEPA's error limits for Method 9. These encouraging results indicate that DOM has the potential to serve as an alternative method to Method 9 to determine the opacity of plumes for regulatory compliance of stationary sources.

Keywords Opacity · Method 9 · Digital photography · Nighttime · Visible emission

1 Introduction

Standards have been developed by the United States Environmental Protection Agency (USEPA) and local authorities to regulate particulate matter (PM) emissions from anthropogenic sources to protect the environment. These emissions can be regulated with an opacity-based standard and/or a mass based standard. Opacity of a plume can be described as the degree to which emissions reduce the transmission of light and obscure the view of an object in the background. Opacity standards are used widely in compliance auditing because opacity is comparatively easier to monitor than mass concentration of PM in a plume. It has been reported that opacity based standards are more lenient than mass emission based standards and thus a violation of an opacity standard is an indication of a violation of the mass emission standard (Conner and McElhoe 1982). Method 9 is the most common way to quantify the opacity of plumes that are emitted from stationary sources because of USEPA regulatory requirements. However, Method 9 does not provide an archival photograph as a record and has been questioned for its subjectivity (Haythorne and

K. Du and M.J. Rood (✉)
Department of Civil and Environmental Engineering, University of Illinois, 205 N. Mathews Ave. Urbana, IL, USA 61801
e-mail: kedu75@gmail.com; mrood@illinois.edu

Y.J. Kim et al. (eds.), *Atmospheric and Biological Environmental Monitoring*,
DOI 10.1007/978-1-4020-9674-7_3, © Springer Science+Business Media B.V. 2009

Rankin 1974). Also, it was estimated that the digital camera opacity method has the potential to save at least $200 million (USD) per year in the United States (OAQPS 2006). In addition, Method 9 was developed for daytime conditions with limitations for the sun's orientation with respect to the observer and the plume (i.e., the plume shall be observed with the sun located in the 140° sector to the back of the observer). However, industrial sources can have emissions during nighttime. Therefore, plumes emitted during nighttime should also be readily characterized for opacity too. Several State regulatory agencies have recognized the need of nighttime quantification of plume opacity. For example, protocols to quantify plume opacity through visual observations during nighttime have been proposed in Colorado State (Air Pollution Control Division 2001). Hawaii and Alaska use a non-EPA-approved method to measure residential wood smoke chimney opacity at night (Intergovernmental Wood Smoke Committee 2002). California Air Resources Board uses a Method 9-based technique for nighttime conditions to read opacity for plume from stationary sources, ships and wood smoke fireplaces.

Other USEPA-approved methods for determining plume opacity include in-stack transmissometry and lidar (LIght Detection And Ranging). Both methods use active light sources and detectors, which make them capable for nighttime operation. However, the in-stack transmissometer method requires instrument installation for each monitored source, which makes it not as flexible as Method 9 and the measurement occurs in the stack instead of the ambient environment. In addition, in-stack transmissometers usually cost > $10 K USD and lidars can cost > $100 K USD, which are much more expensive than digital cameras (e.g. $200 USD). Despite its subjectivity and high cost, Method 9 remains the preferred method to measure plume opacity in the atmosphere. To overcome the aforementioned drawbacks, techniques have been developed to use digital cameras to quantify plume opacity in the past five years. For example, Digital Opacity Compliance System (DOSC) was developed to quantify plume opacity using a specific digital camera that self-calibrates for clear-sky backgrounds (McFarland et al. 2003). DOCS was tested with clear skies at a high mountain desert (McFarland et al. 2003), cloudy skies with mild temperature and moderate wind, and overcast skies with freezing temperature and light rain that was mixed with snow (McFarland et al. 2004). Most recently, DOCS was tested using a range of commercially available cameras in lieu of a specific digital camera that was required to be used for the previous field campaigns (McFarland et al. 2006). In this study, Digital Optical Method (DOM) was developed as the first method to quantify plume opacity during the daytime and nighttime with use of digital photography that uses low-cost (e.g. $200) commercially available off-the-shelf digital cameras. The principles of DOM and results from using DOM during daytime and nighttime field campaigns are described below.

2 Digital Optical Method

A digital camera is used to take photographs of the plume and its background. The background can be a contrasting object behind and next to the plume or a homogeneous sky that is in contrast with the plume. The light from the background and from the plume is detected by the light sensitive device within the digital camera (i.e., a Charge-Coupled Device (CCD)), and recorded as pixel values. A camera response curve can be determined empirically for each individual camera (i.e., camera calibration Du 2007) so that the pixel values can be used to quantify the radiance ratio from two selected areas in the digital photograph.

DOM consists of the contrast model and transmission model (Du et al. 2007b, a, 2006), which were developed for different background conditions and previously described in detail, but are described here briefly for clarity. The contrast model quantifies plume opacity by viewing the plume in front of and next to the background that has areas in contrast with each other (e.g. bright and dark areas). Plume opacity is then determined based on the observed change in contrast of the background areas that are behind and next to the plume. The contrast value between two specified areas is determined from the ratio of radiances coming from those areas. The transmission model is used to quantify the opacity of a plume with homogeneous sky background that is in contrast with the plume, e.g., a black or white plume against a clear sky, or a black plume against white overcast sky. The transmission model quantifies plume opacity from the radiance ratios between the plume and its contrasting sky background. The first principles of contrast and transmission model for daytime and nighttime conditions are described below in detail.

2.1 Principles of DOM for Daytime Conditions

During daytime, the radiance coming from the plume and detected by the camera consists of two parts: (1) the direct radiance that comes from the plume's background (i.e., sky or contrasting background) and is attenuated by extinction of the plume and the atmosphere in the path; (2) the diffusive radiance (i.e., "path radiance") that comes from the sky and the sun and is scattered into the line of sight of the digital camera.

2.1.1 Contrast Model for Daytime

Figure 1A depicts the principles of the contrast model for daytime conditions. x_1 is the distance between the plume and the camera, x_2 is the thickness of the plume,

and x_3 is the distance between the contrasting background and the plume. N_{w0} and N_{b0} are radiance values that originate from the bright and dark parts of the backgrounds, respectively. N_{w01} and N_{b01} are radiance values from the bright and dark areas that reach to the plume after attenuation by the ambient atmosphere in path x_3. $N_{dir,w}$ and $N_{dir,b}$ are the attenuated radiance values that result from N_{w01} and N_{b01} after the light is scattered and/or absorbed by the plume, respectively. $N_{diff,w}$ and $N_{diff,b}$ are diffusive radiance values caused by other sources of light than the background (e.g., sunlight back-scattered from the plume and directed into the camera's line of sight toward the plume). N_{wt} is the radiance value resulting from $N_{dir,w}$ and $N_{diff,w}$, and N_{bt} is the radiance value resulting from $N_{dir,b}$ and $N_{diff,b}$. N_{wp} and N_{bp} are the radiances actually received by the camera corresponding to the bright and dark part of the contrasting background behind the plume. N_{wp} and N_{bp} result from the attenuation of N_{wt} and N_{bt} by the atmosphere between the camera and plume plus path radiance.

The radiances received by the digital camera are expressed as:

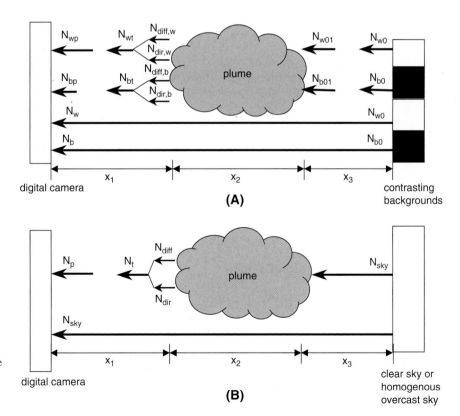

Fig. 1 Schematic describing the contrast model (**A**) and the transmission model (**B**) to determine plume opacity during daytime

$$N_{wp} = [(N_{w0} \times T_3^* + N_3^*) \times T_p^* \times T_2^* + N_{diff,w}]$$
$$\times T_1^* + N_1^* \qquad (1)$$

$$N_{bp} = [(N_{b0} \times T_3^* + N_3^*) \times T_p^* \times T_2^* + N_{diff,b}]$$
$$\times T_1^* + N_1^* \qquad (2)$$

$$N_w = [(N_{w0} \times T_3^* + N_3^*) \times T_2^* + N_2^*]$$
$$\times T_1^* + N_1^* \qquad (3)$$

$$N_b = [(N_{b0} \times T_3^* + N_3^*) \times T_2^* + N_2^*]$$
$$\times T_1^* + N_1^* \qquad (4)$$

where T_1^*, T_2^*, T_3^*, and T_p^* are transmittance values of the atmosphere along paths x_1, x_2, and x_3, and of the plume, respectively. N_1^*, N_2^*, and N_3^* (which are not included in the schematics for concision) are path radiance values of the atmosphere along paths x_1, x_2, and x_3, respectively, and $N_{diff,w}$ and $N_{diff,b}$ are the path radiance values of the plume in front of the bright and dark backgrounds. From Eqs. (1) and (2):

$$N_{wp} - N_{bp} = (N_{w0} - N_{b0}) \times T_1^* \times T_2^* \times T_3^* \times T_p^* \qquad (5)$$

From Eqs. (3) and (4):

$$N_w - N_b = (N_{w0} - N_{b0}) \times T_1^* \times T_2^* \times T_3^* \qquad (6)$$

From Eqs. (5) and (6):

$$T_p^* = \frac{N_{wp} - N_{bp}}{N_w - N_b} \qquad (7)$$

According to the definition of opacity (O), $O \equiv 1 - T_p^*$, the contrast model describes plume opacity by:

$$O = 1 - \frac{N_{wp} - N_{bp}}{N_w - N_b} \qquad (8)$$

2.1.2 Transmission Model for Daytime

As previously mentioned, the transmission model determines the plume's opacity based on the radiance values from the plume and the plume's background. N_{sky} is the radiance value from the uniform contrasting background that is located behind and next to the plume at the same height (e.g. uniform clear or overcast sky) (Fig. 1B), N_{dir} is the direct radiance value

from N_{sky} but attenuated by absorption and/or scattering by the plume, and N_{diff} is the path radiance value from the plume's particles and the background atmosphere within the plume; N_t is the radiance value that originated from the plume, which is the sum of N_{dir} and N_{diff}. x_1 is the distance between the plume and the camera; x_2 is the thickness of the plume, and x_3 is the distance between the plume and its background. The radiance from the plume is expressed as:

$$N_p = N_t \times T_1^* + N_{sky}(1 - T_1^*)$$
$$= (N_{dir} + N_{diff})T_1^* + N_{sky}(1 - T_1^*) \qquad (9)$$

According to the definition of opacity, attenuated radiance from the sky background is described by:

$$N_{dir} = N_{sky} \times (1 - O_t) = N_{sky} \times T_p^* \times T_2^* \qquad (10)$$

where O_t is the opacity of the plume and the background atmosphere, and T_p^* is the transmittance of the plume only. Therefore, the opacity of the plume only, O_p, can be expressed by N_p, N_{sky}, and N_{diff} in a more general form through substitution of Eq. (10) into (9):

$$O_p = 1 - T_p^* = 1 - \frac{N_p - N_{sky}(1 - T_1^*) - N_{diff} \times T_1^*}{N_{sky} \times T_1^* \times T_2^*} \qquad (11)$$

Typically, the extinction values of the background atmosphere along x_1 and x_2 are negligible compared to the plumes if the plume is <1,000 m away (Du 2007) from the digital camera. Therefore, it is reasonable to assume $T_1^* = 1$ and $T_2^* = 1$, and Eq. (11) becomes

$$O_p = O_t = 1 - \frac{N_p - N_{diff}}{N_{sky}} \qquad (12)$$

N_{diff} is proportional to the product of sky background radiance, N_{sky}, and O_t, such that $N_{diff} = KN_{sky}O_t$, where K can be quantified for typical atmospheric conditions and optical properties of the plume particles (Du 2007). Therefore, the plume opacity can be quantified from the radiance ratio between the plume and the sky and the K value by:

$$O_p = \frac{1 - \dfrac{N_p}{N_{sky}}}{1 - K} \qquad (13)$$

The parameter K describes the ability of the plume to scatter light into the line of sight of the camera. A larger value of K indicates a brighter plume.

The method for determining K for typical plumes and atmospheric conditions is described in Du (2007). K values of 0.16 and 1.4 were used in DOM for black and white plumes during daytime conditions.

2.2 Principles of DOM for Nighttime Conditions

When quantifying plume opacity during nighttime, a light source is required to illuminate either the contrasting background (for contrast model, Fig. 2A) or the plume (for transmission model, Fig. 2B).

2.2.1 Contrast Model for Nighttime

The nighttime-based contrast model requires that the contrasting background is set behind and next to the plume and sufficient light is available to illuminate the background for a digital still camera to obtain a digital image of the plume and its background over a reasonable time period (e.g., 1/8 s). The contrast model considers the transmission of light from the illuminated background towards the camera as it passes through the plume (path length x_2) and through the ambient atmosphere (path lengths $x_1 + x_3$), before detection of that light by the camera (Fig. 2A).

The derivation of contrast model for nighttime conditions is similar to that for daytime conditions except that the diffusive scattering terms caused by the plume (i.e. $N_{diff,w}$ and $N_{diff,b}$) are considered insignificant. Therefore, the radiance values received by the camera are expressed as:

$$N_{wp} = [(N_{w0} \times T_3{}^* + N_3{}^*) \times T_p{}^* \times T_2{}^*] \times T_1{}^* + N_1{}^* \tag{14}$$

$$N_{bp} = [(N_{b0} \times T_3{}^* + N_3{}^*) \times T_p{}^* \times T_2{}^*] \times T_1{}^* + N_1{}^* \tag{15}$$

$$N_w = [(N_{w0} \times T_3{}^* + N_3{}^*) \times T_2{}^* + N_2{}^*] \times T_1{}^* + N_1{}^* \tag{16}$$

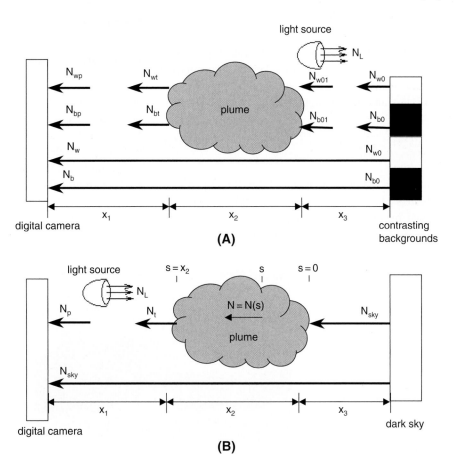

Fig. 2 Schematic describing the contrast model (**A**) and the transmission model (**B**) to determine plume opacity during nighttime

$$N_b = [(N_{b0} \times T_3{}^* + N_3{}^*) \times T_2{}^* + N_2{}^*] \times T_1{}^* + N_1{}^* \quad (17)$$

Therefore, the final equation to determine plume opacity during nighttime is the same as that for daytime:

$$O_p = 1 - \frac{N_{wp} - N_{bp}}{N_w - N_b} \quad (18)$$

2.2.2 Transmission Model for Nighttime

The nighttime-based transmission model requires the dark sky as the background and a light source that is used to directly illuminate the plume. N_L is the radiance value from the light source. The resulting equation for the nighttime transmission model has the same form as the daytime transmission model (Du 2007):

$$O_p = \frac{1 - \dfrac{N_p}{N_{sky}}}{1 - K} \quad (19)$$

However, the K value for nighttime transmission model is different from that of daytime transmission model. Instead, K is a function of N_L, plume type $(P(\theta), \omega)$, and ambient lighting (N_{sky}) (Du 2007):

$$K = \frac{N_L}{N_{sky}} \frac{P(\theta)}{4\pi} \omega \quad (20)$$

where P is the phase function of the plume particles, which is a function of the scattering angle, θ. θ is determined by the relative orientation among the plume, light source and the camera. ω is the single scattering

albedo of the plume particles. Thus, K is specific to the field setting and is not a specific value for dark or bright plumes that exist when using the transmission model during daytime conditions. Therefore, for nighttime conditions, K needs to be determined empirically from a photograph of a plume with known opacity during typical field settings. For example, with a photograph of a plume that is 50% opaque, the radiance ratio between the plume and sky, $N_{p50\%}/N_{sky}$ can be determined by means of the camera response function (Du 2007). Then the calibrated value for K can be determined by inverting Eq. (19) and then used to determine the opacity of the plume at different opacity values:

$$K = \frac{1}{50\%} \left(\frac{N_{p50\%}}{N_{sky}} - 1 \right) + 1 \quad (21)$$

The term 1/50% in Eq. (21) is from the derivation of inverting Eq. (19) for a plume with 50% opacity.

3 Field Tests

3.1 Daytime Tests of DOM

DOM was initially tested during a sunny day, with low wind and clear sky (Fig. 3 and Table 1) (Daytime, IEPA smoke generator). DOM was evaluated using the contrast model with black and white contrasting backgrounds and the transmission model with a clear-sky background. Two cameras (Canon® Powershot® G3, which is a manual exposure control camera, and Sony® DSC-S30, which is an automatic exposure control camera) were used simultaneously to quantify the

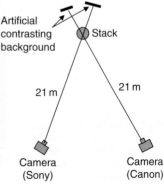

Fig. 3 Site photograph and experimental set-up during daytime test of DOM with IEPA smoke generator and in-stack transmissometer

Table 1 Meteorological conditions during the field tests of DOM

Meteorological condition	Daytime		Nighttime
	Test with transmissometer (IEPA smoke generator)	Test with method 9 (Industrial sources)	Test with transmissometer (IEPA smoke generator)
Mean temperature (°C)	24	25	10
Average wind speed (km/h)	5.6	9.5	14
Sky conditions	Clear to overcast	Clear to overcast	Clear
Visibility (km)	16.1	NA	16.1
Sun elevation angle (°)	33–70	NA	NA
Relative humidity (%)	47–78	21–95	50–55

NA = not available.

plume opacities. Black and white plumes were generated by a smoke generator, which was operated by Illinois Environmental Protection Agency (IEPA). The tests started at 0% opacity and then increased to 100% opacity at 14 levels for black plumes first, and then white plumes with the same sequence. Each camera took one photograph every 15 s for a total of 24 photographs for each plume. The sun was oriented within the 208° sector to the back of the cameras for the entire duration of the test.

DOM was also implemented in the field to monitor plume opacity at two industrial stationary point sources at Fort Hood, which is located at Killeen TX. A Method 9 qualified observer estimated the plumes' opacities using Method 9 while also determining the plumes' opacities using DOM. Opacity measurements from the qualified observer and DOM were then compared to evaluate the consistency of those opacity values. The plumes were emitted from an exhaust tack for three diesel-fired 1,320 KW Caterpillar electrical generators (Facility No. 1001) and an exhaust stack for a diesel-fired 100 KW Cummins electrical generator (Facility No. 91012). Backgrounds for the plumes were either a clear-sky or a building (Fig. 4). The meteorological conditions during the test are summarized in Table 1 (Industrial sources).

During each measurement, the qualified observer made 24 observations within 6 min with the sun located to the back of the observer, and the results of the 24 observations were averaged to provide the final average opacity value for that plume. At the same time, 24 photographs were taken for that plume and were analyzed using DOM software to determine the plume's opacity. Results from DOM for the 24 photographs were averaged to provide the final average opacity value for that plume. Plume opacity was quantified for both sites with one site during each individual day for eleven days between September and December 2004.

3.2 Nighttime Tests of DOM

The nighttime tests with the smoke generator were completed during two consecutive nights to test DOM's contrast and transmission models. When testing the contrast model, two 500 W light sources (Model PAR56LB, Sound Division LC) was located between the stack and the contrasting background. The light sources were directed toward the contrasting background but away from the plume (Fig. 5). Two digital cameras (Canon® Powershot® G3 and Sony®

Fig. 4 Exhaust stacks for industrial sources and backgrounds when monitoring plume opacity using DOM (contrast (*right*) and transmission (*left*) models) and Method 9 at Ft. Hood, TX

Fig. 5 Site photographs (**A**) and experimental set-ups for the nighttime tests of the contrast model (**B**) and transmission model (**C**) with the in-stack transmissometer

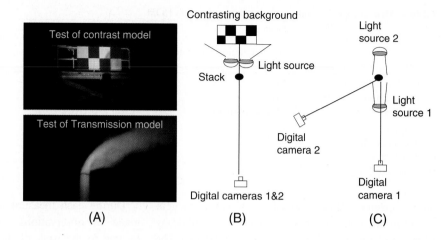

(A)   (B)   (C)

Cybershot® P100) were used to take photographs of the plumes simultaneously. The cameras were placed together when testing the contrast model. When testing the transmission model, the two 500 W light sources were setup in front of and behind the stack and directed toward the plume. During each test, only one light source was powered on to evaluate the orientation of the light source on the DOM's method to determine plume opacity. The cameras were setup at two orientations with respect to the stack when testing the transmission model. The cameras took one photograph every 15 s and a total of 6–12 photographs were taken at each opacity level for the black and then the white plumes. The results from the 6 to 12 photographs were averaged to provide the DOM-derived opacity value for that plume. Opacity values obtained from the two digital cameras were then compared to each other and to the opacity values from the in-stack transmissometer. The test started at 0% opacity and then increased to 100% opacity at 10 levels for the black plumes and then white plumes.

4 Results and Discussion

USEPA set criterion for the smoke school participants by requiring the individual opacity error (IOE) $\leq 15\%$ and the average opacity error (AOE) $\leq 7.5\%$ for both groups of black plumes and white plumes during one smoke school test. The smoke school participants who achieve those two standards are certified as a "Method 9 observer". IOE with unit of percent is defined as the absolute difference between an opac-

ity value, $O_{1,i}$, that is obtained by a human or a digital camera observation and a corresponding opacity value, $O_{2,i}$, that was measured by a reference in-stack transmissometer, as described by:

$$IOE \equiv d_i = \left| O_{1,i} - O_{2,i} \right| \cdot 100 \qquad (22)$$

where subscript i represents each corresponding observation and measurement. AOE is defined as follows:

$$AOE \equiv \bar{d} = \frac{1}{N} \cdot \sum_{i=1}^{N} d_i \qquad (23)$$

where N is the total number of corresponding observations and measurements for a particular test.

4.1 Results from Daytime Tests

During the initial testing of DOM (Fig. 3), the sky was clear in the morning when testing DOM with black plumes. However, clouds developed during the afternoon when testing DOM with white plumes, which prevented the testing of the transmission model to analyze white plumes because the plumes could not be distinguished from their white cloudy backgrounds. Thus, only the digital photographs of the black plumes were analyzed using the transmission model, but all of the photographs of the black and white plumes were analyzed to determine plume opacities using the contrast model.

The results for the contrast model for black and white plumes, and results for the transmission model for black plumes are compared to the simultane-

Fig. 6 Comparison of daytime DOM results to in-stack transmissometer measurements (contrast model for *black* plumes (**A**), contrast model for *white* plumes (**B**), and transmission model for *black* plumes (**C**))

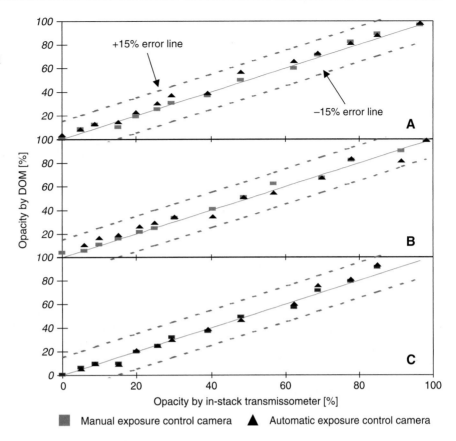

ous opacity values that were provided by the in-stack transmissometer in Fig. 6A, B, and C, respectively. The solid line represents a perfect 1:1 correspondence between modeled opacity values and the transmissometer measurement. The bold dashed line is USEPA's acceptable 15% limit for the individual opacity errors.

All of these DOM measurements have IOEs ≤15%. Therefore, the errors associated with the opacity values obtained with the contrast model and the transmission model (excluding white plumes with a white cloudy background) satisfy the individual error limits specified in Method 9. The results from both models have good linearity with R² values >0.97 for all linear regressions.

The AOE values obtained with the contrast and transmission model for black and white plumes are compared to USEPA's acceptable AOE values (Fig. 7). All of the AOEs for the contrast and transmission model results were <7.5%. AOEs of 2.3 and 3.8% were obtained for the contrast model when using the manual-exposure and automatic-exposure con-

trolled cameras, respectively. AOE values of 3.2 and 3.3% were obtained for the transmission model when using the manual-exposure and automatic-exposure controlled cameras, respectively. These results indicate that DOM quantifies plume opacity well within USEPA's acceptable error limits for the previously described conditions. The results obtained with the two cameras were also compared to each other. The average absolute difference for all results obtained by

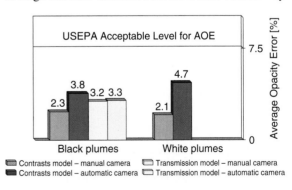

Fig. 7 Average opacity errors for DOM during the field test with the IEPA smoke generator and in-stack transmissometer

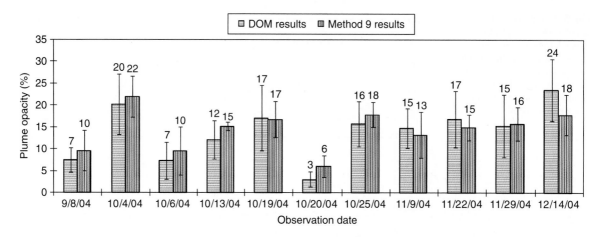

Fig. 8 Mean values (*vertical bars*) and their standard deviations (*vertical lines*) for plume opacity values obtained with DOM and a Method 9 observer

the two cameras is 3%, which suggests that DOM is able to be implemented using different calibrated digital still cameras under these conditions. The 3% difference accounts for the difference between hard-ware (digital camera) and plume-camera orientations tested here.

Both dark and light colored plumes were observed during the daytime tests at Ft. Hood. Figure 8 shows

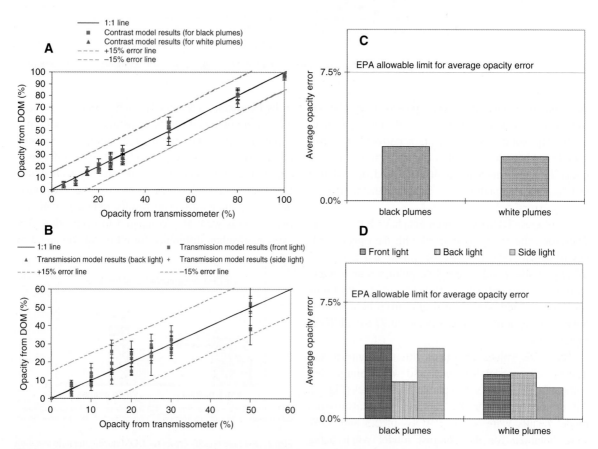

Fig. 9 Comparison of nighttime DOM results to in-stack transmissometer measurements for contrast model (**A**) and transmission model (**B**), and the AOEs for contrast model (**C**) and transmission model (**D**)

opacity values and their standard deviations for both sources when using DOM and Method 9 to determine plume opacity values. The mean difference between the results from DOM and Method 9 for all plumes is 2.2%. A paired t-test, at a confidence level of 99% (i.e., level of significance $= 1 - 0.99 = 0.01$), demonstrates that the results obtained by DOM are not significantly different from those by Method 9. This suggests that DOM has the capability to yield results that are consistent with Method 9.

4.2 Results from Nighttime Tests

Individual and average opacity errors for the contrast and transmission models during the nighttime opacity tests are described in Fig. 9. Opacity results from the contrast model for opacity values ranging from 0 to 100% and the transmission model for opacity values ranging from 0 to 50% compare well to the corresponding opacity results from the in-stack transmissometer with all results meeting USEPA's Method 9 IOE requirement of $\leq 15\%$ (Fig. 9A, B). The AOE values for DOM's contrast model for opacity values from 0 to 100% and the transmission model for opacity values from 0 to 50% were well below the 7.5% acceptable error limit required by USEPA for the black and white plumes, and for all tested camera/stack/light source configurations (Fig. 9C, D). However, the transmission model underestimated opacity values at opacity values $>50\%$. The underestimation resulted from the one-dimensional illumination and large extinction of the plume. When the light illuminated high opacity plumes from one direction, the other side of the plume became dark because of the light extinction of the plume. Thus, the radiance ratio between the plume and the sky no longer increases with the corresponding increasing opacity values as observed with the plumes with lower opacity values.

The limitations of using DOM during nighttime include: (1) an external light source was needed because the light source from the camera was not sufficient to illuminate the plume, (2) contrasting backgrounds are needed for the contrast model, (3) the backgrounds should be close to the plume so that they can by readily illuminated, (4) negative bias increases with the opacity value for the transmission model because of the plume's extinction and non-uniformly illumination, and (5) that bias is sensitive to the relative orientation of the light source and the plume.

5 Summary

The Digital Optical Method (DOM) was tested under both daytime and nighttime conditions. Field results demonstrated that DOM met Method 9's requirements during both daytime and nighttime. DOM also provided consistent results with Method 9 observations when monitoring plumes from industrials sources. Such results indicate that DOM has the potential to be used as an alternative to Method 9. Nighttime field tests also demonstrated DOM's ability in quantifying plume opacity during nighttime, although some limitations exist.

Future areas of research include the development of an automatic, handheld device that integrates DOM with other environmental data to provide more comprehensive real-time opacity monitoring and recording of environmental conditions, development of DOM to quantify plume opacity during daytime for bright plumes with cloudy backgrounds, development of DOM's transmission method to quantify plume opacity for values $> 50\%$. However, even with these constraints, DOM has been clearly shown to meet USEPA Method 9 requirements over a wide range of conditions and has the potential to be used as an alternative to Method 9 while reducing cost, reducing subjectivity, and providing digital photographs of the plumes and their environments.

Abbreviations

AOE	average opacity error
CCD	Charge-Coupled Device
DOCS	Digital Opacity Compliance System
DOM	Digital Optical Method
IEPA	Illinois Environmental Protection Agency
IOE	individual opacity error
PM	particulate matter
USEPA	United States Environmental Protection Agency

References

Air Pollution Control Division (2001) Protocol for Nighttime Opacity Observations Colorado Operating Permit.

Conner WD, McElhoe HB (1982) Comparison of opacity measurements by trained observer and in-stack transmissometer. J Air Pollut Control Assoc 32: 943–946.

Du K (2007) Optical Remote Sensing of Airborne Particulate Matter to Quantify Opacity and Mass Emissions. Dissertation, University of Illinois at Urbana-Champaign, Urbana.

Du K, Rood MJ, Kim BJ et al. (2006) Field Testing of the Digital Opacity Method to Quantify Plume Opacity during Nighttime, 99th Annual Meeting of the Air & Waste Management Association, New Orleans, LA, 10.

Du K, Rood MJ, Kim BJ et al. (2007a) Quantification of plume opacity by digital photography. Environ Sci Technol 41: 928–935.

Du K, Rood MJ, Kim BJ et al. (2007b) Field evaluation of digital optical method to quantify the visual opacity of plumes. J Air Waste Manage Assoc 57: 836–844.

Haythorne RE, Rankin JW (1974) Visual plume readings – too crude for clean air laws. Nat Resour Lawyer 7: 457–477.

Intergovernmental Wood Smoke Committee (2002) Wood smoke program review for city of fort collins, Natural Resources Department.

McFarland MJ, Rasmussen SL, Stone DA et al. (2006) Field Validation of Visible Opacity Photographic Systems, the 99th Annual Meeting of the Air & Waste Management Association, New Orleans, LA, pp. 18.

McFarland MJ, Terry SH, Calidonna MJ et al. (2004) Measuring visual opacity using digital imaging technology. J Air Waste Manage Assoc 54: 296–306.

McFarland MJ, Terry SH, Stone DA et al. (2003) Evaluation of the digital opacity compliance system in high mountain desert environments. J Air Waste Manage Assoc 53: 724–730.

OAQPS (2006) New Digital Camera Approach for Opacity. Memorandum by Stephen D Page from OAQPS of USEPA.

Scanning Infrared Remote Sensing System for Detection, Identification and Visualization of Airborne Pollutants

Ulrich Klenk, Eberhard Schmidt and Andreas Beil

Abstract The use of passive Fourier Transformation Infrared Spectrometry (FT/IR) is a way for the fast recognition and identification of airborne pollutants. In studies done by the Department of Safety Engineering/Environmental Protection at the University of Wuppertal a passive FT/IR was used to show the possibilities of such a system. The fast identification of accidentally released substances is a very important hint in finding leaks and stopping the release. This gives not only the chance of a recognition of leaks; it although shortens the time to find the point of release and sealing the leak. At least the released mass of a substance is reduced. The study shows, that a passive FT/IR is an easy to deploy system for the fast detection, identification and visualization of airborne pollutants. It can be used for the continuous monitoring of plants, dumpsites, production lines or other points of interest.

Keywords Detection · Visualization · Quantification · Remote sensing · FTIR · Passive infrared spectrometry · Toxic cloud imaging

1 Introduction

The use of passive Fourier Transformation Infrared Spectrometry (FTIR) is a way for the fast recognition and identification of airborne pollutants.

In studies done by the Department of Safety Engineering/Environmental Protection at the University of Wuppertal in cooperation with the municipal fire brigade of Ludwigshafen on the Rhine, the Institute of Measurement Technology at the Technical University of Hamburg-Harburg and the company Bruker Daltonik, a passive FTIR was used to show the capabilities of such a system. Different case studies were done, for example:

- Identification and visualisation of accidentally released substances
- Tracing exhausts of chimneys
- The use for civil protection during soccer world championship

These three case studies are only examples for a very wide range of applications for passive Fourier Transformation Infrared Spectrometry. The fast identification and visualization of accidentally or premeditated released substances is a very important hint in finding the source and stopping the release as soon as possible. This gives the chance to reduce the totally released mass of substances by shortening the time of the release, to evacuate affected areas or to supervise the prescriptive limits. The identification of the released substances also allows a propagation calculation with correct parameters like the density of the now known substance. For line emission sources and diffuse emission sources it is a possibility to measure the concentration not only depending on a few points (measurements at certain points) but on the complete lines or areas.

The range of substances that could be identified by classical FTIR is well described in literature. But there is a limit for passive FTIR used for remote sensing: The

U. Klenk (✉)
Department of Safety Engineering/Environmental Protection, University of Wuppertal, D-42097 Wuppertal, North Rhine-Westphalia, Germany
e-mail: klenk@uni-wuppertal.de

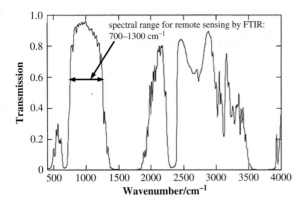

Fig. 1 Transmission of the atmosphere for wavenumbers between 500 and 4000 cm^{-1} (Harig et al. 2006)

2 Theory of Passive FTIR

2.1 Measurement Technique

The remote sensor system includes an infrared emission spectrometer (FTIR), which primarily detects the natural infrared radiation (infrared spectrum) of the environment. Likewise an infrared camera the detector systems determines the radiation contrast between the cloud of a chemical pollutant and the natural background (soil, buildings, forest, sea etc). The measured radiation contrast is spectral resolved and not spatial resolved as for a standard infrared camera. Spatial resolution can be achieved by scanning the observation direction of the remote sensor, where the spatial resolution is limited by the Field of View (FOV) of the sensor system.

The chemical compounds in a gas cloud interact with the infrared radiation of the environment. If a chemical agent is inside the FOV of the sensor, very characteristic spectral structures appear in the measured radiation contrast. These structures which are very specific for each chemical compound, are representing the IR transmission spectrum of the compound. This signatures can be separated from the predictable background radiation and compared to a specific IR library.

The intensity of the radiation contrast allows quantitative values being retrieved about the passive IR signal intensity as the product $c \, d \, \Delta T$, where c is the pollutant concentration, d the thickness of the pollutant cloud and ΔT the brightness temperature difference between background and target cloud.

The sensitivity of the measurement technique depends rather on the brightness temperature difference ΔT. Typical values for temperature differences (fluctuations) in the environment are 1–5 K.

There is a specific spectral contrast for the radiation (feature of the absorption coefficient) of each chemical compound. Chemical compounds, which will give a weak spectral contrast can hardly be detected. Some few molecules do not show a detectable radiation contrast in the accessible spectral range of passive remote detection (e.g. hydrogen chloride, chlorine), so that they cannot be detected. For the majority of vaporized toxic chemical compounds passive remote detection is possible.

useable wavenumber (reciprocal to the wavelength) of the infrared spectra is limited by the transmission of the ambient air and the signal to noise ratio. As shown in Fig. 1 for wavenumbers between 500 and 4000 cm^{-1} and transmissions between 0 and 1, the useable spectral range for FTIR are between 700 and 1300 cm^{-1}. Best results are achieved at wavenumbers between 800 and 1200 cm^{-1}. In this case the transmission is about 95%; additionally the signal to noise ratio is at a maximum (Griffith 1996). Substances which have IR-lines or significant parts of them within this range could be identified best. Identifications at other wavelengths are possible with a higher analytical effort. The analysis range of the FTIR systems used for the studies are restricted to wavenumbers between 700 and 1300 cm^{-1}.

Passive FTIR remote sensing is based on the spectral analysis of ambient infrared radiation. Due to the principle of measurement the probability of detection p_d does not only depend on the concentration c of an airborne pollutant but also on the optical path length d through the cloud of the substance and the temperature difference ΔT between the pollutant and the background (as shown in Eq. (1)).

$$p_d \approx c \cdot d \cdot \Delta T \qquad (1)$$

The theory of passive FTIR, following in Chapter 2, was published by Dr. Andreas Beil during the 5th Joint Conference on Standoff Detection for Chemical and Biological Defense at Williamsburg, Virgina, 24–28th September 2001 (Beil 2001).

2.2 Calibration and Transformation of the Infrared Spectra

From the measured spectrum the spectral radiance is calculated using the linear approximation of the inverse response function of the instrument

$$L(\nu, T) \cong \frac{S(\nu, T) - b(\nu)}{m(\nu)}, \qquad (2)$$

where $S(\nu, T)$ is the measured spectrum, $m(\nu)$ the spectral response factor and $b(\nu)$ the spectral offset. The instrumental response function parameter is determined by a two point calibration with the integrated reference source of the RAPID.

For further analysis of the spectra it is convenient to transform the radiance to the corresponding equivalent radiation temperature (*Brightness Temperature*) $T(\nu)$ using Planck's Radiation Law (Beil et al. 1998; Flanigan 1996).

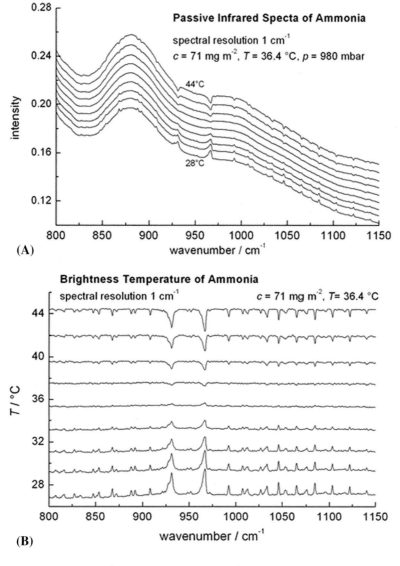

(A)

(B)

Fig. 2 (**A**): Passive IR spectra of 71 mg m^{-2} Ammonia gas at 36.4 °C and background temperatures of $T_b = 28$–44 °C (total pressure $p = 980$ mbar, nitrogen purge gas, spectral resolution 1 cm^{-1}). The intensity difference of each spectrum is related to a temperature difference of approximately 2 K. (**B**): Corresponding temperature spectra T(ν) of (**A**)

$$T(v) = \frac{h\,c\,v}{\ln\left[\dfrac{L(v) + 2\,h\,c^2 v^3}{L(v)}\right] k} \qquad (3)$$

The following interpretation of $T(v)$ uses the temperature as a helpful quantity value for the analysis of the spectra and not as a classical thermodynamic temperature.

As an example of the described spectra evaluation Fig. 2 shows (A) the raw spectra $S(v)$ of a passive measurement of Ammonia vs. the wavenumber and (B) the corresponding temperature spectra $T(v)$ vs. the wavenumber. The Ammonia was measured in a gas cell at 36 °C were the background temperature T_b was varied between 27 and 44 °C. Figure 2 shows that the application of Eqs. (2) and (3) corresponds to a baseline correction of the measured spectra $S(v)$. For background temperatures T_b, which are lower than the temperature of the Ammonia T_t, i.e. $T_b < T_t$, emission lines of Ammonia are observed in the spectra. For reverse temperature conditions, were $T_b > T_t$, absorption lines of Ammonia are observed. The intensities of the spectral Ammonia lines increase nearly proportional with the temperature difference $\Delta T = T_t - T_b$.

The example of Fig. 2A, B clearly illustrates that the infrared emission or absorption from chemical compounds in the target cloud will appear as a characteristic deviation $\Delta T(v) = T(v) - T_b(v)$ from the background temperature T_b.

2.3 Evaluation Model

For the evaluation of the passive detected infrared radiation a simple effective 2-layer model is applied. A target cloud (t) and a background (b) are predefined. Equation (4) shows this 2-layer model:

$$L = L_t(1 - \tau_t) - L_b\,\tau_t, \qquad (4)$$

where L_t is the radiance of a black body at the temperature of the target cloud; L_b is the radiance of the background (sky, forest etc.), and τ_t the effective transmittance of the target cloud.

The measurements have shown that this simple model can be used very well as an effective model for the analysis of the measured radiance in an open environment.

The atmospheric absorption from water and Carbon Dioxide are effectively included in the radiance $L_b(v)$ and the transmittance $\tau_t(v)$. Following Eq. (4) the transmittance of the target can be calculated as follows:

$$\tau_t = \frac{L - L_t}{L_b - L_t} = \exp(-\varepsilon\,c\,d), \qquad (5)$$

where ε is the Naperian absorption coefficient, c the concentration, and d the optical pathlength.

For moderate temperature differences ($\Delta T < 10\,\mathrm{K}$ at 300 K), ΔT is approximately proportional to ΔL, therefore Eq. (5) can be rewritten in terms of the brightness temperatures as follows:

$$\Delta T(v) \cong \Delta T\,\{1 - \exp[-\varepsilon(v)\,c\,d]\}. \qquad (6)$$

For small values of $\varepsilon \times c \times d < 0.1$, Eq. (6) can be further simplified to

$$\Delta T(v) \approx \Delta T\,\varepsilon(v)\,c\,d. \qquad (7)$$

The advantage of the spectra analysis using Eq. (7) is the linear relation between the measured spectral temperature deviation $\Delta T(v)$ and the $c \cdot d$ value of the target compounds. This allows the implementation of a simple and fast algorithm for the analysis of the spectra and identification of the target cloud.

For high values of $\varepsilon\,c\,d\,(>1)$ it is possible to determine $c \cdot d$ and ΔT by a non-linear fit of $\Delta T(v)$ applying Eq. (6). Such an analysis will be more difficult and is not as universal as the analysis using Eq. (7).

2.4 Identification and Quantification

In order to indicate the characteristic spectral signature of a chemical agent in the target cloud the correlation $|r|$ (scalar product) of the brightness temperature spectrum $\Delta T(v)$ with the absorption coefficient spectrum of the chemical agent $\varepsilon(v)$ is calculated using Eq. (8):

$$r_i = \frac{\int \Delta T(v)\,\varepsilon(v)\,dv}{\sqrt{\left(\int \Delta T^2(v)\,dv\right)\left(\int \varepsilon^2(v)\,dv\right)}}. \qquad (8)$$

If the value of $|r_i|$ reaches a certain limit (typically $|r_i| > 0.97$) the chemical compound i is identified with high probability. The actual detection threshold of each compound i has to be determined from the statistics of the noise equivalent $|r_i|$ values of different environments.

As shown in Eq. (9), quantitative information of the chemical compounds can be obtained from the comparison of the spectral band (line) integrals of the $T(\nu)$ spectrum and the spectrum of the target compound $\varepsilon_i(\nu)$.

$$(c \, d \, \Delta T)_i \approx \frac{\int_{\text{band}} \Delta T(\nu) \, d\nu}{\int_{\text{band}} \varepsilon_i(\nu) \, d\nu} \qquad (9)$$

Depending on the units of ε [m^2 mg^{-1}, cm^2 mol^{-1}, ppm^{-1} m^{-1}] the resulting units for the quantification value $c \, d \, \Delta T$ are [mg m^{-2} K], [mol cm^{-2} K] or [ppm m K].

Generally the structure of chemical compounds which appears in the measured $L(\nu)$ spectra corresponds to the structure of a transmittance spectrum $\tau(\nu)$, see Eq. (5), and not to an absorbance spectrum $A(\nu)$. Equation (10) shows the non-linear relation between the absorbance A and the transmittance τ.

$$A = \varepsilon \, c \, d = -\ln(\tau) \cong 1 - \tau - \dots \qquad (10)$$

Only for small absorbency values < 0.1, a linear relation of the $c \, d$ values and the measured line intensities of $L(\nu)$ can be assumed. For high absorbency values, i.e. high concentrations, a non-linearity correction has to be applied.

In case of the Ammonia molecule, where $\varepsilon(\nu)$ has a significant spectral line structure and measurements with limited spectral resolution, even the absorbency line intensity has a non-linear relation to the $c \, d$ value.

The mole fraction x (i.e. volume fractions) is calculated within good approximation by the ideal gas law following Eq. (11):

$$\left(\frac{x}{\text{ppm}}\right) = 1000 \left(\frac{c}{\text{mg m}^{-3}}\right) \frac{R \, T}{p_0 \, M} = \left(\frac{c}{\text{mg m}^{-3}}\right) \frac{24.46}{M}, \qquad (11)$$

with the concentration c, the gas constant R, the temperature T ($= 298$ K), the total pressure p_0 ($= 1013$ mbar), and the molar weight M [g mol^{-1}].

3 The Scanning Infrared Remote Sensing System

The FTIR used in these studies has a military background (RAPID, Bruker Daltonik). It was originally designed for the fast detection of chemical warfare agents (CWAs) and some industrial toxic substances (TICs). The field of view (30 mrad) can be fixed to a single point or scanned over a wide area up to a 360° view. The vertical range can be between -10 and 50°. The spectral range of the rapid is between 700 and 1300 cm^{-1} with a spectral resolution of 4 cm^{-1}. Scans are possible during driving, so that air pollutants can be persecuted (Table 1).

The main advantage of a passive FTIR is the possibility to use it as a scanning device. Large areas, the field of regard, can be scanned very quickly. The additional results are the azimuth and the elevation of the pollutant cloud. Changing the position of the FTIR or using a second FTIR allows overlaying the angles of the azimuth and the elevation and showing the position of the vapour cloud in three dimensions.

Figure 3 shows a schematic view of the RAPID. The IR radiation goes through the entrance window (2) and is transferred by the scanner optics (1) to the interferometer and IR detector unit (3). The reference source (5) is used for the internal radiometric calibration of the RAPID. The RAPID is mounted on a shock frame (4).

There are different ways to use the RAPID FTIR. First way is to use it as a stand-alone device, identifying TICs and CWAs that are in an onboard database, giving the azimuth and elevation angles where a substance is detected and the name of the substance.

Table 1 Technical data of the RAPID

Sensor module	
Spectral range	700–1300 cm^{-1}
Spectral resolution	4 cm^{-1}
Sensitivity	< 0.06 K*
Measurement speed	20 measurements/second
Scanner module	
Scanner speed	Azimuth: max. 120°/second
	Elevation: max 20°/second
Field of regard (FOR)	360° azimuth
	-10–50° elevation
Field of view	30 mrad

* Black body at 30 °C, wavenumber of 1000 cm^{-1}, 0.05 s measurement time

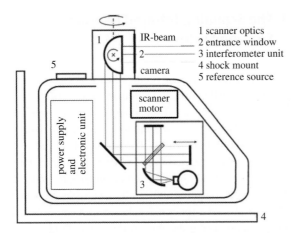

Fig. 3 Scheme of the RAPID with the scanner optics (1), the entrance window (2) and the interferometer unit (3). It is mounted on a shock frame (4). For the calibration an internal reference source (5) is used

Second way is the use of military (Bruker Daltonics 2005a) or civil (Institute of Measurement Technology at the Technical University of Hamburg-Harburg 2006) software on a laptop, giving the angles in degrees, a picture or video of the scanned-area and the names of the identified substances from larger databases on the laptop. Third way is the use of academic software (Bruker Daltonics 2005b), giving the raw data of the recorded interferograms for further use with scientific analyzing software and a video of the point of view.

Other scanning FTIRs were developed at the Technical University of Hamburg-Harburg. The SIGIS and SIGIS 2 are based on the Bruker OPAG 22 and EM 27 (Bruker Optics) with a telescope and a system of mirrors (http://www.et1.tu-harburg.de/ftir/).

4 Laboratory and Field Measurements

4.1 Laboratory Measurements with a Gas-Cell

For the use of passive FTIR in areas with high background concentrations like chemical plants it should be shown, that passive FTIR can be used to identify pollutant clouds through other clouds. Therefore a simple gas-cell with three compartments was built (Kraugmann 2007). Figure 4 shows a scheme of the gas-cell with the inlets and blowoffs, the instrument openings, the heatable black background and the Polyethylene foil (PE foil) between the compartments.

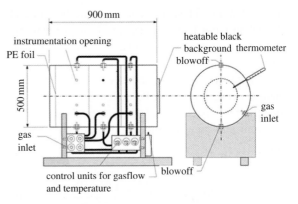

Fig. 4 Scheme of the gas-cell with three compartments, the control unit, the instrument openings and the inlets and blowoffs for the test substances

Before using the gas-cell with different compounds, the influence of the PE foil between the compartments have to be examined. Figure 5 shows the change of the temperature depending on the wavenumbers with and without PE foils. Due to the PE foils the wavenumbers between 700 and 760 cm^{-1} can not be used for identification of pollutants in the gas-cell. Other wavenumbers in the atmospheric spectral window are not affected. The use of PE foils as windows between the compartments in the gas-cell is possible for pollutants with wavenumbers of their characteristic IR-lines higher than 760 cm^{-1}.

The following three tests were done with Ammonia and Ethanol at different concentrations. The first compartment of the gas-cell was filled with ambient air and Ammonia at concentrations of 50, 125 and 524 ppm. The second compartment of the gas-cell was

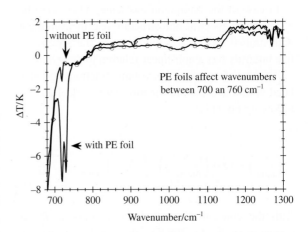

Fig. 5 Effect of PE foils between the compartments of the gas-cell

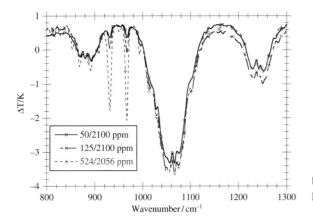

Fig. 6 Ammonia and Ethanol at different concentrations in the gas-cell

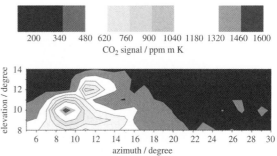

Fig. 7 CO_2 identified over a chimney of a waste-fuel power plant

filled with ambient air and Ethanol at concentrations of 2100, 2100 and 2056 ppm. The third compartment with the heatable background was not used. The temperature of the black background was 41.6 °C. Figure 6 shows, that the significant IR-lines of Ethanol between wavenumbers 1000 and 1100 cm^{-1} can be identified. Although two significant IR-lines of Ammonia between wavenumbers 920 and 980 cm^{-1} are identified, depending on the concentration. Due to the PE foils, wavenumbers lower than 800 cm^{-1} were cut off in the diagram.

Depending on the used software algorithms, the identification of those two pollutants are carried out automatically because the significant IR-lines are detectable. In general it is possible to identify single pollutants through other pollutant clouds or at high background concentrations, if the significant IR-lines can be found. In some cases, the IR-lines are too close together. Therefore it must be possible to switch the used software from "clear identification" to "possible identification". Because of the small amount of wavenumbers that can be used for passive FTIR as a scanning measurement system it could be useful, to identify the pollutant groups and not single pollutants. Another way could be to visualize possible target clouds without the identification of single substances.

4.2 Tracing Exhausts of Chimneys

For another field of use, some measurements were made at chimneys of power plants. The identified sub-stances are Sulfur Dioxide, Carbon Dioxide and traces of Ammonia. Identification and visualization worked well, especially for CO_2 and SO_2. The result for CO_2 of one measurement is shown in Fig. 7. We have analyzed the intensity of the weak IR emission band of CO_2 at 960 cm^{-1} applying Eq. (9), with prior spectral compensation of water and ozone interferences in the measured $T(v)$ spectrum. The noise signal of the sky radiation background was below 350 ppm m K. The image is generated by linearly interpolation of the scanned intensities. The distance between the passive FTIR and the chimney was about 1.7 km. The CO_2 signal was calculated to be between 200 and 1600 ppm m K. The maximum is located at the top of the chimney at an azimuth of 9° and an elevation of 10°. The CO_2 cloud is drifting in direction of the wind from the left side of the picture to the right side with a decreasing CO_2 concentration.

Figure 8 is an overlay of the CO_2-signal with a picture of the chimney. Depending on the used software, the operator is able to see those pictures in real time during the measurement. But there is no information about the third dimension, until the position of the FTIR is changed or a second FTIR is used. In case of Fig. 8 there are no other possible sources between the FTIR and the chimney or beyond the chimney. The CO_2-source is identified surely without the need of overlaying the results of a second measurement from another direction. The box in Fig. 8 shows the scanned area (FOR).

The signal evaluation includes several approximation like the effective radiation model, Eq. (4), the signal linearity given by Eqs. (6) and (7). However, the detection result includes the location, a contour and a signal strength for the plume. The measured signal values $c \cdot d \ \Delta T$ are only a rough linear estimation of the

Fig. 8 Overlay of the CO_2-signal with a picture of the chimney

true values. Without a further detailed non-linear spectral analysis, the transmittance (column density) of the plume can't be determined. Such an sophisticated analysis is not the subject of this work. Our major interest is fast identification and localisation of chemical clouds. Thus we can not give any error estimation for the determined cloud signals, but for sure we have determined the correct order of magnitude.

4.3 Civil Protection

During summer 2006 large amounts of German civil protection units, police, and military were in state of emergency. The soccer world championship took place in different stadiums, which are ideal aims for terroristic attacks. Additionally to the well known chemical, biological, radiological, and nuclear (CBRN) detection devices and vehicles, different kinds of scanning infrared sensors were used in the stadiums and at the fan-parties to detect airborne pollutants as soon as possible.

One passive FTIR RAPID was mounted in the stadium of Leipzig/Germany and operated by the staff of Bruker Daltonics on behalf of the Fire Department of the city of Leipzig. Of course, in the narrow area of a soccer stadium a remote sensor cannot detect the chemical cloud long before the civilians are potentially affected. In case of a quick evaporation of a chemical agent, it is very unlikely that there will be enough time to fully evacuate a stadium safely. Nevertheless, the RAPID is able to identify the release point and the agent type of such a chemical attack within 30 s. The information of chemical remote detection gives fast warning to the protection personnel and allows to guide them close to the release point for the verification with classical chemical detection. Countermeasure can be started earlier, and valuable information are given for the first aid of injured people. Furthermore it is possible to report spectral information of the scenario, which is very useful for the post-analysis of the attack. It gives forensic evidence of the released chemicals, their concentration, distribution, and the progression of contamination.

As shown in Fig. 9 the RAPID was installed on a tripod near the display panel of the stadium. Local power supply (220 V AC) and local area network (LAN) capability was used to operate the system from the control office of the fire brigade with a Laptop PC and the RapidControl software (Bruker Daltonics 2005a).

Fig. 9 RAPID mounted beside the display panel of the soccer stadium at Leipzig/Germany

In order to achieve an optimal coverage of the stadium from the elevated position, the instrument was tilted down by approximately 30°. Figure 10 shows a panoramic view from the RAPID together with the scan lines for azimuth scan (−90° to +90°) and different elevations (−10° to +50°).

The stadium was filled close to its maximum capacity of 45000 people (game: Angola/Iran started at 4:00 p.m. on June-21-2006). During the game the sun came

from the right side of Fig. 10 (West). As mentioned above, the radiation contrast (i.e. contrast of brightness temperature) between the chemical cloud and the background is essential for the detection limits. The pseudo color image of the background radiance in the stadium at 7:40 p.m. (close to the end of the game) is shown in Fig. 11. This image was scanned during the game and corresponds to an IR image similar to typical pictures measured with a monochromatic IR camera in the

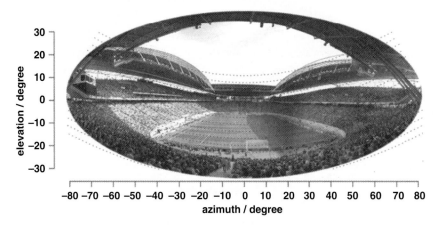

Fig. 10 Panoramic view of the scanned area from the detection point at the stadium of Leipzig (view from NNW to SSW, June 2006). The IR entrance optic of the RAPID scanned an azimuth angle of −90° to +90°. The tilt of the instrument by −30° leads to bended scan lines of the azimuth scans (*dotted lines*). With the maximum scanning speed of 120°/s the whole stadium was scanned within about 30 s. The panorama was photographed at 4:15 p.m. shortly after the beginning of the match

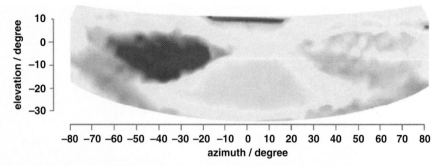

Fig. 11 IR pseudo-color image of the background radiance measured at 7:40 p.m. (close to the end of the match): The *dark* areas on the *left* corresponds to background temperatures of max. 50 °C, where the *dark* area *top middle* (sky) corresponds to −50 °C. The *green* of the field in the center is colder and does also reflect radiation from the cold sky. The shadow covered major areas of the field, only some smaller parts on the *left* side are not inside the *shadow*. The hottest background radiation was emitted from the seats on the *lower left* side of the stadium. These seats were strongly heated by direct solar irradiation

mid IR. The image resolution is approximately 2500 pixel, each covering a FOV of 2°. After interpolation and color shading the contour of the stadium and the soccer field is well visible in Fig. 11.

The IR image of Fig. 11 shows, that the temperature differences are high (up to 20 K). This provides low concentration limits for chemical detection. Furthermore the image allows to control the alignment and scan range of the RAPID compared to the panorama image shown in Fig. 10: The sky section, the left tiers and the border of the field can be well indicated.

In order to test the system, 9.5 grams (1.51 at 25 °C) of Sulfur Hexafluoride (SF_6) were released at a distance of 180 m from the remote sensor. The small plume of this simulant gas was well detected by the RAPID. Figure 12 shows the panorama of the scan range with an overlay of the pseudo color image of the detected SF_6 cloud. Correlation values for SF_6 were determined for the most intensive IR band at 947 cm^{-1}. The interference of atmospheric water was spectrally compensated prior to the signal analysis by Eq. (8). The pseudo colours image of Fig. 12 was generated by linear interpolation of spatially adjacent correlation values. It shows the detected correlation value for SF_6 (dark = 0.8 to light = 1.0).

Correlation = 1 means 100% identity between the effective IR signal $\Delta T(\nu)$ and the library spectrum $\varepsilon(\nu)$

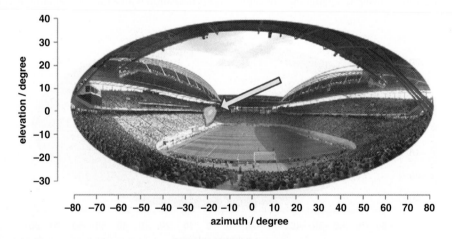

Fig. 12 Panorama view of the scan range with an overlay of the pseudo-color image of the detected SF_6 cloud. The signal was generated from 9 g SF_6, which was released on a distance of 180 m in the *left* corner of the stadium. The colors represents the correlation value of the measured IR structure and the IR band of SF_6 (dark = 0.8, light = 1.0). The color contours corresponds to the contour of the invisible SF_6 cloud. A correlation of 1 means 100% identity. The maximum value of the measured correlation for SF_6 was 0.990. The imaged cloud at the distance of 180 m had a max. size of 20–30 m

of SF_6, and a correlation value equal 0.8 is the upper noise level of SF_6 in the given scan range. The maximum correlation level for the detected SF_6 was 0.99, which is a very clear identification. This high correlation was detected within an azimuth angle of $5°$. For the distance of 180 m this area of max. correlation corresponds to a cloud diameter of approximately 16 m.

The radiance spectrum for the detection with the strongest signal is shown in Fig. 13. The small signal from the SF_6 cloud is indicated by the arrow. The dynamic range for the SF_6 signal band is limited by the Planck radiance for the background temperature (upper solid line, T = 309.1 K) and the Planck radiance for the air/cloud temperature (lower solid line, T = 299.5 K). The difference of the brightness temperature for the detection is almost 10 K. From the analysis of the radiance spectrum the transmittance can be calculated: The signal strength for the detection corresponds to a column density of 21 mg m^{-2} ($\Delta T = 9.5$ K). If one estimates a cloud thickness of 5–10 m, the local concentration in the cloud is 2–4 mg m^{-3} (0.3–0.6 ppm).

Further the signal intensities of the cloud image of Fig. 12 were analyzed. The detection limit for the column density of SF_6 is approximately 5 mg m^{-2}. From the integration of all signals of the cloud, a total mass of less than 5 g can be estimated. These are 50% of the released amount. The difference may be explained by the uncertainty of the estimation, the missing contribution of column densities below the detection limit, and the obstruction of the cloud signal by the crowd of spectators and superstructures of the stadium. Nevertheless it can be clearly verified from the spectral analysis, that the order of magnitude of released SF_6 were several grams.

The results for the detection of SF_6 can not be compared directly with the detection performance for a live agent, because SF_6 is a very strong IR absorber, and forms immediately a detectable vapor cloud (vapor pressure at 20 °C is 21.6 bar). From the comparison of IR-band strengths, simulations and field test results, it can be estimated, that a release of 100 g of a nerve agent with local vapor concentrations of 1–10 ppm, can be detected by the RAPID under the given conditions.

A second RAPID was used at the stadium of Kaiserslautern/Germany to scan the ambient air in the streets around the stadium during the games. The measurements took place in the wider surrounding of the Betzenberg, where the stadium is located. Figure 14 shows the RAPID FTIR mounted on a vehicle of the Fire Department of the city Ludwigshafen at the Rhine.

Fortunately there were no results of the measurements to be reported. During this time period of the field measurements a lot of positive experiences were made. The operator of the FTIR was a firefighter of the municipal fire brigade of the city Ludwigshafen at the Rhine without analytical background. After half a day of training he was able to operate the FTIR, to generate ROTA-reports (Releases other than Attack), and to transmit the data to the situation room, where an

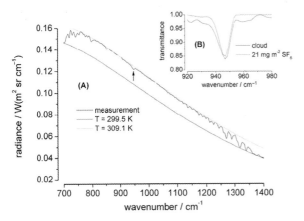

Fig. 13 Infrared spectra for the SF_6 detection in the soccer stadium. (**A**) Measured radiance spectrum: The small SF_6 signal band is indicated by an *arrow*. The Plank radiance for the temperature of 36 °C (background), and 26.4 °C (air/cloud) indicates the *lower* and *upper* limit for the dynamic range of the cloud signal. (**B**) *Upper right* corner: Transmittance of the measured SF_6 cloud, together with the transmittance of a reference spectrum for a SF_6 with a column density of 21 mg m^{-2}. The detection limit is about 5 mg m^{-2}

Fig. 14 RAPID mounted on a vehicle of the fire department

expert advisor for chemicals could made recommendations for the incident command.

5 Summary

Using passive FTIR for the detection, identification and visualization of airborne pollutants is very helpful for different questions on air monitoring. According to the different kinds of software used with passive FTIR it can be shown, that academic use is although possible like the use for civil protection, for air monitoring through government agencies, and for fast recognition of leaks in chemical industries. Especially for the monitoring of diffuse emission sources or line emission sources, passive FTIR is a good choice. The software must be adjusted to the different kinds of questions on advanced air monitoring.

References

Beil A.: Real time Remote Detection and Cloud Imaging of CWA and TIC using high speed FTIR Systems. 5th Joint Conference on Standoff Detection for Chemical and Biological Defense at Williamsburg, Virgina, 24–28th September 2001

Beil A., Daum R., Matz G., Harig R.: "Remote sensing of atmospheric pollution by passive FTIR spectrometry" in Spectroscopic Atmospheric Environmental Monitoring Techniques, Klaus Schäfer, Editor, Proceedings of SPIE Vol. 3493 1998, 32–43

Bruker Daltonics; RAPID Control Software Version 2.0; Leipzig/Germany, 2005a

Bruker Daltonics; Spectra Acquisition Tool Version 1.2; Leipzig/Germany, 2005b

Flanigan D. F.: "Prediction of the limits of detection of hazardous vapors by passive infrared spectroscopy with the use of modtran", Applied Optics 35, 1996, 6090–6098

Griffith, H.: Fourier Transform Infrared Spectroscopy. John Wiley & sons; New York/USA, 1996

Harig R., Matz R., Rusch P.: Infrarot-Fernerkundungssystem für die chemische Gefahrenabwehr. Bundesamt für Bevölkerungsschutz und Katastrophenhilfe Zivilschutz-Forschung Band 48, Bonn/Germany 2006

http://www.et1.tu-harburg.de/ftir/

Institute of Measurement Technology at the Technical University of Hamburg-Harburg; GeDetekt – Experimental Rapid Version; Hamburg/Germany, 2006

Kraugmann, F.: Einsatz passiver FT/IR-Spektrometrie zur Frueherkennung von Stofffreisetzungen; Diplomarbeit, Wuppertal/Germany, 2007

Remote Sensing of Tropospheric Trace Gases (NO$_2$ and SO$_2$) from SCIAMACHY

Chulkyu Lee, Randall V. Martin, Aaron van Donkelaar, Andreas Richter, John P. Burrows and Young J. Kim

Abstract Atmospheric trace gases can be measured by remote sensing of scattered sunlight from space, using its unique absorption features in the ultraviolet region. The satellite remote sensing approach associated with the spectral fit technique has been successfully employed for measurements of tropospheric trace gases on global and regional scales. Here we present the retrievals of tropospheric traces gases (NO$_2$ and SO$_2$) from SCIAMACHY (Scanning Imaging Absorption Spectrometer for Atmsopheric Chartography) onboard the ENVISAT satellite and the calculation of their air mass factor (AMF) used to convert slant columns to vertical columns. The AMF used here is calculated from the integral of the relative vertical distribution of the trace gases from a global 3-D model of tropospheric chemistry (GEOS-Chem), weighted by altitude-dependent scattering weights computed with a radiative transfer model (Linearized Discrete Ordinate Radiative Transfer), and accounts for cloud scattering using cloud fraction and cloud top pressure. The results demonstrate a high sensitivity of the SCIAMACHY instrument to NO$_2$ and SO$_2$ concentrations, and the possibility to retrieve them in the boundary layer.

Keywords Remote sensing · SCIAMACHY · NO$_2$ · SO$_2$ · Air mass factor (AMF)

1 Introduction

Over the last decade, solar backscatter satellite measurements of chemical species in the atmosphere have become an integral and in many cases complementary development to existing ground-based and airborne measurements. Development of algorithms used to retrieve data from satellite observations has been of primary interest. Satellite remote sensing of tropospheric trace gases began in 1978 with the launch of the Total Ozone Mapping Spectrometer (TOMS) instrument onboard the Nimbus-7 satellite. The TOMS instrument was first aimed at determining global knowledge on stratospheric O$_3$. It was later recognized that this instrument also produces information about volcanic SO$_2$ (Krueger 1983; Krueger et al. 1995; Carn et al. 2004) and tropospheric O$_3$ (Fishman et al. 1990). Since then, numerous satellite-based instruments have provided important measurements from earth observing platforms in low-earth orbit giving nearly global coverage of several key trace gases (Burrows et al. 1999; Bovensmann et al. 1999; Levelt et al. 2006; Fishman et al. 2008; Martin 2008). Table 1 shows a summary of major UV-Vis nadir satellite instruments designed for remote sensing of trace gases in the troposphere. These instruments are in Sun-synchronous, polar orbits.

Satellite remote sensing of tropospheric trace gases effectively began with the launch of the Global Ozone Monitoring Experiment (GOME) (Burrows et al. 1999) onboard the ERS-2 satellite. GOME made global measurements from July 1995 to June 2003, and in particular tropospheric NO$_2$, SO$_2$, HCHO, BrO, and O$_3$ were retrieved from GOME (Chance 1998; Eisinger and Burrows 1998; Wagner and Platt 1998; Burrows et al. 1999; Chance et al. 2000; Palmer et al. 2001; Martin et al. 2002; Richter and Burrows 2002; Beirle et al. 2003; Ladstätter-Weißenmazer et al. 2003; Boersma et al. 2004; Irie et al. 2005; Khokhar et al. 2005; Liu et al. 2005; Richter et al. 2005; Liu et al. 2007; Uno

C. Lee (✉)
Department of Physics and Atmospheric Sciences,
Dalhousie University, Halifax, Nova Scotia, Canada
e-mail: chulkyu.lee@dal.ca

Y.J. Kim et al. (eds.), *Atmospheric and Biological Environmental Monitoring*,
DOI 10.1007/978-1-4020-9674-7_5, ⓒ Springer Science+Business Media B.V. 2009

63

Table 1 Major satellite instruments for remote sensing of atmospheric trace gases

Instrument	Satellite platform	Measurement period	Equator crossing time (local time)	Spectral range (μm)	Spectral resolution (nm)	Spatial resolution, Nadir (km)	Global coverage (days)
GOME[a]	ERS-2	1995–2003[b]	10:30	0.23–0.79	0.2–0.4	320×40	3
SCIAMACHY[c]	Envisat	2002	10:00	0.23–2.3	0.25–0.4	60×30	6
OMI[d]	EOS-Aura	2004	13:30	0.27–0.50	0.5	24×13	1
GOME-2	MetOp	2006	09:30	0.24–0.79	0.26–0.51	80×40	1.5

[a]Global Ozone Monitoring Experiment.
[b]Operating at reduced coverage since June 2003.
[c]SCanning Imaging Absorption SpectroMeter for Atmospheric CHartographY.
[d]Ozone Monitoring Instrument.

et al. 2007). Validation of GOME observations with in situ measurements is complicated by its large spatial footprint and sparse temporal coverage. However the limited validations of the retrievals with aircraft measurements for HCHO (Ladstätter-Weißenmayer et al. 2003; Martin et al. 2004) and NO_2 (Heland et al. 2002; Ladstätter-Weißenmayer et al. 2003; Martin et al. 2004) are consistent with the expected uncertainty.

The SCanning Imaging Absorption SpectroMeter for Atmospheric CHartography (SCIAMACHY) is an extended version of GOME, and measures back-scattered solar radiation upwelling from the atmosphere, alternately in nadir and limb viewing geometry (Bovensmann et al. 1999). Retrieval algorithms for tropospheric trace gases (e.g., BrO, IO, CO, NO_2, CHOCHO, HCHO, SO_2, H_2O, O_3, etc.) and their expected uncertainty for SCIAMACHY are similar to those for GOME for the UV and visible channels (Afe et al. 2004; Buchwitz et al. 2005; Frankenberg et al. 2005; Richter et al. 2005; Martin et al. 2006; Wittrock et al. 2006; Lee et al. 2008a,b; Schönhardt et al. 2008). Validation of the retrieved tropospheric NO_2 columns with aircraft measurements yields mean agreement within 20% of the retrieved column (Heue et al. 2005; Martin et al. 2006).

The Ozone Monitoring Instrument (OMI) is a nadir-viewing, imaging spectrometer that uses two-dimensional CCD detectors to measure the solar radiation backscattered by the Earth's atmosphere and surface over 270–500 nm with a spectral resolution of 0.5 nm (Levelt et al. 2006). The spatial resolution of OMI is up to $13 \times 24 \, km^2$ at nadir with coarser resolution at larger viewing angles. Retrieval algorithms of tropospheric trace gases (e.g., HCHO, NO_2, SO_2, and O_3) and their expected uncertainty for OMI are similar to those for GOME and SCIAMACHY (Krotkov et al. 2006; Bucsela et al. 2006; Boersma et al. 2007). Validation of the retrieved columns with ground-based remote sensing measurements (Celarier et al. 2008; Wenig et al. 2008) and aircraft measurements (Boersma et al. 2008; Bucsela et al. 2008) yields mean agreement to within 30%.

Another instrument currently in orbit is GOME-2 onboard the MetOp satellite launched in October 2006, the first operational meteorological platform to have instrumentation dedicated to making tropospheric trace gas measurements. Retrievals of tropospheric trace gases from GOME-2 are similar to those from GOME, SCIAMACHY, and OMI.

The satellite remote sensing approach associated with spectral fitting has been successfully employed for measurements of atmospheric trace gases on global and regional scales (e.g., Platt 1994; Chance et al. 1997; Eisinger and Burrows 1998; Wagner and Platt 1998; Burrows et al. 1999; Martin et al. 2002; Palmer et al. 2001; Afe et al. 2004; Richter et al. 2005; Wittrock et al. 2006; Lee et al. 2008b). Here we present the retrieval of NO_2 and SO_2 slant columns from SCIAMACHY using the spectral fitting and AMF calculation to convert the slant columns to vertical columns. Section 2 gives a description of the principle of the spectral fit, and the AMF formulation is shown in Section 3. The retrieval of tropospheric vertical columns is presented in Section 4.

2 Principle of Spectral Fitting

Spectral fitting has been applied to measure trace-gas concentrations in the troposphere and stratosphere (Solomon et al. 1987; Platt 1994). A large number of other molecules absorbing the light in the UV and the visible wavelength region, e.g. BrO, CS_2, HCHO, IO, NO_2, NO, NH_3, O_3, OClO, SO_2, etc, have also been detected (e.g., Platt et al. 1979; Sanders et al. 1993; Platt 1994; Chance 1998; Eisinger and Burrows 1998;

Wagner and Platt 1998; Alicke et al. 1999; Martin et al. 2002; Palmer et al. 2003; Afe et al. 2004; Richter et al. 2005; Wittrock et al. 2006; Lee et al. 2008b). Spectral fitting is a very sensitive measurement technique for these trace gases since they exhibit strong and highly structured absorption cross sections in the UV and visible spectral regions. Because spectral fitting is capable of inferring trace species at low concentrations in the atmosphere it is especially useful in the detection of highly reactive species, such as the free radicals OH, NO_3, halogen oxides (ClO, BrO, IO, etc.). The simultaneous determination of the concentration of several trace gases, by analyzing the sum of their absorptions in one wavelength interval, reduces measurement time and allows analysis of the chemical composition of the observed air mass at high temporal resolution.

Spectral fitting is based on the Beer-Lambert Law, which describes the decrease of the light intensity I while passing through an infinitesimally thin layer dl of an absorbing trace gas. The change in the light intensity is directly proportional to the intensity of light before the absorbing matter I_0, the concentration of the trace gas that might not be constant over the absorbing path $C(l)$, the temperature-dependent absorption cross section of the trace gas $\alpha(\lambda, T)$ with wavelength λ and the length of the absorbing layer dl.

$$I(\lambda) = I_0(\lambda) \cdot e^{-\alpha(\lambda,T) \int C(l)dl}. \quad (1)$$

One of the many problems in atmospheric measurements is that it is difficult to determine the true intensity $I_0(\lambda)$ that would be received from the light source without any extinction. The fundamental concept of spectral fitting is to determine the 'differential' absorption of trace gases by dividing the total temperature-dependent absorption cross sections $\alpha_i(\lambda, T)$ of a single trace gas i down into a contribution from the structure portion: $\alpha'_i(\lambda, T)$ that shows rapid variations in wavelength and $\alpha_{i0}(\lambda, T)$, that varies slowly in wavelength.

$$\alpha_i(\lambda, T) = \alpha_{i0}(\lambda, T) - \alpha'_i(\lambda, T) \quad (2)$$

$\alpha_{i0}(\lambda, T)$ describes the general slope caused by Rayleigh or Mie scattering, while $\alpha'_i(\lambda, T)$ indicates the rapid variations in λ due to absorption lines or bands. 'Rapid' and 'slow' variations of the absorption cross sections are in fact a function of both the observed wavelength interval and the width of the absorption bands that needs to be detected.

$$I(\lambda) = \left\{ I_0(\lambda) \cdot A(\lambda) \cdot e^{-\sum(\alpha_{i0}(\lambda,T) \int C_i(l)dl) + \varepsilon_{Ray}(\lambda) + \varepsilon_{Mie}(\lambda)} \right\}$$
$$\cdot e^{-\sum(\alpha'_i(\lambda,T) \int C_i(l)dl)} \quad (3)$$

where ε_{Ray} and ε_{Mie} are extinction coefficients of Rayleigh and Mie scattering, respectively.

The exponential function behind the bracket describes the rapid variation of the spectrum, the so-called *differential absorption*, while the first exponential function considers the broadband variations of the cross sections and the effects of Rayleigh and Mie scattering. The transmission of the optical system, which is varying slowly with wavelength, is summarized in the attenuation factor $A(\lambda)$. The bracket of (3) can be defined as the quantity I_0'.

$$I(\lambda) = I_0'(\lambda) \cdot e^{-\sum(\alpha'_i(\lambda) \int C_i(l)dl)} \quad (4)$$

Usually this analysis in scattered sunlight algorithms yields the so called slant columns (SC), which is defined as the trace gas concentration integrated along the light path, and in the case where only one absorber is present can be given by:

$$SC = \int C_i(l) dl = \frac{1}{\alpha'_i(\lambda, T)} \ln\left(\frac{I_0'(\lambda)}{I(\lambda)}\right). \quad (5)$$

3 Air Mass Factor (AMF)

For a single SC measurement the individual photons registered in the detector may have traveled different paths through the atmosphere before being detected. Since the SC depends on the observation geometry and the meteorological conditions, it is usually converted to a vertical column (VC), which is defined as the trace gas concentration $C(z)$ integrated along the vertical path through the atmosphere.

$$VC = \int C(z) dz \quad (6)$$

Since the VC only depends on the trace gas profile, it is independent of the viewing geometry and the trajectories along which the light traveled through the atmosphere before reaching the instrument. The air mass factor (AMF) is defined as the ratio of SC of the absorber (i.e. that viewed by the satellite in the measured radiance spectrum) to the vertical column (VC):

$$AMF = \frac{SC}{VC} \qquad (7)$$

The AMF calculation applied here is based on the formulation of Palmer et al. (2001) and Martin et al. (2002) for the retrieval of HCHO and NO_2 from GOME measurements. Using the Reference Sector Method (RSM) (Martin et al. 2002; Richter and Burrows 2002), the stratospheric component of the slant column is subtracted prior to the application of the AMF to the tropospheric component. The AMF is sensitive to the relative vertical distribution of the absorber due to Rayleigh scattering and to Mie scattering by clouds. A vertical shape factor $S(\sigma)$ (dimensionless) representing a normalized vertical profile of mixing ratio over the sigma (σ) vertical coordinate is determined by a global 3-D model of tropospheric chemistry, GEOS-Chem (Bey et al. 2001), for each observation pixel

$$S(\sigma) = C(\sigma)\frac{\tau_{air}}{\tau_{absorber}}, \qquad (8)$$

where $C(\sigma)$ is the mixing ratio of the absorber, and τ_{air} and $\tau_{absorber}$ are the tropospheric vertical columns of air and absorber from surface pressure to the pressure of the upper boundary layer of the model. Scattering weights $\omega(\sigma)$ to describe the radiance I observed by SCIAMACHY to the abundance of the absorber at each level σ can be given (Palmer et al. 2001):

$$\omega(\sigma) = -\frac{1}{AMF_G}\frac{\alpha(\sigma)}{\alpha_e}\frac{\partial(\ln I/I_0)}{\partial \tau}, \qquad (9)$$

where $\alpha(\sigma)$ is the temperature-dependent absorption cross section, α_e is the effective absorption cross-section representing an average cross section weighted by the vertical distribution of the absorber in the tropospheric column, I_0 is a normalization constant of value unity, and $\partial \tau$ is the optical depth increment of the absorber as a function of σ. A geometric air mass factor AMF_G is a simple function of the solar zenith angle θ_s and the SCIAMACHY viewing angle θ_v

$$AMF_G = \sec(\theta_s) - \sec(\theta_v). \qquad (10)$$

The AMF is then written as (Palmer et al. 2001)

$$AMF = AMF_G \int_{\sigma_T}^{1} \omega(\sigma)S(\sigma)d\sigma \qquad (11)$$

where the integral is taken here from the model tropopause σ_T to the surface.

The AMF formulation accounts for cloud-contaminated pixels of SCIAMACHY, which are typical in satellite measurements, using the cloud information (e.g. cloud fraction and cloud top height) for the AMF calculation (Martin et al. 2002) (the cloud fraction and cloud top height for this work were determined from SCIAMACHY measurements using the FRESCO algorithm (Koelemeijer et al. 2001), available from the Tropospheric Emission Monitoring Internet Service (http://www.temis.nl/)). Scattering weights for both the clear-sky and cloudy fractions of the pixel at all levels in the troposphere are calculated using Mie scattering by clouds included in the radiative transfer model (Linearized Discrete Ordinate Radiative Transfer, LIDORT (Spurr et al. 2001; Spurr 2002)). Assuming the same shape factor in each pixel, AMFs are calculated for the clear-sky and cloudy fractions for the pixel, AMF_a and AMF_c. The combined AMF (accounting for the cloud-contaminated pixel) is then given by Martin et al. (2002).

$$AMF = \frac{AMF_a \cdot R_a(1-f) - AMF_c \cdot R_c \cdot f}{R_a(1-f) + R_c f}, \qquad (12)$$

where f is the cloud fraction ($0 \leq f \leq 1$). The reflectivity of the clear- and cloudy-sky R_a and R_c is defined as

$$R_a = \frac{\pi \cdot I_a}{E_0 \cdot \cos(\theta_s)} \qquad (13a)$$

$$R_c = \frac{\pi \cdot I_c}{E_0 \cdot \cos(\theta_S)} \qquad (13b)$$

where E_0 is the solar irradiance at the top of the atmosphere perpendicular to the direction of incident sunlight, and the irradiance observed by SCIAMACHY is decomposed into the contributions for the clear-sky (I_a) and cloudy fractions (I_c), respectively. Each reflectivity includes contributions from the surface, Rayleigh scattering, and also cloud scattering for the cloudy pixel. Values of R_a and R_c are obtained from a radiative transfer model.

Figure 1 shows the mean AMFs of SCIAMACHY tropospheric NO_2 and SO_2 for 2005 and 2006, respectively. AMFs tend to be largest over ocean and decrease over polluted regions. In polluted regions, SO_2 AMFs are generally smaller than NO_2 AMFs, reflecting the

Fig. 1 Air mass factor (AMF) for conversion of tropospheric slant columns to tropospheric vertical columns in the SCIAMACHY measurements. AMF values of tropospheric NO_2 (*top*) and SO_2 (*bottom*) are yearly means for 2005 and 2006, respectively. The AMF accounts for aerosol effects, following Martin et al. (2003)

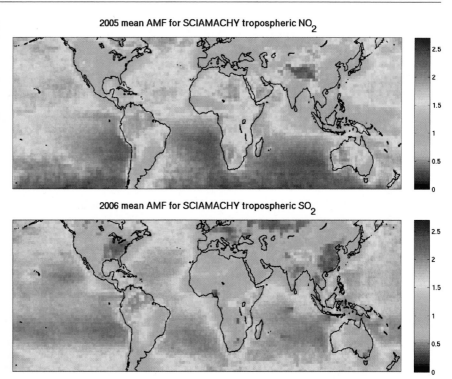

increased molecular scattering at short SO_2 wavelengths. In the SO_2 AMF calculation, volcanic emissions of SO_2 from continuously active volcanoes are included from the database of Andres and Kasgnoc (1998) in the global 3-D model.

4 Retrieval of Tropospheric Columns

4.1 NO_2

The retrieval of tropospheric NO_2 columns presented here is based on the algorithms of Martin et al. (2002, 2006). Total slant columns of NO_2 were determined by directly fitting backscattered radiance spectra observed by SCIAMACHY. The specification of the NO_2 retrieval is summarized in Table 2.

The spectral fit is optimized for SCIAMACHY over the wavelength region 426–452 nm using measured absorption cross sections for NO_2, O_3, H_2O, $O_2 - O_2$, and the Ring effect (Chance and Spurr 1997). The solar spectra for each orbit are from the elevation scan mirror on the SCIAMACHY instrument. NO_2 slant columns retrieved from SCIAMACHY for 2005 are shown in the upper of Fig. 2.

Table 2 Specifications of NO_2 and SO_2 retrievals for SCIAMACHY data

Species	Wavelength range (nm)	Polynomial order	Cross-sections included in the fit
NO_2	426–450	3	NO_2 at 243 K (Vandaele et al. 2002)
			O_3 at 223 and 243 K (Bogumil et al. 2003)
			H_2O at 296 K (Rothman et al. 2005)
			O_2-O_2 at 296 K (Greenblatt et al. 1990)
			Ring (Chance and Spurr 1997)
SO_2	315–327	4	SO_2 at 295 K (Vandaele et al. 1994)
			O_3 at 223 and 243 K (Bogumil et al. 2003)
			Ring (Vountas et al. 1998)
			Undersampling
			Polarization dependency

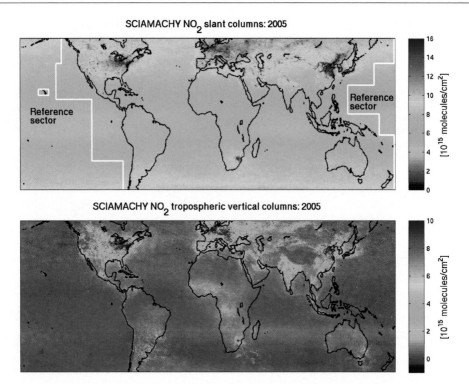

Fig. 2 NO$_2$ slant column (*upper*) and tropospheric vertical columns (*bottom*) retrieved from SCIAMACHY. The *white lines* of the *upper panel* bound the reference sector for the tropospheric NO$_2$ over the central Pacific region, which is used to determine tropospheric columns. The columns averaged over the region bounded by the *white line* represent a latitudinal bias and non-tropospheric amounts of NO$_2$. The tropospheric vertical columns (VC) are determined by dividing tropospheric slant columns (SC) with the air mass factor (AMF) (VC = SC/AMF)

The stratospheric NO$_2$ column and instrument biases are removed by the Reference Sector Method (RSM) (Martin et al. 2002; Richter and Burrows 2002) to obtain the tropospheric slant columns (see the upper of Fig. 2). The RSM used here is done by subtracting the corresponding columns from the ensemble of SCIAMACHY observations for the appropriate latitude, after correcting for the small amount of tropospheric NO$_2$ over the central Pacific, and using a stratospheric assimilation to account for zonal variability (Boersma et al. 2004).

Figure 2 shows tropospheric vertical columns of NO$_2$ retrieved from SCIAMACHY for 2005. Pronounced enhancements of NO$_2$ are visible over major industrial and metropolitan areas like the United States East coast, Western Europe, East China, the Persian Gulf, South Africa, or Hongkong. Tropospheric NO$_2$ columns are closely related to surface NOx emissions (Leue et al. 2001; Martin et al. 2003). Weaker enhancements are found in central Africa and Brazil from biomass burning (Richter and Burrows 2002) and observed over northern equatorial Africa with a maximum at the beginning of the rainy season that could be attributed to rain-induced soil NOx emissions (Jaeglé et al. 2004). There are weak enhancements along ocean ship tracks near Kuala Lumpur (Beirle et al. 2004a, b; Richter et al. 2004).

The error budget of satellite measurements of tropospheric NO$_2$ columns has been discussed in detail in several publications (Martin et al. 2002; Richter and Burrows 2002; Boersma et al. 2004). The main contributions to the error are random fitting uncertainties, uncertainties related to the subtraction of the stratospheric contributions, uncertainties from residual clouds, and the AMF. The total uncertainties in the retrieval of tropospheric NO$_2$ columns over continental source regions is largely determined by the AMF calculation due to surface reflectivity, clouds, aerosols, and the trace gas profile. An overall assessment of errors leads to 5×10^{14}–1×10^{15} molecules/cm^2 for the absolute error and 40–60% relative errors for monthly averages (Richter and Burrows 2002;

Fig. 3 SO$_2$ slant column (*upper*) and tropospheric vertical columns (*bottom*) retrieved from SCIAMACHY. The *white lines* of the *upper panel* bound the reference sector for the tropospheric SO$_2$ over the central Pacific region. The columns averaged over the region bounded by the *white line* represent a latitudinal bias and non-tropospheric amounts of SO$_2$. In the area of the Southern Atlantic Anomaly (SAA), large scatter in SCIAMACHY SO$_2$ results from exposure of the instrument to radiation and particles

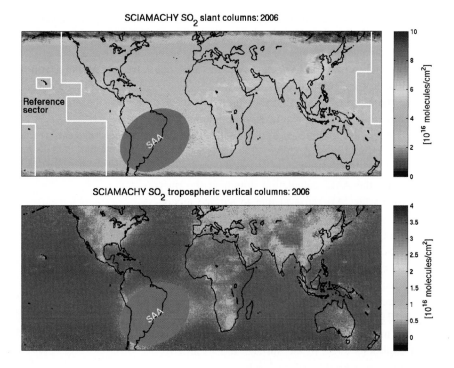

SCIAMACHY SO$_2$ slant columns: 2006

SCIAMACHY SO$_2$ tropospheric vertical columns: 2006

Martin et al. 2003; Richter et al. 2005; Boersma et al. 2004; Martin et al. 2006).

4.2 SO$_2$

The SO$_2$ analysis for SCIAMACHY is based on the algorithm of Eisinger and Burrows (1998), Afe et al. (2004), Richter et al. (2006), and Lee et al. (2008a, b). The wavelength range of 315–327 nm was used for the SO$_2$ fit as the differential absorptions are large and interference by other species is relatively small. In addition to the SO$_2$ cross-section, two ozone cross-sections, a synthetic Ring spectrum, an undersampling correction, and the polarization dependency of the SCIAMACHY instrument are included in the fit (see Table 2). Daily solar irradiance measurements taken with the ASM diffuser are used as background spectrum. Slant columns of SO$_2$ retrieved from SCIAMACHY for 2006 are shown in the upper of Fig. 3. There is a latitude-dependent offset in the SO$_2$ slant column, related to the strong interference with ozone and an imperfect correction of Ring effect (Van Roozendael et al. 2002; Khokhar et al. 2005; Lee et al. 2008b). This latitude-dependent offset was removed by subtracting monthly mean of the columns

taken over the Pacific from the total column by the reference sector method (Martin et al. 2002; Richter and Burrows 2002; Khokhar et al. 2005). The reference sector representing an area with presumably little SO$_2$ over the Pacific is shown in the upper of Fig. 3.

Figure 3 shows tropospheric vertical columns of SO$_2$ retrieved from SCIAMACHY for 2006. Enhanced anthropogenic SO$_2$ is seen over metropolitan areas (e.g., East China, Hongkong in China, or Chongqing in China, or Northeast United States) and industrial areas (e.g., refineries in the Persian Gulf or the Highveld region in South Africa). Volcanic SO$_2$ is observed, such as around Nyamuragira in the Democratic Republic of Congo, El Reventador in South America, Yasur in Vanuatu. In the region of South America and the Southern Atlantic Ocean, large scatter results from exposure of the instrument to radiation and particles, and the SO$_2$ data in this region is screened in Fig. 3. This is a doughnut-shaped region of high-energy charged particles trapped by the earth's magnetic field (Van Allen radiation belts) and named South Atlantic Anomaly (SAA). When low-orbiting satellites (e.g., ERS-2 and Envisat) pass through this radiation belt in the SAA, the charged particles in this region may cause the detector to be exposed to higher-than-normal irradiance, decreasing the quality of the measurements, notably in UV. This reduction of the signal to noise also affects the SO_2 retrieval in the SAA region and results in a large scatter in SO_2 slant columns. NO_2 retrieval is less sensitive to the SAA due to its longer wavelength range.

The main errors of satellite measurements of tropospheric SO_2 columns come from random fitting uncertainties, uncertainties from residual clouds, and the AMF (Khokhar et al. 2005; Thomas et al. 2005; Richter et al. 2006; Krotkov et al. 2008). The uncertainties by the AMF calculation dominate the overall error of the SO_2 retrieval due to surface albedo, aerosol loading and most importantly the vertical profile of SO_2 in the atmosphere (Khokhar et al. 2005; Richter et al. 2006). In addition, the sensitivity also changes with wavelength as Rayleigh scattering increases to shorter wavelengths. If a simplified volcanic profile (e.g., with a maximum between 9 and 12 km) is used globally, this could lead to systematic underestimation of SO_2 columns in regions with emissions close to the surface and to an overestimation over bright surfaces such as ice or clouds. The overall error for the monthly mean uncertainties is 1×10^{16}–2×10^{16} molecules/cm^2 for the absolute error and 100–200% relative errors.

5 Conclusions

Global mapping of trace gases from space provides critical information for constraining their emissions and improving our understanding of tropospheric chemistry. Satellite measurements of trace gases in the troposphere (e.g., NO_2 and SO_2) are being applied for these purposes. We have presented the retrievals of NO_2 and SO_2 in the troposphere and the AMF calculation to convert slant columns to vertical columns. The AMF has been calculated from the relative vertical distribution determined from a global 3-D model of tropospheric chemistry (GEOS-Chem), weighted by altitude-dependent scattering weights computed with a radiative transfer model (LIDORT).

Satellite remote sensing has been successfully applied to measurements of trace gases in the troposphere. Continued development of algorithms to reduce the uncertainty in the retrievals of trace gases in the troposphere and boundary layer is still of primary interest. Satellite measurements of the composition of the boundary layer over land, which is of direct relevance for surface air quality, is a rapidly advancing area (Lamsal et al. 2008). Satellite measurements of trace gases in the atmosphere are a crucial step forward for real time monitoring of air quality and forecasting such as hazard warning on a global scale.

Acknowledgments The SCIAMACHY SO_2 slant column data were obtained from the Institute of Environmental Physics and Remote Sensing (IUP/IFE), University of Bremen, Germany.

Abbreviations

AMF	Air Mass Factor
CCD	Charge Coupled Device
ENVISAT	Environment Satellite
FRESCO	Fast Retrieval Scheme for Clouds from the Oxygen A-band
GOME	Global Ozone Monitoring Experiment
LIDORT	Linearized Discrete Ordinate Radiative Transfer
OMI	Ozone Monitoring Instrument
RSM	Reference Sector Method
SC	Slant Column
SCIAMACHY	Scanning Imaging Absorption Spectrometer for Atmsopheric Chartography
VC	Vertical Column

References

Afe O. T., Richter A., Sierk B. et al. (2004) BrO emission from volcanoes: A survey using GOME and SCIAMACHY measurements. Geophys Res Lett. doi:10.1029/2004GL020994

Alicke B., Hebestreit K., Stutz J. et al. (1999) Detection of iodine oxide in the marine boundary layer. Nature. 397:572–573

Andres R. J., Kasgnoc A. D. (1998) A time-averaged inventory of subaerial volcanic sulfur emissions. J Geophys Res. 103:25251–25261

Beirle S., Platt U., Wenig M. et al. (2003) Weekly cycle of NO_2 by GOME measurements: A signature of anthropogenic sources. Atmos Chem Phys. 3:2225–2232

Beirle S., Platt U., Wenig M. et al. (2004a) NOx production by lightning estimated with GOME. Adv Space Res. 34:793–797

Beirle S., Platt U., von Glasow R. et al. (2004b) Estimate of nitrogen oxide emissions from shipping by satellite remote sensing. Geophys Res Lett. doi:10.1029/2004GL020312

Bey I., Jacob D. J., Yantosca R. M. et al. (2001) Global modeling of tropospheric chemistry with assimilated meteorology: Model description and evaluation. J Geophys Res. 106:23073–23096

Boersma K. F., Eskes H. J., Brinksma E. J. (2004) Error analysis for tropospheric NO_2 retrieval from space. J Geophys Res. doi:10.1029/2003JD003962

Boersma K. F., Eskes H. J., Veefkind J. P. et al. (2007) Near-real time retrieval of tropospheric NO_2 from OMI. Atmos Chem Phys. 7:2103–2118

Boersma K.F., Jacob D.J., Bucsela E.J. et al. (2008) Validation of OMI tropospheric NO_2 observations during INTEX-B and application to constrain NOx emissions over the eastern United States and Mexico. Atmos Environ. doi:10.1016/j.atmosenv.2008.02.004

Bogumil K., Orphal J., Homann T. et al. (2003) Measurements of molecular absorption spectra with the SCIAMACHY Pre-Flight Model: Instrument characterization and reference data for atmospheric remote sensing in the 230–2380 nm region. J Photoch Photobio A. 157:167–184

Bovensmann H., Burrows J.P., Buchwitz M. et al. (1999) SCIAMACHY: Mission objectives and measurements modes. J Atmos Sci. 56:127–150

Buchwitz M., de Beek R., Noël S. et al. (2005) Carbon monoxide, methane and carbon dioxide columns retrieved from SCIAMACHY by WFM-DOAS: Year 2003 initial data set. Atmos Chem Phys. 5:3313–3329

Bucsela E. J., Celarier E. A., Wenig M. O. et al. (2006) Algorithm for NO_2 vertical column retrieval from the Ozone Monitoring Instrument. IEEE Transac Geosci Remote Sensing. 44:1245–1258

Bucsela E. J., Perring A. E., Cohen R. C. et al. (2008) Comparison of NO_2 in situ aircraft measurements with data from the Ozone Monitoring Instrument. J Geophys Res. doi:10.1029/2007JD008838

Burrows J. P, Weber M., Buchwitz M. et al. (1999) The global ozone monitoring experiment (GOME): Mission concept and first scientific results. J Atmos Sci. 56: 151–175

Carn, S. A., Krotkov N. A., Gray M. A. et al. (2004) Fire at Iraqi sulfur plant emits SO_2 clouds detected by Earth Probe TOMS. Geophys Res Lett. doi: 10.1029/2004GL020719

Celarier E. A., Brinksma E. J., Gleason J. F. et al. (2008) Validation of Ozone Monitoring Instrument nitrogen dioxide columns. J Geophys Res. doi:10.1029/2007JD008908

Chance K. (1998), Analysis of BrO measurements from the global ozone monitoring experiment. Geophys Res Lett. 25:3335–3338

Chance K. V., Burrows J. P., Perner D. et al. (1997) Satellite measurements of atmospheric ozone profiles including tropospheric ozone, from ultraviolet/visible measurements in the nadir geometry: A potential method to retrieve tropospheric ozone. J Quant Spectrosc Radiat Transfer. 57: 467–476

Chance K., Palmer P., Spurr R. J. D. et al. (2000) Satellite observations of formaldehyde over North America from GOME. Geophys Res Lett. 27:3461–3464

Chance K. V., Spurr R. J. D. (1997) Ring effect studies: Rayleigh scattering, including molecular parameters for rotational Raman scattering, and the Fraunhofer spectrum. Appl Opt. 36:5224–5230

Eisinger M., Burrows J. P. (1998) Tropospheric sulfur dioxide observed by the ERS-2 GOME instrument. Geophys Res Lett. 25:4177–4180

Fishman J., Bowman K. W., Burrows J. P. et al. (2008) Remote sensing of tropospheric pollution from space. Bull Am Meteorol Soc. doi:10.1175/2008BAMS2526.1

Fishman J., Watson C. E., Larsen J. C. et al. (1990) Distribution of tropospheric ozone determined from satellite data. J Geophys Res. 95:3599–3617

Frankenberg C., Platt U., Wagner T. (2005) Retrieval of CO from SCIAMACHY onboard ENVISAT: Detection of strongly polluted areas and seasonal patterns in global CO abundances. Atmos Chem Phys. 5:1639–1644

Greenblatt G. D., Orlando J. J, Burkholder J. B. et al. (1990) Absorption measurements of oxygen between 330 and 1140 nm. J Geophys Res. 95:18577–18585

Heland J., Schlager H., Richter A. et al. (2002) First comparison of tropospheric NO_2 column densities retrieved from GOME measurements and in situ aircraft profile measurements. Geophys Res Lett. doi:10.1029/2002GL015528

Heue K.-P., Richter A., Bruns M. et al. (2005) Validation of SCIAMACHY tropospheric NO_2-columns with AMAXS-DOAS measurements. Atmos Chem Phys. 5:1039–1051

Irie H., Sudo K., Akimoto H. et al. (2005) Evaluation of long-term tropospheric NO_2 data obtained by GOME over East Asia in 1996-2002. Geophys Res Lett. doi:10.1029/2005GL022770

Jaeglé L., Martin R. V., Chance K. et al. (2004) Satellite mapping of rain-induced nitric oxide emissions from soils. J Geophys Res. doi:10.1029/2004JD004787

Khokhar M. F., Frankenberg C., Van Roozendael M. et al. (2005) Satellite observation of atmospheric SO_2 from volcanic eruptions during the time-period of 1996–2002. Adv Space Res. 36:879–887

Koelemeijer R. B. A., Stammes P., Hovenier J. W. et al. (2001) A fast method for retrieval of cloud parameters using oxygen A band measurements from the Global Ozone Monitoring Experiment. J Geophys Res. 106:3475–3490

Krotkov N. A., Carn S. A., Krueger A. J. et al. (2006) Band residual difference algorithm for retrieval of SO_2 from the Aura Ozone Monitoring Instrument (OMI). IEEE Trans Geosci Remote Sensing. 44:1259–1266

Krotkov N. A., McClure A. B., Dickerson R. R. et al. (2008) Validation of SO_2 retrievals from the Ozone Monitoring Instrument (OMI) over NE China. J Geophys Res. doi:10.1029/2007JD008818

Krueger A. (1983) Sighting of El Chichon sulfur dioxide clouds with the Nimbus 7 total ozone mapping spectrometer. Sci. 220:1377–1379

Krueger A., Walter L., Bhartia P. et al. (1995) Volcanic sulfur dioxide measurements from the total ozone mapping spectrometer instruments. J Geophys Res. 100(D7):14057–14076

Ladstätter-Weißenmayer A., Heland J., Kormann R. et al. (2003) Transport and build-up of tropospheric trace gases during the MINOS campaign: Comparision of GOME, in situ aircraft measurements and MATCH-MPIC-data. Atmos Chem Phys. 3:1887–1902

Lamsal L. N., Martin R. V., van Donkelaar A. et al. (2008) Ground-level nitrogen dioxide concentrations inferred from the satellite-borne Ozone Monitoring Instrument. J Geophys Res. doi:10.1029/2007JD009235

Lee C., Richter A., Burrows J. P. et al. (2008a) Impact of transport of sulfur dioxide from the Asian continent on the air quality over Korea during May 2005. Atmos Environ. doi: 10.1016/j.atmosenv.2007.11.006

Lee C., Richter A., Weber M. et al. (2008b) SO_2 retrieval from SCIAMACHY using the Weighting Function DOAS (WFDOAS) technique: Comparison with standard DOAS retrieval. Atmos Chem Phys Diss. 8:10817–10839

Leue C., Wenig M., Wagner T. et al. (2001) Quantitative analysis of NO_x emissions from GOME satellite image sequence. J Geophys Res. 106:5493–5505

Levelt P. F., van den Oord G. H. J., Dobber M. R. et al. (2006) The Ozone Monitoring Instrument. IEEE Trans Geosci Remote Sens. doi:10.1109/TGRS.2006.872333

Liu X, Chance K., Sioris C. E. et al. (2005) Ozone profile and tropospheric ozone retrievals from the Global Ozone Monitoring Experiment: Algorithm description and validation. J Geophys Res. doi:10.1029/2005JD006240

Liu X., Chance K., Sioris C. E. et al. (2007) Impact of using different ozone cross sections on ozone profile retrievals from Global Ozone Monitoring Experiment (GOME) ultraviolet measurements. Atmos Chem Phys. 7:3571–3578

Martin R.V. (2008) Satellite remote sensing of surface air quality. Atmos Environ. 42:7823–1902

Martin R. V., Chance K., Jacob D. J. et al. (2002) An improved retrieval of tropospheric nitrogen dioxide from GOME. J Geophys Res. doi:10.1029/2001JD001027

Martin R. V., Jacob D. J., Chance K. et al. (2003) Global inventory of nitrogen oxide emissions constrained by space-based observations of NO_2 columns. J Geophys Res. 108:4537–4548

Martin R. V., Parrish D. D., Ryerson T. B. et al. (2004) Evaluation of GOME satellite measurements of tropospheric NO_2 and HCHO using regional data from aircraft campaigns in the southeastern United States. J Geophys Res. doi:10.1029/2004JD004869

Martin R. V., Sioris C. E., Chance K. et al. (2006) Evaluation of space-based constraints on nitrogen oxide emissions with regional aircraft measurements over and downwind of eastern North America. J Geophys Res. doi:10.1029/2005JD006680

Palmer P. I., Jacob D., Chance K. et al. (2001) Air mass factor formulation for spectroscopic measurements from satellites: Application to formaldehyde retrievals from the Global Ozone Monitoring Experiment. J Geophys Res. 106: 14539–14550

Palmer P. I., Jacob D. J., Fiore A. M. et al. (2003) Mapping isoprene emissions over North America using formaldehyde column observations from space. J Geophys Res. doi:10.1029/2002JD002153

Platt U. (1994) Differential optical absorption spectroscopy (DOAS). In: Sigrist M W (ed) Air Monitoring by Spectrometric Techniques. John Wiley & Sons, New York

Platt U., Perner D., Pätz H. W. (1979) Simultaneous Measurement of Atmospheric CH_2O, O_3 and NO_2 by differential optical absorption. J Geophys Res. 84:6329–6335

Richter A., Burrows J. P. (2002) Retrieval of Tropospheric NO_2 from GOME measurements. Adv Space Res. 29: 16673–1683

Richter A., Burrows J. P., Nüß H. et al. (2005) Increase in tropospheric nitrogen dioxide over China observed from space. Nature. 437:129–132

Richter A., Eyring V., Burrows J. P. et al. (2004) Satellite measurements of NO_2 from international shipping emissions. Geophys Res Lett. doi:10.1029/2004GL020822

Richter A., Wittrock F., Burrows J. P. (2006) SO_2 Measurements with SCIAMACHY, In Proc. Atmospheric Science Conference, 8–12 May 2006, ESRIN, Frascati, ESA publication SP-628

Rothman L. S. et al. (2005) The HITRAN 2004 molecular spectroscopic database. J Quant Spectrosc Radiat Transfer. 96:139–204

Sanders R. W., Solomon S., Smith J. P. et al. (1993) Visible and near-ultraviolet spectroscopy at McMurdo station Antarctica, 9. Observations of OClO from April to October 1991. J Geophys Res. 98:7219–7228

Schönhardt A., Richter A., Wittrock F. et al. (2008) Observations of iodine monoxide (IO) columns from satellite. Atmos Chem Phys. 8:637–653

Solomon S., Schmeltekopf A. L, Sanders R. W. (1987) On the interpretation of zenith sky absorption measurements. J Geophys Res. 92:8311–8319

Spurr R. J. D. (2002) Simultaneous derivation of intensities and weighting functions in a general pseudo-spherical discrete ordinate radiative transfer treatment. J Quant Specrosc Radiat Transfer. 75:129–175

Spurr R. J. D., Kurosu T. P., Chance K. V. (2001) A linearized discrete ordinate radiative transfer model for atmospheric remote-sensing retrieval. J Quant Spectrosc Radiat Transfer. 68:689–735

Thomas W., Erbertseder T., Ruppert T. et al. (2005) On the retrieval of volcanic sufur dioxide emissions from GOME backscatter measurements. J Atmos Chem. 50:295–320

Uno I., He Y., Ohara T. et al. (2007) Systematic analysis of interannual and seasonal variations of model-simulated tropospheric NO_2 in Asia and comparison with GOME-satellite data. Atmos Chem Phys. 7:1671–1681

Vandaele A. C., Hermans C., Fally S. et al. (2002) High-resolution Fourier transform measurement of the NO_2 visible and near-infrared absorption cross section: Temperature and pressure effects. J Geophys Res. doi:10.1029/2001JD000971

Vandaele A. C., Simon P. C., Guilmot J. M. et al. (1994) SO_2 absorption cross section measurement in the UV using a Fourier transform spectrometer. J Geophy Res. 99:25599–25605

Van Roozendael M., Soevijanta V., Fayt C. et al. (2002) Investigation of DOAS issues affecting the accuracy of the GDP version 3.0 total ozone product, In: ERS-2 GOME GDP 3.0 Implementation and Delta Validation, ERSE-DTEX-EOAD-TN-02-0006, ESA/ESRIN, Frascati, Italy, pp. 97–129

Vountas M., Rozanov V. V., Burrows J. P. (1998) Ring effect: Impact of rotational Raman scattering on radiative transfer in earth's atmosphere. J Quant Spectrosc Rad Transfer. 60:943–961

Wagner T., Platt U. (1998) Satellite mapping of enhanced BrO concentrations in the troposphere. Nature. 395:486–490

Wenig M. O., Cede A. M., Bucsela E. J. et al. (2008) Validation of OMI tropospheric NO_2 column densities using direct-sun mode Brewer measurements at NASA Goddard Space Flight Center. J Geophys Res. doi:10.1029/2007JD008988

Wittrock F., Richter A., Oetjen H. et al. (2006) Simultaneous global observations of glyoxal and formaldehyde from space. Geophys Res Lett. doi: 10.1029/2006GL026310

An Advanced Test Method for Measuring Fugitive Dust Emissions Using a Hybrid System of Optical Remote Sensing and Point Monitor Techniques

Ram A. Hashmonay, Robert H. Kagann, Mark J. Rood, Byung J. Kim, Michael R. Kemme and Jack Gillies

Abstract A new test method for measuring fugitive dust emissions has been developed. This method includes one open path laser transmissometer (OP-LT) extended to a path of several hundred meters to measure ground-level extinction coefficients across an entire plume combined with one tower with at least two vertically distributed and time-resolved dust monitors (DM) (in the middle of the OP-LT path) to measure vertical gradients of PM_{10} and $PM_{2.5}$ concentration. At least two wind monitors are mounted on the tower at the same elevation as the DM instruments to measure wind speed and wind direction for input into the PM flux calculations. The extinction data from the OP-LT (from a specific dust source) are calibrated to the PM_{10} concentration data from calibrated DM instruments. We found that such calibration is mainly a function of dust type and its typical airborne particle size distribution. The performance of this method is demonstrated through comparison to a more traditional upwind-downwind method that deploys three towers with five DM instruments on each tower to define the flux plane with multiple measurements. It is shown that the new hybrid method (one tower with two or three DM instruments and OP-LT) provides comparable flux calculation to the traditional method.

Keywords Fugitive dust emissions · Particulate matter · Dust emission factors · Isokinetic dust samplers · Exposure profiling · Optical remote sensing · Open path laser transmissometer · Dust monitors

1 Introduction

Particulate Matter (PM) emissions from fugitive sources are a major concern because of their contribution to degradation of air quality and visibility in many areas of the world (Watson 2002). Past studies have shown that PM with aerodynamic diameter $\leq 10\,\mu m$ (PM_{10}) and PM with aerodynamic diameter $\leq 2.5\,\mu m$ ($PM_{2.5}$) have adverse effects on human health in the areas surrounding PM sources (Sheppard et al. 1999; Norris et al. 1999; Zanobetti et al. 2000). This paper describes a new and innovative test method that determines PM concentrations and mass fluxes in plumes generated by fugitive dust events, from which emission factors can be calculated. This test method combines traditional PM point monitors and optical remote sensing (ORS) techniques to allow reliable measurement at a fraction of the cost of traditional methods.

Cowherd et al. (1974) developed one of the first methodological approaches for estimating fugitive PM emissions and emission factors. Their original methodology required two lines of isokinetic dust samplers in two crosswind locations downwind of the source area or dust emitting activity. At least three samplers were situated in each of the crosswind lines to profile the plume mass concentrations in the horizontal crosswind direction. The dilution between the two downwind lines provided the vertical profiling of the PM concentrations, and the source emission rate was estimated by

R.A. Hashmonay (✉)
ARCADIS, 4915 Prospectus Drive, Suite F Durham, NC 27713, USA
e-mail: rhashmonay@arcadis-us.com

using inverse dispersion modeling. Later, another measurement technique, termed exposure profiling, offered distinct advantages for source-specific quantification of fugitive emissions from open dust sources (Cowherd and Engelhart 1984, 1985; Cowherd and Kinsey 1986; Pyle and McCaln 1986). This method uses the isokinetic profiling concept that is the basis for conventional (ducted) source testing. The passage of airborne pollutants immediately downwind of the source is measured directly by means of simultaneous multipoint sampling over the effective cross section of the fugitive emissions plume. This technique uses a mass-balance calculation scheme similar to EPA Method 5 stack testing, rather than requiring indirect calculation through the application of a generalized atmospheric dispersion model. Typically, three towers with several dust samplers were considered adequate to directly provide the vertical and horizontal crosswind profiling. This method, applied throughout the 1980s and 1990s, provided integrated information on dust emission factors. Recently, Etyemezian et al. (2004) and Gillies et al. (2005) have replaced the time integrating mass filters of earlier methods with time resolved dust monitors (DM). This allows for the isolation of short-term events and better understanding of the relationships between the dust generating mechanisms and the measured flux downwind, leading to a significant improvement in development of fugitive dust emission factors. This improved method has been applied in fugitive dust measurement studies that have focused on characterizing military fugitive dust sources (Gillies et al. 2005, 2007).

Researchers at the University of Washington (Hashmonay and Yost 1999) proposed another approach for estimating fugitive dust emission factors based on the use of ORS techniques. Varma et al. (2007) demonstrated this approach (ORS method) for fugitive dust emissions from vehicle activity on unpaved surfaces. These two methods have recently been merged for a series of concurrent studies to characterize fugitive dust emissions from unique military sources under the auspices of the Strategic Environmental Research and Development Program (SERDP) projects, SI-1399 and SI-1400 (see www.serdp.org). The new test method was developed as a subset of the two approaches (time resolved exposure profiling and ORS) and tested in 2007 at the Yuma Proving Grounds (YPG), Yuma AZ, a desert location in the southwestern U.S., where dust plumes were generated from a rotary-wing military aircraft.

2 The Method

The new test method includes one open path laser transmissometer (OP-LT) operating in the visible wavelength, (at least) two time-resolved DM instruments, and (at least) two wind monitors. Each DM should be capable of measuring PM_{10} and $PM_{2.5}$ concentrations and may consist of a co-located pair of separate monitors for PM_{10} and $PM_{2.5}$ concentrations (DM_{10} and $DM_{2.5}$, respectively). Figure 1 shows a schematic of the setup configuration with three DM instruments mounted on the tower at different elevations. Preferably, these DM instruments would be calibrated to integrated mass filters measurements in a laboratory setup using a gravimetric method (Gillies et al. 2007). The OP-LT beam path distance from the sensor to the retroreflector can vary according to the expected plume width and possible trajectories. Typically, the path distance will be on the order of several hundred meters. The simplest configuration would use one tower, located at the expected plume center line. The two (or three) DM pairs and the two wind monitors would be distributed evenly at different heights, the lowest at the same height as the OP-LT beam.

The concept behind this emission measurement method is to use the OP-LT to determine the entire cross-plume, plume-averaged mass concentration of PM_{10} dust at ground level, $\overline{C_y^{PM_{10}}}$, and use the vertically distributed PM_{10} DM instruments to determine the vertical distribution of the dust mass concentration. These two parameters can be used to determine the plane-integrated PM_{10} mass concentration ($PIMC_{10}$), which is the mass concentration integrated over the vertical downwind measurement plane:

$$PIMC_{10} = \iint C^{PM_{10}}(y, z)\,dy\,dz \qquad (1)$$

In order to determine $\overline{C_y^{PM_{10}}}$ from the OP-LT measurements, the OP-LT must be calibrated against the DM_{10} instruments using the procedure described in the next section.

The $PIMC_{10}$ is calculated by the expression in Eq. (2):

$$PIMC_{10} \cong \frac{\overline{C_y^{PM_{10}}}}{C_{z_0}^{DM_{10}}} \cdot \overline{C_z^{DM_{10}}} \cdot A \qquad (2)$$

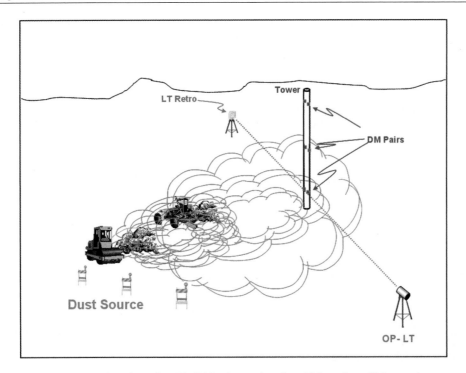

Fig. 1 Schematic of the dust method configuration. The DM pairs consist of one PM_{10} and one $PM_{2.5}$ monitor

where $C_{z_0}^{DM_{10}}$ is the measurement by the DM_{10} that was installed at the same height as the OP-LT beam. $\overline{C_z^{DM_{10}}}$ is the average of all of the measurements by the vertically distributed DM_{10} monitors on the tower, and A is the area of the plane defined by the product of the OP-LT path distance, L, and the height of the top DM on the tower, H. The values for $PIMC_{10}$ are in units of g/m. The ratio $\overline{C_y^{PM_{10}}} \big/ C_{z_0}^{DM_{10}}$ is a correction factor for the horizontal capture of the plume by the single tower. This factor depends on the position and width of the plume. If the ratio has a value of 1, the plume is distributed uniformly over the OP-LT beam path. The correction factor ratio would be smaller than 1 when a narrow plume is centered on the tower and the PM_{10} concentrations measured at the tower location overestimate the plane average concentration. This ratio would be larger than 1 when a narrow plume misses the tower and the PM_{10} concentrations measured at the tower location underestimate the plane average concentration.

The PM_{10} flux, $\varphi_{PM_{10}}^A$, through the measurement area is given by:

$$\varphi_{PM_{10}}^A = PIMC_{10} \cdot \overline{U}_x \qquad (3)$$

where $\overline{U_x}$ is the normal wind component averaged over the two or three vertically distributed anemometers. If the entire plume is encompassed by the measurement area, $\varphi_{PM_{10}}^A$ is equal to the total mass emission rate for PM_{10} released from the target fugitive source.

The $PM_{2.5}$ flux is calculated by:

$$\varphi_{PM_{2.5}}^A \cong \frac{\overline{C_y^{PM_{10}}}}{C_{z_0}^{DM_{10}}} \cdot \overline{C_z^{DM_{2.5}}} \cdot A\overline{U_x} \qquad (4)$$

where $\overline{C_z^{DM_{2.5}}}$ is the average concentration determination from the two or three $PM_{2.5}$ DM instruments on the tower. Using Eqs. (3) and (4), one can determine the emission flux for both PM_{10} and $PM_{2.5}$ using one OP-LT, (at least) two $PM_{10}/PM_{2.5}$ DM instruments, and two or three anemometers.

3 Method Demonstration and Validation

3.1 Calibration of the OP-LT

The OP-LT was calibrated, and the validity of the new dust monitoring method was demonstrated with data collected during a field campaign at YPG, in May 2007. This campaign provided us with an abundance of OP-LT and DM measurement data, with which we tested and validated the new method. The goal of the campaign was to develop PM_{10} and $PM_{2.5}$ emission factors for rotary-wing aircraft operations in a desert environment. The measurements were event based with each dust plume originating from a rotary-wing aircraft travelling close to the desert surface (rotor height $\sim$ 7 m). The PM and extinction data were recorded as the dust plume passed through the measurement plane.

Figure 2 shows the relevant measurement configuration at YPG. The setup includes three towers, each with five vertically distributed pairs of DustTraks (DT) dust monitors (Model 8520, TSI, Inc., Minnesota, USA). One of each pair was fitted with a PM_{10} size selective inlet and the other fitted for $PM_{2.5}$. The OP-LT was setup near one tower, and its retroreflector array was setup near the third tower at a distance of $\sim$ 105 m. The DT monitors and OP-LT measurements were made at 1-s intervals.

The subset configuration of instruments that was used for calibration of the OP-LT to the three ground level DT monitors is shown in Fig. 3. Such calibration configuration can be undertaken as a separate effort or can be combined with measurements by adding the two DM instruments, $DM_{10}1$ and $DM_{10}3$ in Fig. 3, to the measurement configuration shown in Fig. 1.

The OP-LT measures the path-integrated extinction given by:

$$PIE^{LT} = \int_0^L \varepsilon(y)dy \qquad (5)$$

where $\varepsilon(y)$ is the extinction value at a location y along the beam path; L is the beam-path distance from the OP-LT sensor to the retroreflector. For fixed particle composition and size distribution, the path-averaged extinction, $PAE = PIE^{LT}/L$, is proportional to the path-averaged dust concentration.

The OP-LT is calibrated by performing a linear regression fit of the path-averaged extinction to the average value of the PM_{10} measurements by the DM instruments, distributed along the cross-plume OP-LT beam path $\overline{C_y^{DM_{10}}}$, to compute the calibration factor, F^{DM-LT},

$$\overline{C_y^{DM_{10}}} = F^{DM-LT} \cdot PAE^{LT} \qquad (6)$$

The total measurement set at YPG consisted of 1840 1-s dust-plume measurements over a 1.8-h period, for 32 dust plume events. The 1-s PM_{10} concentration data

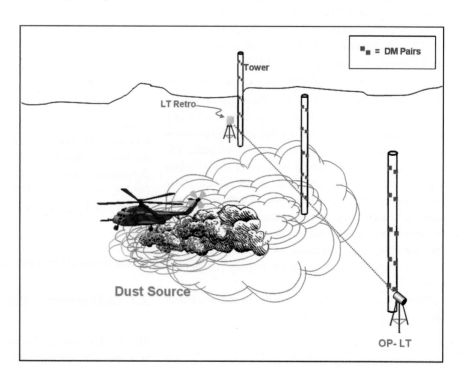

Fig. 2 The measurement setup at YPG. The DT pairs consist of one PM10 and one PM2.5 monitor

Fig. 3 Schematic of the OP-LT calibration configuration using the minimum of 3 DMs

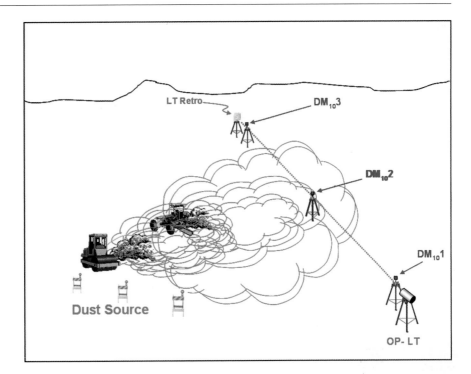

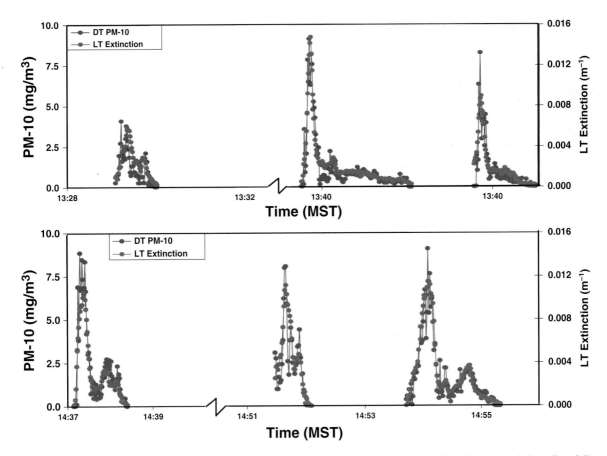

Fig. 4 OP-LT Calibration measurements made at yuma proving grounds – The six dust events with high dust correlations (R > 0.9)

Fig. 5 Plot of the OP-LT and
DM PM$_{10}$ calibration
measurements. (**a**) The full
1840-point data set. (**b**) The
six high-correlation events
(508-point) data set

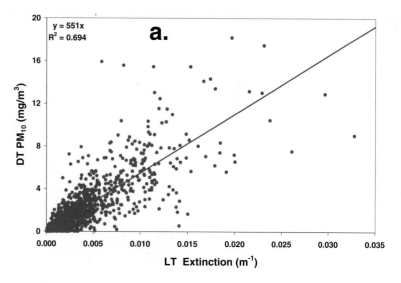

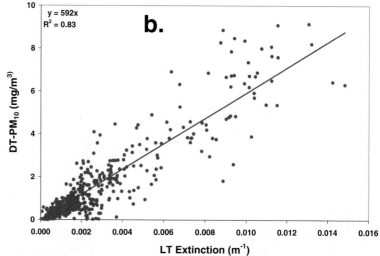

(calibrated to gravimetric filter-based methods) from
the three lowest DT monitors (situated along the OP-
LT beam path) were spatially averaged and synchro-
nized against the path-averaged extinction. The Pearson correlation was calculated for each event and six
events out the total 32 had a Pearson correlation higher
than 0.9 (R > 0.9). The other 26 events had Pearson
correlation values between 0.6 and 0.9, which indicates a less homogeneous plume along the LT path
length. Figure 4 shows an example of the calibration
data collected at YPG for the six events with the highest correlation between the three DTs averaged PM$_{10}$
concentration and the OP-LT path-averaged extinction.
The double ordinates were adjusted to overlap the two

traces to show the strong correlation between the DM
and OP-LT measurements.

Plots of $\overline{C_y^{DT_{10}}}$ measurements versus the path-
averaged OP-LT extinction measurements are shown
in Fig. 5. Figure 5a shows the plot of the entire 1840-
point data set (32 events) and Fig. 5b shows the plot of
the measurements made during the six events shown in
Fig. 4. The slope of this linear plot for the full dataset,
which is the calibration factor, F^{DT-LT}, is 551 mg/m^2
and the R^2, the square of the Pearson correlation coefficient, has a value of 0.69. As one may expect, reducing
the data set to the six high-correlation events results
in much less scatter in the linear plot, as shown in
Fig. 5b. The linear plot of the six events resulted in

Fig. 6 Comparison of PIMC
determined with the new
method to the PIMC
determined from the grid of
DT monitors. *Top*: 5 DT
monitors in new method,
middle: 3 DT monitors in new
method, *bottom*: 2 DT pairs in
new method

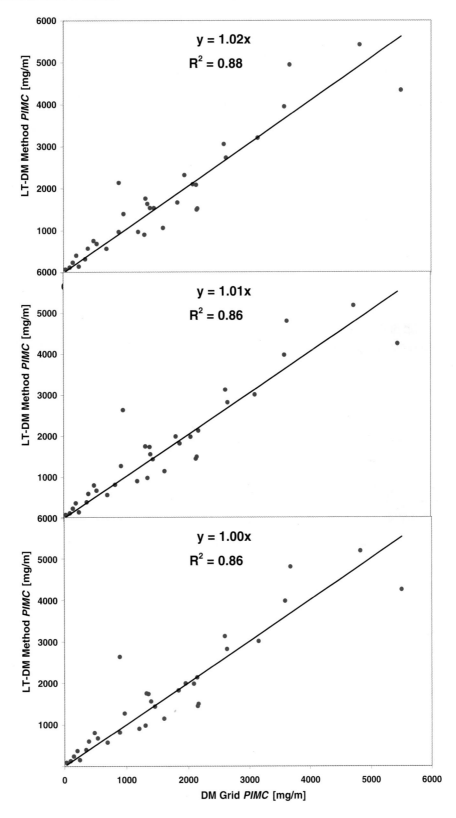

the improved R^2 value of 0.83 and a calibration factor of 592 g/m^2. The six highly-correlated events represent the case when the plume is more evenly distributed along the OP-LT beam path and conditions are favorable for such calibration. It is not necessary to use any preset number of DM monitors along the OP-LT beam path as long as we have a large number of data sets with Pearson correlation coefficient larger than 0.9, and an overall R^2 > 0.8 in the linear regression calibration curve.

3.2 Proof-of-Concept of the New Method

The fifteen PM$_{10}$ dust monitors formed a 3-by-5 grid that defines the measurement area, A, as a 105-m wide by 10-m high plane. The PIMC$_{10}$ is calculated as the product of the mean value of the 15 DM$_{10}$ on the three towers and the area A:

$$PIMC_{10}^{Grid} \cong \overline{DM_{10}^{Grid}} A \qquad (7)$$

The configuration used at YPG (Fig. 2) has the elements required for the new method (Fig. 1) if we ignore Towers 1 and 3. We compared the new method calculation of $PIMC_{10}$ (using Eq. (2)) to the current method calculation of $PIMC_{10}^{Grid}$ (using Eq. (7)) by performing a linear regression for all 32 helicopter-generated plumes. We then tested the effect on the emission estimate of reducing the number of DM pairs on the tower (center tower) from five to three and then to two. The results are shown in Fig. 6 for three cases: five, three, and two DT monitors in the calculation of $PIMC_{10}$. These results indicate that the new method (OP-LT and DM hybrid) calculations of PIMC (equivalent to flux for the same wind conditions) values are comparable (slope close to 1 with R^2 > 0.85) to the calculations of the PIMC of the current method (DM grid). Also, it is shown that the use of five DM instruments on the tower for the new method is not required, as similar results can be obtained by using only two or three DM instruments, which simplifies the instrumentation set up and logistics of deployment, making this method an attractive alternative to the more costly emission profiling method.

4 Conclusion

We compared the results for emission flux determinations using this simple hybrid setup versus a more elaborate setup involving multiple (three) towers and (30) DM instruments, to determine the feasibility of the new method and to demonstrate the equivalence of the results.

Acceptance of the new test method would provide accurate, cost-effective, and more easily acquired fugitive dust emission measurement data for the development of emission factors. The new test method brings closure to an issue that has hindered the development of other ORS techniques used to define fugitive dust emission flux, which is that the relationship between open path extinction (or opacity) and mass concentration for specific dust plumes has previously been highly uncertain. This new method provides a calibration relationship between ORS and in situ PM measurements and will allow for the development and eventual deployment of other open path extinction measurement tools such as LIDAR or digital cameras, which can be used to develop fugitive PM emission factors.

Acknowledgments This research was sponsored by the Strategic Environmental Research and Development Program grants number CP-1399 and CP-1400

Abbreviations

OP-LT	Open path laser transmissometer
DM	Dust monitors
PM	Particulate matter
PM$_{10}$	Particulate matter with aerodynamic diameter $\leq 10\,\mu$m
PM$_{2.5}$	Particulate matter with aerodynamic diameter $\leq 2.5\,\mu$m
ORS	Optical remote sensing
SERDP	Strategic Environmental Research and Development Program
YPG	Yuma Proving Grounds
PIMC$_{10}$	Plane-integrated PM$_{10}$ mass concentration
DT	DustTraks (proprietary dust monitor)

References

Cowherd, C., Axetell, K., Guenther, C.H., and Jutze, G.A., *Development Of Emission Factors For Fugitive Dust Sources*, EPA-450/3-74-037, U. S. Environmental Protection Agency, Research Triangle Park, NC, June 1974.

Cowherd, C., and Englehart, P.J., *Paved Road Particulate Emissions*, EPA-600/7-84-077, U. S. Environmental Protection Agency, Cincinnati, OH, July 1984.

Cowherd, C., and Englehart, P.J., *Size Specific Particulate Emission Factors For Industrial And Rural Roads*, EPA-600/7-85-038, U. S. Environmental Protection Agency, Cincinnati, OH, September 1985.

Cowherd, C., Jr. and Kinsey, J.S., *Identification, Assessment, and Control of Fugitive Particulate Emissions*. EPA-600/8-86-023, U. S. Environmental Protection Agency, Research Triangle Park, NC, August 1986.

Etyemezian, V., Ahonen, S., Nikolic, D., Gillies, J., Kuhns, H., Gillette, D., and Veranth, J., Deposition and removal of fugitive dust in the arid southwestern United States: measurements and model results. *J. Air Waste Manage. Assoc.*, 54(9):1099–111, 2004.

Gillies, J. A., Etyemezian, V., Kuhns, H., Nikolic, D., and Gillette, D.A., Effect of vehicle characteristics on unpaved road dust emissions. *Atmos. Environ.*, *39*(13): 2341–2347, 2005.

Gillies, J.A., Kuhns, H., Engelbrecht, J.P., Uppapalli, S., Etyemezian, V., and Nikolich, G., Particulate emissions from U. S. department of defense artillery backblast testing. *J. Air Waste Manage. Assoc.*, 57(5):551–60, 2007.

Hashmonay, R.A., and Yost, M.G., On the application of OP-FTIR spectroscopy to measure aerosols: observation of water droplets. *Environ. Sci. Technol.*, 33(7):1141–1144, 1999.

Norris, G., Young Pong, S.N., Koenig, J.Q., Larson, T.V., Sheppard L., and Stout, J.W., An association between fine particles and asthma emergency department visits for children in Seattle. *Environ Health Perspect*, 107(6): 489–493, June 1999.

Pyle, B.E., and McCaln, J.D., *Critical Review of Open Source Particulate Emission Measurements, Part II – Field Comparison*. EPA-60012-86-072, U. S. Environmental Protection Agency, Research Triangle, NC, 1986.

Sheppard, L., Levy, D., Norris, G., Larson, T.V., and Koenig J.Q., Effects of ambient air pollution on nonelderly asthma hospital admissions in Seattle, Washington, 1987–1994 *Epidemiology*, 10(1): 23–30, 1999.

Varma, R.M., Hashmonay, R.A., Du K., Rood, M.J., Kim, B.J., and Kemme M.R.A Novel Method to Quantify Fugitive Dust Emissions Using Optical Remote Sensing, in Advanced Environmental Monitoring (ed. Kim, Y.J., and Platt. U.) 143–154 (Springer Netherlands, 2007).

Watson, J.G., Visibility: science and regulation. *J. Air Waste Manage. Assoc.* 52: 628–713, 2002.

Zanobetti, A., Schwartz, J., and Dockery, D.W., Airborne particles are a risk factor for hospital admissions for heart and lung disease. *Environ Health Perspect.*, 108(11): 1071–1077, November 2000.

Aerosol Sampling Efficiency Evaluation Methods at the US Army Edgewood Chemical Biological Center

Jana Kesavan* and Edward Stuebing

Abstract This chapter presents information on test aerosols, generation methods, and analysis techniques that are used in the Aerosol Sciences Laboratories, US Army Edgewood Chemical Biological Center (ECBC) to quantitatively characterize the performance of aerosol samplers. The Sampling Efficiency results of three aerosol samplers characterized at ECBC are also presented in this chapter solely for the purpose of illustrating the application of these methods. Solid, liquid, and biological aerosols may have different transmission and collection efficiencies. Solid particles can bounce when they impact onto internal surfaces, or they can be re-entrained into the airflow after deposition; however, liquid particles are permanently captured upon impact. Biological particles fall in between solid and liquid particles with respect to their characteristics of bounce. Bioparticles may also be tacky and stick to tubing and walls which can seriously affect recovery by elution and aerosol sampler performance. Therefore, the test methodology should be carefully selected to answer the research question. In addition, Sampling Efficiency tests can be conducted either by filling a chamber with aerosols and conducting tests or by delivering the test aerosol to the inlet of the sampler and the reference filter using the Ink Jet Aerosol Generator (IJAG). Using the IJAG and delivering the aerosol to the inlet allows many tests to be conducted in a single day.

Keywords Aerosol generation methods · Aerosol sampling · Aerosol samplers

1 Introduction

Air samplers are important in the war against terrorism and on the battlefield to detect the presence of chemical, biological, and nuclear aerosols. Knowledge and use of efficient air samplers enhance the ability to protect soldiers, first responders, and the general public from airborne agents. Samplers and detection systems must be tested and their performance efficiencies determined so that suitable samplers and detectors can be matched appropriately for various challenges. There are specialized aerosol samplers made for sampling chemical, biological, and nuclear aerosols. This chapter will focus on bioaerosol samplers. Usually, these samplers are high-volume samplers that sample the air continuously to find a few organisms in the air. Commonly, the high-volume air samplers used for bioaerosol sampling are designed to collect $1-10\,\mu m$ particles. Each air sampler has multiple components such as: an inlet, transmission tubes, a pre-separator skimmer to reject large particles, aerosol concentrating stages, and a collector such as an impactor. An exploded illustration of an aerosol sampler is shown in Fig. 1.

Each of these components can be as simple as a filter or as complicated as a multistage virtual impactor aerosol concentrator. The performance of an aerosol sampler, the Sampling Efficiency, is the overall

J. Kesavan (✉)
US ARMY Edgewood Chemical Biological Center,
AMSRD-ECB-RT-TA E5951, 5183 Blackhawk Road,
Aberdeen Proving Ground, MD 21010, USA
e-mail: Jana.Kesavan@US.ARMY.MIL

* The use of either trade or manufactures' names in this report does not constitute an official endorsement of any commercial products.

Fig. 1 Illustration of an aerosol sampler

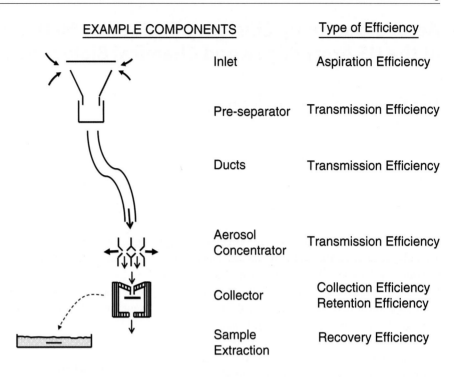

EXAMPLE COMPONENTS	Type of Efficiency
Inlet	Aspiration Efficiency
Pre-separator	Transmission Efficiency
Ducts	Transmission Efficiency
Aerosol Concentrator	Transmission Efficiency
Collector	Collection Efficiency Retention Efficiency
Sample Extraction	Recovery Efficiency

end-to-end ratio of the amount of aerosol contained in the sample produced by the sampler to the amount of aerosol contained in the volume of ambient air sampled by the system's inlet. In a well-designed, well-fabricated, well-assembled system it is the product of the performance efficiencies of the sampler's individual components, variously: aspiration, transmission, and collection efficiencies.

The aspiration efficiency of a sampler's inlet describes the efficiency with which particles are extracted from the air and transmitted through the sampler inlet and is dependent on particle aerodynamic size and wind speed. Inlets intended for use in calm air or indoors can be characterized in an aerosol chamber, however inlets to be used outdoors, on moving platforms, or in ducts of moving air must be characterized in an aerosol wind tunnel using the appropriate range of wind speeds. The other components can be characterized in an aerosol chamber.

Transmission efficiency describes the efficiency with which particles are transported from the intake of a component to its outlet, and the collection efficiency describes the efficiency with which particles are captured by the component. Retention efficiency indi-

cates how efficiently particles are retained by the sampler during a long sampling time, for example, in an impinger or in a wetted cyclone that stores the collected particles in the active collection fluid throughout the collection time. Particles in the collection fluid can escape into the air (reaerosolization) and be ejected with the exhaust. The collected particles are recovered for analysis and the efficiency with which they are recovered is indicated by the recovery efficiency. We include this step in the Sampling Efficiency analysis because it is not until recovery has been accomplished that we have a sample, and there are instances of systems with inefficient recovery steps making this step an important determinant of end-to-end Sampling Efficiency. All efficiencies described above can depend on particle size, density, charge, composition, and biofactors.

There has been ambiguity in the use of these efficiency terms in the literature. In particular, the term collection efficiency is used to mean a variety of different experimental observations, e.g., (1) how much aerosol comes out in the sample or (2) how much of the aerosol that enters a sampler inlet does not come out in the exhaust in which case the "collected" aerosol may be in the sample or it may be stuck on the walls inside the sampler. In reading the literature and in discussions it is therefore critical to clarify the context of a term's use to understand its meaning, and in our experience this can lead to misunderstandings. We suggest that Sampling Efficiency be reserved to refer to the overall, end-to-end efficiency from the amount of aerosol available in the air outside the inlet to the amount of aerosol contained in the final sample to be analyzed or preserved. Furthermore, the other efficiency terms should always be used in connection with a component or subset of components. Confusion arises when an aerosol sampler is called an aerosol collector and its performance is then referred to its collection efficiency. The term collection efficiency can not even be assumed to only apply to the performance of the collector element of a sampler. For example, a simple duct has a transmission efficiency and that is its purpose, however, a duct also has a collection efficiency that measures what fraction of aerosol that enters the duct is captured and retained. This applies to losses at all components of a sampler, and it is an important measurement when doing a diagnostic analysis of the contribution of components to the overall performance of a sampler. The goal is to develop an understanding of the sampler and this is best accomplished if a mass balance is achieved for each component by measuring the amount of aerosol collected, transmitted, and in the case of split flows like virtual impactors, the amount rejected. When done properly, the result is a Sampling Efficiency that matches the product of all the relevant component efficiencies in the sampling train unless there are problems in the connections between components such as changes in duct diameter, offsets in mating the components, or leaks. Real world instances of all three of these have been seen at ECBC resulting in losses of as much as 95% of the aerosol depending on particle aerodynamic diameter.

Bioorganisms have two additional issues: survival fraction and culturable fraction. Survival of an organism can be measured by flow cytometry using different dyes that reveal viable vs. non-viable organisms, and by other life function measures such as ATP, and culturability is determined by plating (Rule et al., 2007). These are reported as fractions rather than efficiencies because they are characteristics of the aerosol in the sample not the amount of aerosol in the sample. For samplers that affect the viability of organisms in the aerosol, it is important to make this measurement when the use to be made of the sample requires viability or culturability of organisms.

Many inlets are simple with just a tube sampling the air; however, other inlets are complicated and may have a rain hat, bug screens, and pre-separator. A rain hat prevents rain from entering the inlet and a pre-separator removes unwanted large particles. The transmission region can be simple with a tube connecting the inlet to the collection site, or it can be complicated with aerosol concentrators, bends, constrictions, and expansions. The collection site uses one or more of the four main collection mechanisms: impaction, interception, diffusion, and collection using electric forces. Some samplers have no collection site, for example, delivering an aerosol stream for optical interrogation.

Each component of the sampler is characterized differently using various facilities. Sampler inlets are characterized in aerosol wind tunnels to determine the effect of wind speed on the aspiration efficiencies of different size particles. Tests are conducted in flow-through cells to determine sampler component efficiencies for different size particles. Sampling Efficiency tests at calm air conditions of components other than the inlet are conducted in chambers using several methods.

Many samplers have been characterized at the US ARMY ECBC in the past 15 years as new samplers are developed and existing samplers are improved. Sampler characterization methods have been modified and improved over time to achieve faster, more effective sampler characterization. This chapter describes methods used in chambers at ECBC to characterize aerosol sampler components at calm air conditions. These methods have been developed under a formal quality assurance program at the Good Laboratory Practices (GLP) level, including a large library of internal operating procedures (IOPs) which are reviewed, approved, and audited by professional quality management personnel. The safety of these methods is governed by formalized Standard Operating Procedures (SOPs) that are reviewed, approved, and witnessed in pre-operation runs by safety professionals. The SOP procedures protect people and the environment by ensuring the use of safe aerosol substances and engineering controls including containment. Diagnostic testing of aerosol samplers is also conducted in addition to end-to-end performance testing. Location of aerosol losses due to undesirable deposition inside the device and mass balance studies are conducted with fluorescing aerosols.

UV lights are used to observe the location of deposition losses. Sampler components are also washed to recover the deposited aerosols for quantification.

The sampler characterization and diagnostic test methods used at ECBC are:

(1) Monodisperse fluorescent and non-fluorescent polystyrene latex (PSL) microspheres with fluorometric analysis or Coulter Multisizer analysis (Kesavan and Hottell, 2005),

(2) Polydisperse solid aluminum oxide particles with Coulter Multisizer analysis (Kesavan and Doherty, 2001),

(3) Fluorescent oleic acid particles with fluorometric analysis (Kesavan et al., 2002),

(4) Bioparticles with Coulter Multisizer, culturing, enzyme-linked immunosorbent assays (ELISA), polymerase chain reaction (PCR), and Aerodynamic Particle Sizer (APS) analyses (Kesavan and Hottell, 2005).

The aerosols are generated using many methods. Table 1 lists the materials, generation methods, resultant size distribution, and analysis methods that are used at ECBC.

Some samplers, such as wetted wall cyclones or large format filters, can collect the aerosol from the high volume flow, however many samplers have aerosol concentrators followed by an aerosol collection stage in a much lower volume flow. Aerosol concentrators concentrate aerosol particles from a larger volume of air into a smaller volume of air. Some of the aerosol

collection methods used by the aerosol sampler are: collection on filters, impactors, impingers, and electrostatic collection.

Filters are used in many aerosol samplers to collect particles (Lippmann, 1995). There are many advantages in using filters to collect particles. Most of the filters have a particle collection efficiency of 100%. The sampling time can be very long because there is no liquid to evaporate. The sampler is very simple and there are fewer things to go wrong. However, there are many disadvantages in using a filter collection system. The recovery of particles from filters may not be 100%. Bioparticles sampled onto filters may desiccate over time and lose their culturability.

Impaction is used as the collection mechanism in some samplers (Hering, 1995). Impaction can be onto a dry surface, wet surface, fan blades, rotating surface, or an agar plate. In traditional impactors, an aerosol stream is first accelerated and then its direction is suddenly changed 90 degrees so that the centrifugal force on a particle will cause it to move perpendicularly to the direction of the turning air stream and impact on a surface. Larger particles will impact onto the surface and smaller particles will follow the air and escape. Impactors can be operated in series, referred to as cascade impactors (Hinds, 1999), and with increasing acceleration can collect smaller and smaller particles on the downstream stages.

Impingers use the same principle as real impactors; however, the collection surface of the impinger is a liquid or a solid (e.g., glass) immersed in liquid (Hering, 1995). Particles larger than $\approx 1\ \mu m$ are captured by

Table 1 Aerosol, generation methods, size distributions, and analysis methods used in sampler characterization tests

Aerosol type	Generation method	Size distribution	Analysis method
PSL	Nebulizers, Sonic nozzle, IJAG, Puffers	Monodisperse 0.5–6 μm	Microscopy, fluorometry, coulter counter
Fluorescent oleic acid droplets	Vibrating Orifice Aerosol Generator Spinning top, IJAG	Monodisperse 3–20 μm	Fluorometry
Test dust Al$_2$O$_3$	Sonic nozzle	Polydisperse 0.5 to ≲20 μm	Gravimetric, Coulter Counter
Biosimulants	Nebulizers, Puffers, IJAG, Sonic nozzle, Bubblers	Poly, mono, or narrowly dispersed 1 to ≈20 μm	Culturing, coulter counter, APS

the inertial mechanisms in the liquid. Particles are collected in the same liquid volume over time (i.e., the collection is on a batch basis, and the particles are thereby concentrated in the liquid).

Electrostatic force is used in many samplers to collect particles (Swift and Lippmann, 1995). Particles are charged in the first region of the sampler and then collected in the second region of the sampler. This is a more efficient collection mechanism for small (<1 μm) particles. This is also a gentle mechanism for collecting particles.

Virtual impactors are used to concentrate the aerosols and operate under the same principle as traditional impactors. Virtual impactors can be operated in series, where the minor flow from a first stage becomes the inflow to a second stage, which can provide much higher concentrations than a single stage alone. With a traditional impactor, the particles are collected on a real surface placed transverse to the initial airflow direction; however, in a virtual impactor, instead of a real surface there is a port into which the larger particles are driven. About 10% of the air stream is drawn into the port to transport the large particles away from the fractionation zone, while the remaining 90%, which is devoid of large particles, is exhausted away from the port. The two streams are referred to as the minor (concentrated aerosol) and major flows, respectively (Hinds, 1999).

A variety of aerosol generation and analysis methods are used at ECBC to characterize aerosol samplers. Solid, liquid, and biological particles are used as test aerosols. Solid particles can bounce when they impact onto internal surfaces, or they can be re-entrained into the airflow after deposition; however, liquid particles are permanently captured upon impact. Biological particles (microorganisms and biological substances such as proteins) fall in between solid and liquid particles with respect to their characteristics of bounce. Bioparticles may also be tacky and stick to tubing and walls which can seriously affect recovery by elution of the sample from the collector as well as impacting the aerosol collector's overall performance. Some bioparticles are hydrophilic, others are hydrophobic, and this may determine how easily they can be washed from surfaces. Some of our tests have shown that bioparticles stick to plastics and tubing compared to inert particles. Therefore, a suitable aerosol needs to be selected for each test. The test aerosol can fill the whole chamber or it can be delivered to the inlet of the sampler and the reference unit. There are many advantages to both methods. However, using an IJAG to deliver the aerosol to the inlet of the sampler and reference unit allows many tests to be conducted in a single day and minimizes the amount of expensive materials to aerosolize.

2 Background

The performance of aerosol samplers as a function of particle size is determined using inert and biological particles. Aerosolization of biological materials requires Biosafety Level 1+ (BSL1+) or higher containment and practices, and workers must be medically cleared for certain operations. Accommodating these requirements can be expensive and time consuming, and the results can be inherently less precise than the physical quantification of inert aerosols (e.g., fluorescence). In addition, biomaterial quantification techniques such as PCR, culturing, and flow cytometry can be expensive and time consuming. As a result, it is simpler to use inert particles such as solid PSL microspheres, solid aluminum oxide particles, and liquid fluorescent oleic acid droplets for the aerosol sampler characterization. These tests can be rapidly, precisely, and accurately conducted at many particle sizes to get a detailed assessment of sampler performance. If an aerosol sampler meets performance requirements or expectations, then confirmatory studies with bioaerosols at a few particle sizes can be conducted.

Sampler characterization studies have been reported by others using non-biological, inert particles. Six samplers were evaluated by Li et al. (2000) using solid, monodisperse ammonium fluorescein test particles. These samplers were also tested to determine how well they matched the inhalable convention. The particles were generated with a vibrating orifice aerosol generator and the fluorescence was quantified by a fluorometer. This study showed that the Sampling Efficiency depends on the particle size, wind speed, wind direction, and the stickiness of the particles. John and Kreisberg (1999) characterized samplers with dry polystyrene beads that were generated with a fluidized bed and analyzed with an APS. The use of a fluidized bed aerosol generator minimized the generation of small particles. Therefore, an APS can be used for the analysis of this small number of particles. Maynard

(1999) and Maynard et al. (1999) developed a system to rapidly measure sampler performance using polydisperse glass microspheres and an APS analysis method. A rotating brush generator was used to generate the glass microspheres. McFarland et al. (1991) used liquid fluorescent oleic acid with fluorometer analysis in their sampler characterization tests. A vibrating jet atomizer was used to generate the monodisperse fluorescent oleic acid particles. Gao et al. (1997) used a fluidized bed aerosol generator to generate ceramic and polystyrene beads for characterizing samplers. An APS was used to measure the particle size and concentration. Particles collected on filters were also examined by bright field microscopy and the particles were counted. Aizenberg et al. (2000a) and Witschger et al. (1998) used aluminum oxide particles with gravimetric analysis to characterize sampler performance. The aerosol was generated with a rotating brush generation system. Willeke et al. (1998) conducted tests with PSL aerosols generated with a Collison nebulizer and analysis was conducted with an Aerosizer.

Viable bioaerosol particles have been used by others in a number of sampler performance studies. The overall Sampling Efficiency depends on the physical collection efficiency, and if the analysis is by culturing, it also depends on the stresses imposed on the test organism. However, even dead organisms can be quantified by PCR or similar techniques, although the sensitivity and repeatability of such methods may not be particularly good. Li et al. (1999) conducted tests with *Escherichia coli (E. coli)* and *Bacillus subtilis (B. subtilis)* to evaluate sampling methods: AGI-30 impingers and filters. The results showed that sensitive bacteria are killed due to dehydration effects on filters as the sampling time is increased. The authors believe that impingers could perform much better than filter sampling methods for sampling bioaerosols. Jensen et al. (1992) evaluated eight bioaerosol samplers. *B. subtilis* and *E. coli* were used as the test aerosols and the authors reported viability losses due to stress to the organisms.

A few studies have evaluated the Sampling Efficiency of aerosol samplers using inert and biological aerosols. Aizenberg et al. (2000b) characterized three aerosol samplers (Air-O-Cell, Burkard, and Button Samplers) using PSL and microorganisms (*B. atrophaeus, Penicillium melinii, Penicillium brevicompactum, Cladosporium cladosporioides, and Streptomyces albus*). The aerodynamic diameter (AD) of the microorganisms was ≤3.1 μm; the AD of the PSL

microspheres was <5.1 μm. The Sampling Efficiency results showed that there were no significant differences between the efficiency of PSL microspheres and the microorganisms. A Swirling Aerosol Collector (SAC) was characterized using PSL beads with Aerosizer analysis (Willeke et al., 1998) and bioaerosols with culturing (Lin et al., 2000). *Bacillus subtilis var. niger* (BG) (now redesignated Bacillus atrophaeus) and *Pseudomonas fluorescens* were used as the test bioaerosols. For particles ≤2 μm AD, similar efficiencies were obtained for PSL microspheres and bioparticles.

At ECBC, the Sampling Efficiency tests in aerosol chambers are conducted by generating the aerosol for a specified time, mixing it in a chamber to achieve a uniform aerosol concentration, and then sampling from the chamber using the samplers and reference filters. The samples and reference filters are analyzed and the Sampling Efficiency is determined by comparing the particles in the sample to the particles on the reference filters. The air flow rates and liquid volumes of the reference filters and samplers are taken into account in the Sampling Efficiency calculations. For example, in a test using fluorescent particles, the Sampling Efficiency is calculated using the following equation.

$$Sampling\ Efficiency = \frac{\left[\dfrac{F_s \times SV_s}{Q_s}\right]}{\left[\dfrac{F_{RS} \times SV_{RS}}{Q_{RS}}\right]} \times 100 \quad (1)$$

where

F_s = Fluorescence concentration (ml^{-1}) measurements of the sample from the tested sampler

SV_s = Sample liquid output volume of the tested sampler

Q_s = Air flow rate of the tested sampler

F_{RS} = Fluorescence concentration (ml^{-1}) measurements of the reference sampler

SV_{RS} = Liquid output volume of the reference sampler

Q_{RS} = Air flow rate of the reference sampler

The samples can be quantified by fluorometry, Coulter Multisizer quantification, or culturing. More information on these methods is given in Section 3 (Chambers and Equipment).

To achieve accurate Sampling Efficiency results, the air flow rates must also be measured carefully. The air flow rates of the reference filters and samplers are measured using a Buck calibrator (A.P. Buck, Inc., Orlando, FL), a Kurz airflow meter (Kurz Instruments, Inc., Monterey, CA), a TSI mass flow meter (TSI Incorporated, St. Paul, MN), or a DryCal (BIOS International Corp., Butler, NJ). All flow meters are calibrated annually by the ECBC Calibration Laboratory. Other characteristics of the samplers are also measured including weight and dimensions. Power usages are measured using a power meter (Extech Instruments, Taiwan).

The Sampling Efficiency tests using the IJAG are conducted slightly differently. The output of the IJAG is delivered directly to the inlet of the sampler. Similarly, the reference samples are made by delivering the IJAG output to a reference filter. Sampling Efficiency is determined by comparing the particles in the sampler to the particles on the reference filters and the air flow rates are not taken into account because the IJAG is used to deliver an identical number of particles to both the sampler and reference filters. Reference filters are taken before and after challenging the sampler.

3 Chambers and Equipment

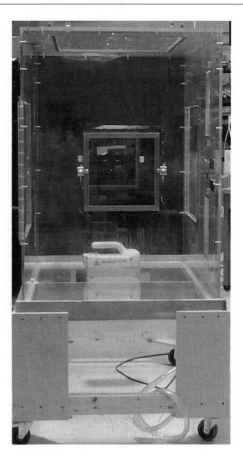

Fig. 2 Small plexiglass chamber (3 ft × 4 ft × 5 ft)

This section describes some of the chambers, aerosol generators and analysis equipment used at ECBC to characterize aerosol samplers. Most of the sampler characterization tests are conducted in a 70 m³ chamber. There are also smaller stainless steel and plexiglass chambers available if very high concentration aerosols are needed for low air flow rate sampler tests. Many aerosol generation methods and analysis techniques are used at ECBC to generate the test aerosol and to quantify the collected samples. The most commonly used aerosol generators and sample analysis equipment are described in this section.

3.1 Small Plexiglass Chamber

A 2 m³ plexiglass chamber, Fig. 2, is used in many low airflow sampler characterization tests. The small volume allows high concentrations of aerosols to be pro-

duced for the test. Reference filters and samplers under test are placed in the chamber, and aerosol generators are attached to the chamber to deliver the test aerosol to the chamber. A fan in the chamber mixes the air to produce uniform concentration of aerosol in the chamber. A High-Efficiency Particulate Air (HEPA) filter unit is attached to evacuate the chamber.

3.2 Stainless Steel Chamber

Sampler characterization tests are conducted in a 70 m³ BSL1+ chamber, Fig. 3. Temperature and humidity of the chamber can be set and maintained easily and accurately by a computer. Power receptacles inside the chamber are also controlled by this computer. HEPA filters are installed at the inlet to filter air entering the chamber to achieve very low background particle

Fig. 3 70 m^3 BSL1+
chamber used at ECBC

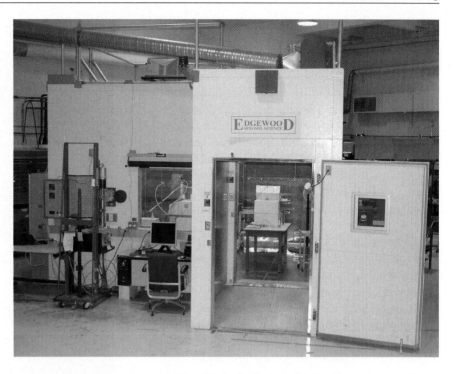

concentrations in the chamber. Similarly, HEPA filters are installed at the exhaust port to filter all particles leaving the chamber. The chamber aerosol is
cleaned by exhausting the chamber air through the
HEPA filters, and by pumping HEPA-filtered air into
the chamber. The maximum amount of airflow that
can be exhausted from the chamber by the exhaust
pump is approximately 2×10^4 L/ min. There is also
a small recirculation system that removes air from the
chamber, passes it through a HEPA filter, and deliv-
ers it back to the chamber. This system is useful when
the aerosol concentration in the chamber needs to be
reduced incrementally.

Aerosols can be generated outside and delivered
to the chamber, or they can be generated inside the
chamber. The chamber air is mixed by fans after
and/or during the aerosol generation to achieve a uniform aerosol concentration in the chamber. Previous
tests showed that mixing the aerosol in the chamber
for 1 min is adequate to achieve a uniform aerosol
concentration.

3.3 Nebulizers

Collison nebulizers (BGI Inc., Waltham, MA), Fig. 4,
are used at ECBC to generate the PSL aerosols. Different jet number Collison nebulizers are available: 4 jet,

6 jet, 24 jet, and 36 jet nebulizers. PSL microspheres
are added to deionized water and this solution is used
in the nebulizer. Low PSL concentration in liquid is
used to minimize doublet and triplet PSL aerosol gen-

Fig. 4 Picture of a 24 jet nebulizer

eration. The Collison nebulizer is connected to compressed air that exits from small holes inside the nebulizer at high velocity. The low pressure created in the exit region causes liquid to be drawn from the bottom of the nebulizer through a second tube by the Bernoulli effect. The liquid exits the tube as a thin filament that is stretched out as it is accelerated in the air stream until it breaks into droplets. The spray stream is directed onto the wall where larger droplets are impacted and removed from the air. The PSL microspheres are carried out of the nebulizer. Particles generated by this method are charged, so they are neutralized by passing them through a Model 3054 Kr-85 radioactive neutralizer (TSI Incorporated, St. Paul, MN).

3.4 Sonic Nozzles

Dry aluminum oxide, dry PSL microspheres, and dry bioparticles are aerosolized using a two-fluid pneumatic sonic nozzle, Fig. 5, that employs two concentric nozzles. One nozzle is connected to the compressed air that exits out through a small annular opening. The other nozzle is located axially in the concentric annular nozzle and is connected to the powder that will be aerosolized. The low pressure created in the exit region due to the air flow causes powder to be pulled through the axial tube at a very low feed rate due to the Bernoulli Effect. The desired air to powder mass ratio is 80–100:1. Because the air flow rate (1100 L/min) and the aerosol generation rate are high, particles generated by this method are highly charged and cannot be neutralized using the Kr-85 neutralizer. The sonic nozzle was initially developed for ECBC under contract by SRI International (Menlo Park, California), but currently, it is built by ECBC.

Aerosol Output

Connect tube to a reservoir of powder that will be aerosolized

Connect to compressed air supply

Fig. 5 Picture of a two-fluid Pneumatic Sonic nozzle

3.5 Puffers

Puffers are metered dose inhalers (MDI) adapted by ECBC for use with biological simulants, Fig. 6. The puffer is a portable, convenient aerosol source and ejects a small reproducible cloud of simulant aerosol particles at the push of a button. A puffer releases a 60 mg spray of its contents into the air when it is activated. The droplets expand and evaporate instantly leaving a small cloud of the loaded particles. It can be used as a quick source of particles in the laboratory and as an operation check in equipment deployed in the field. A puffer can also be used to fill a small chamber for more quantitative experiments. We have used many materials in the puffers such as BG, ovalbumin, and PSL microspheres. A typical formulation is 0.1% by weight of simulant material in pharmaceutical grade 1,1,1,2-tetrafluoroethane (HFA-134a-P) as the propellant. The shot-to-shot mass variation is small, $\approx 10\%$.

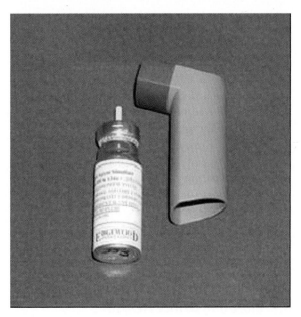

Fig. 6 Picture of a puffer aerosol generator

3.6 Ink Jet Aerosol Generator (IJAG)

The IJAG, Fig. 7, was developed at ECBC for low-concentration aerosol applications; however, it can also be used in some mid- and high-concentration applications (Bottiger and DeLuca, 1999). In aerosol sampler characterization tests, the IJAG is typically used

Fig. 7 Picture of an Ink Jet
Aerosol GENERATOR
(IJAG)

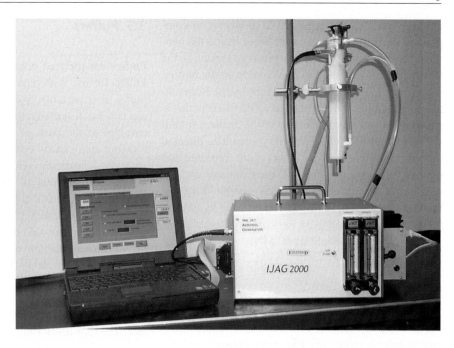

to generate particles directly into the aerosol sampler
inlet and reference unit.

A 12-nozzle ink jet cartridge (HP 51612A), purchased empty, is filled with a slurry of ultra-clean water
and the material of interest. Droplets are fired downward through a heated drying tube where the water
evaporates leaving an aggregate residue particle. Since
the size of the primary ink jet droplet is fixed, $\approx 50\ \mu m$
diameter, the size of the residue particle depends only
on the concentration of the slurry. We have prepared
different concentrations of slurries to produce different size monodisperse particles (3–10 μm). The IJAG
is capable of producing monodisperse particles at a rate
of 1–500 particles/s.

3.7 Vibrating Orifice Aerosol Generator (VOAG)

A model 3450 VOAG (TSI Incorporated, St. Paul,
MN), Fig. 8, is used to generate monodisperse fluorescent oleic acid droplets. The generated droplets pass
through a TSI Model 3054 Kr-85 neutralizer to neu-
tralize the charged particles before entering the chamber. Different concentrations of sodium fluorescein and
oleic acid in isopropanol are used to generate different size particles. The solution to be aerosolized is

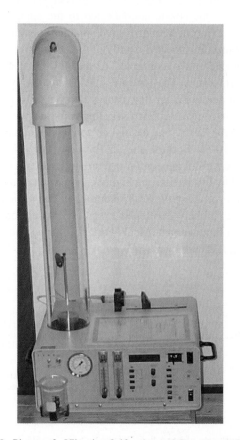

Fig. 8 Picture of a Vibrating Orifice Aerosol Generator (VOAG)

placed in a pressure container. The pressure forces the solution at a constant flow rate from the pressure container through a membrane filter and Teflon tubing into the liquid orifice assembly. The solution is forced through a small orifice to form a liquid jet. A piezoelectric crystal produces mechanical vibration in the liquid orifice disk. The vibration causes the liquid jet to break up into uniform droplets. The alcohol evaporates leaving the fluorescent tagged oleic acid as the final challenge particle. The particle size of the aerosols can be altered by changing the concentration of oleic acid in iso-propanol, the vibration frequency of the liquid orifice disk, and the liquid feed rate.

very small particles (e.g., spores) suspended in the liquid, or may be dissolved in the liquid. The final particle size depends on the concentration of material in the feed liquid and the characteristic size of the Sono-Tek droplet. The rate of aerosol generation depends on the rate at which the mixture fluid is fed to the Sono-Tek nozzle. Five Sono-Tek models are available with number mean diameters ranging from 23 to 70 μm (120 kHz to 25 kHz operating frequencies) and flow rates from a few microliters per second to about 6 gallons per hour. At ECBC, the 25 kHz model, fed by a peristaltic pump, and the 120 kHz micro-bore model, fed by a syringe pump, are used in a laboratory to make controllable aerosol concentrations from about 500–50,000 particles per liter of air.

3.8 Sono-Tek Ultrasonic Atomizing Nozzles

Sono-Tek ultrasonic atomizing nozzles (Sono-Tek Corporation, Milton, N.Y.), Fig. 9, are used at ECBC to produce droplets of liquid/material mixtures that are dried to leave an aerosol of residue material particles. The liquid is usually either water or alcohol in cases where rapid evaporation to form the aerosol particles is desired. The material may either comprise

3.9 Fluorometer

The Turner Model 450 Fluorometer (Barnstead/Thermolyne Corporation, Dubuque, IA), Fig. 10, is used to measure the fluorescence of the collected samples. Appropriate excitation and emission filters are used to detect the blue (Ex: NB360; Em:SC430) and green (Ex:NB460; Em: SC500) PSL microspheres and

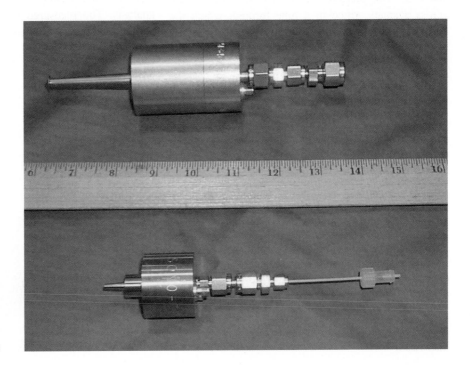

Fig. 9 Picture of Sono-Tek ultrasonic atomizing nozzles

Fig. 10 Picture of a fluorometer

sodium fluorescein (Ex:NB490; Em:SC515). When measuring sodium fluorescein, the pH of the sample must be between 8 and 10 to achieve maximum fluorescence. The temperature also affects the amount of fluorescence; therefore, the samples that are compared should be at the same temperature.

3.10 Coulter Multisizer

The Coulter Multisizer II analyzer (Beckman Coulter, Miami, FL), Fig. 11, is a multichannel particle size analyzer. The analyzer uses electrical impedance across a small orifice in a tube through which the particles travel in a liquid suspension as a method of measurement to provide a particle size distribution analysis. There are different size orifice tubes for measuring different ranges of particle sizes. We generally use the $50\,\mu m$ aperture diameter orifice tube in the Coulter Multisizer that measures $1–30\,\mu m$ diameter particles.

The Coulter Multisizer analysis method requires that the sample be in an electrolyte solution. This is achieved by either using the electrolyte solution as the sample collection liquid in the sampler, or by diluting high concentration samples in electrolyte solution. Filters that were used to collect particles are put in electrolyte solution and vortexed to remove the particles from the filter into the solution for Coulter Multisizer analysis. The measured geometric diameter is converted to AD using the density of the particles and the

Fig. 11 Picture of a Coulter Multisizer II

shape factor of the particles. Disassociation of agglomerated and clustered particles in the electrolytic solution is not desired and should be checked before using the Coulter Multisizer as the analysis method.

3.11 Aerodynamic Particle Sizer (APS)

The APS Model 3320 (TSI Incorporated, St. Paul, MN), Fig. 12, is a high-performance, general-purpose particle spectrometer that measures both AD and light-scattering intensity. The APS provides accurate particle count and size distributions for particles with aerodynamic diameters from 0.5 to 20 µm. It detects light-scattering intensity for particles from 0.3 to 20 µm.

The aerodynamic size is determined by the rate of acceleration. The APS measures the acceleration of aerosol particles in response to the accelerated flow through a jet; the smaller particles accelerate faster and the larger particles slower. As the particles exit the nozzle, the time of flight between two laser beams is recorded and converted to an AD using an internal calibration table. The light-scattering information is also obtained and plotted against the aerodynamic size to gain additional information.

3.12 RBD 3000 Flow Cytometer

The RBD 3000 flow cytometer (Advanced Analytical, Ames, Iowa), Fig. 13, measures the relative particle size, number, and biological state (biomass, live and dead). The microbes are labeled by stains or antibodies and different stains or antibodies are used to distinguish the dead and alive microbes. Up to forty-two samples can be analyzed by the instrument, one at a time, in automatic mode.

The samples are loaded in the instrument and the stain or antibody is added by the instrument to each sample. Each sample is then ejected from a central orifice stream into a sheath fluid flow which is traveling at a much higher velocity and serves to hydrodynamically focus the sample into a very fine stream with single particles. A red diode laser light source is used to excite the fluorescence tagged microbes for analysis. Two photomultiplier tubes collect the emitted signals for fluorescence and side scatter. The side scatter light is used to determine the particle size and the fluorescence is used to select the particles of interest. The flow cytometer is used to determine the number of particles and different stains allow us to determine the alive, dead, and total biomass.

Fig. 12 Picture of an Aerodynamic Particle Sizer (APS)

Fig. 13 Picture of a RBD
3000 flow cytometer

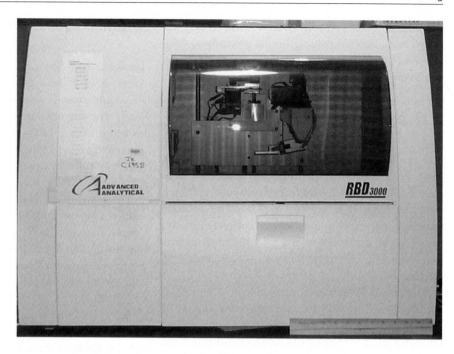

4 Sampler Efficiency Test Methods

4.1 Polystyrene Latex (PSL) Microspheres Tests

Sampling Efficiency tests are conducted with monodisperse fluorescent and non-fluorescent PSL microspheres (Duke Scientific Corp., Palo Alto, CA). The PSL aerosols are generated using a Collison nebulizer, a sonic nozzle, an IJAG, or puffer. The PSL microspheres have to be in powder form for aerosolization using the sonic nozzle. In general, we use the 24 jet collision nebulizer for generating the aerosols because the low air flow rate allows the aerosol to be neutralized using a Kr-85 neutralizer before it enters the chamber. During experiments, aerosols are generated for 10–20 min, and the chamber air is mixed for 1 min to achieve uniform aerosol concentration before sampling using reference filters and the samplers that are being tested.

Polycarbonate membrane filters (Osmonics Inc., Minnetonka, Minnesota) are used as reference filters to collect fluorescent PSL microspheres if the analysis is by fluorometry or by Coulter Multisizer. After sampling, the samples from the system under test and reference filters are collected. Sample liquids are directly

analyzed by a fluorometer or Coulter Multisizer; however, the membrane filters and any dry samples from systems under test are processed to remove microspheres from the filters and surfaces into the liquid for fluorometer or Coulter analysis. The removal procedure consists of placing the membrane filters into 20 mL of filtered deionized water, then shaking on a Multi-Tube Vortexer (VWR Scientific Products, USA) for 10 min to remove the particles from the filter into the liquid. Sampling Efficiency is determined by comparing the samples collected by the test sampler to the samples collected by the reference filter. Sample volumes and air flow rates are taken into account in the Sampling Efficiency calculations.

4.2 Polydisperse Solid Aerosol Tests

Polydisperse aluminum oxide (Al_2O_3) (Saint-Gobain Industrial Ceramics, Worcester, MA) aerosol is generated using the sonic nozzle in the $70\,m^3$ chamber. To get fairly uniform coverage by particle number over the $1–10\,\mu m$ range, a mixture is used (50:50 by weight) of two of the commercially available powders identified by Stokes numbers of $9\,\mu m$ and $14\,\mu m$. One gram of powder takes approximately 10–15 s to be aerosolized. The generated aerosol is mixed in the chamber for

1 min before sampling. Polycarbonate membrane filters (Osmonics Inc., Minnetonka, Minnesota) are used as reference filters as described above. Collected particles on reference filters are removed from the filters as described in PSL tests. Collected samples and reference samples are analyzed using the Coulter Multisizer using an orifice selected according to the size range and resolution best suited to the concerns of each experiment. Usually, the 50 μm orifice tube is used because it covers the aerodynamic size range 1–10 μm, that is of primary interest in biodetection work at ECBC, with more than enough resolution. In fact, size bins are often aggregated when plotting data. The measured geometric diameter of aluminum oxide by the Coulter Multisizer is converted to AD using the density of aluminum oxide (4 g/cm^3) and the shape factor (1.22). Sampling Efficiency is determined as described in the PSL tests bin-by-bin over the size range of interest.

4.3 Sodium Fluorescein Tagged Oleic Acid (Fluorescent Oleic Acid) Tests

Monodisperse fluorescent oleic acid particles are generated using a VOAG (TSI Incorporated, St. Paul, MN). The VOAG can be used to generate monodisperse 3 to >20 μm diameter particles. Sizes of the fluorescent oleic acid particles are determined by sampling the aerosol onto a microscope slide inserted into an impactor and measuring the droplet size using a microscope. A microscopic picture of the collected particles is shown in Fig. 14. The measured fluorescent oleic acid particle diameter is converted to an aerodynamic particle size using a spread factor (Olan-Figueroa et al., 1982) and the density of fluorescent oleic acid. At the end of aerosol generation, the aerosol in the chamber is mixed for 1 min to achieve a uniform aerosol concentration. The samplers and the corresponding reference filters sample the aerosol simultaneously for the same amount of time.

Glass fiber filters (Pall Corporation, Ann Arbor, MI) are used as the reference filters to collect the fluorescent oleic acid particles. After sampling, the filters are removed from filter holders, placed into a fluorescein recovery solution, and shaken on a table rotator (Lab-Line Instruments, Inc., Melrose Park, IL) for one hour. The recovery solution used in the tests is water-alcohol solution with a pH between 8 and 10, obtained by adding a small amount of NH$_4$OH (e.g., 499.5 mL of water, 499.5 mL of isopropanol, and

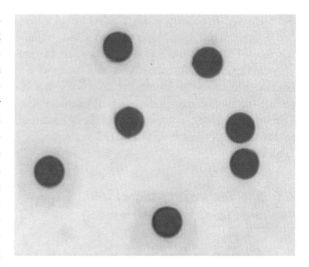

Fig. 14 Microscopic picture of 10 μm AD fluorescent oleic acid droplets

1 mL of 14.8 N NH$_4$OH). Factors that affect fluorescein analysis and the removal of fluorescein from filters are described in detail by Kesavan et al. (2001). The samples from the aerosol samplers are pH corrected by adding NH$_4$OH before the amount of fluorescence is measured by the fluorometer (Barnstead/Thermolyne, Dubuque, IA). All the samples are analyzed on the same day as the experiment or the next day. In appropriate situations, the recovery solution is used as the sample collection liquid. Separate tests did not show any photodegradation of fluorescence with time under normal laboratory lighting in a 12-day period.

4.4 Bioaerosol Tests

Monodisperse clustered bioparticles are generated using an IJAG. Polydisperse or narrowly dispersed biological particles are generated using puffers, nebulizers, and the sonic nozzle. The analysis methods used at ECBC are culturing, PCR, APS, ELISA, and Coulter Multisizer for the organisms; however, a small amount of fluorescein can be added to the solution that is aerosolized and the collected samples can be analyzed by fluorometry. This method assumes that the fluorescein behaves similar to organisms.

In tests where the chamber is filled with aerosols, the aerosol is generated for a certain time, and the chamber air is mixed for 1 min to obtain a uniform aerosol concentration. In tests where the IJAG is used as the aerosol generator, the IJAG output can either go into the inlet of the sampler or into a small chamber for Sampling Efficiency tests. The reference filters, and samplers that are being characterized, sample the aerosol for the same amount of time. At the end of sampling, the reference filters and samples are collected for analysis. Membrane filters are used to collect the bioparticles if the analysis method is by Coulter Multisizer. Samples for analyses by culturing, ELISA, and/or PCR are sent to ECBC's microbiology laboratories.

4.5 IJAG Tests with Solid, Liquid, and Biological Particles

An IJAG could be used to test an aerosol sampler with monodisperse solid, liquid, and biological particles separately. In general, the IJAG output is directed into the inlet of the aerosol sampler and the reference filter. Sodium hydroxide in water with fluorescein is used to generate solid sodium hydroxide monodisperse particles. Tween 80 (a surfactant) in water with fluorescein is used to generate monodisperse fluorescent Tween 80 liquid particles. BG in water with fluorescein is used to generate monodisperse bioclusters. The particle size is adjusted by adjusting the concentration of solute or suspension. The output of the IJAG is placed at the inlet of the sampler to capture all particles exiting the IJAG. Reference samples are generated by capturing the IJAG output on filters. The analysis is conducted by fluorescence for the solid and liquid particle tests. The bioparticles can be analyzed by culturing and/or fluorescence. Because all particles generated by IJAG enter the sampler, the flow rate of the sampler does not need to be known for the Sampling Efficiency calculations. These tests are conducted much faster because the chamber does not need to be filled with aerosols for each change of particle size.

5 Applications: Example Aerosol Samplers and Their Sampling Efficiencies

Many aerosol samplers have been characterized at ECBC using the methods described in Section 4. Three samplers were selected to be presented in this section to illustrate the Sampling Efficiency methods used to characterize these samplers and this is not a com-

prehensive evaluation of the instruments. The results shown here represent a snapshot in time of testing at ECBC and may not actually represent the results of the current models from the manufacturer. The results of the three samplers cover seven different Sampling Efficiency methods and each sampler used three to four different methods: (1) Fluorescent PSL – fluorometry, (2) Fluorescent oleic acid – fluorometry, (3) Aluminum oxide particles – Coulter counting, (4) Clustered bioparticles – APS counting, (5) Single spore bioparticles – culturing, (6) Fluorescent clustered bioparticles – fluorometry, and (7) Clustered bioparticles – culturing.

5.1 BioBadge Samplers

Four BioBadge aerosol samplers (MesoSystem Incorporated, Richland, WA) were tested at ECBC. A picture of the BioBadge aerosol sampler is shown in Fig. 15. The BioBadge is a small battery operated sampler that can be used as a personal sampler. The sampler is designed to sample air at a flow rate of 35 L/min, and it uses MesoSystems' rotating impactor/impeller technology. The impeller is effectively a fan used to move air as well as to collect particles on the blades. The particles are collected dry and then removed into a liquid by sonication. For sonication, the impeller is placed in a zip lock bag with 5 mL of 0.01% of Triton X, a surfactant, and sonicated for 5 min. A BioBadge aerosol sampler is small (L = 5.5 inches, W = 2.75 inches, D = 1.5 inches) and weighs 250 g.

BioBadge aerosol samplers were characterized using 1 and 2.26 μm fluorescent PSL microspheres, 3 and 8 μm fluorescent oleic acid particles, and single spore bioparticles. The fluorescent PSL microspheres were aerosolized using a Collison nebulizer and analyzed by fluorometry. The fluorescent oleic acid particles were aerosolized using a VOAG and analyzed by fluorometry. The single spore bioparticles were aerosolized from dry powder using a sonic nozzle and analyzed by culturing. The Sampling Efficiency results are shown in Fig. 16.

5.2 XMX/2A Aerosol Concentrator

The XMX/2A aerosol concentrator (Dycor Technologies Ltd., Edmonton, Alberta, Canada) is a 3-stage

Fig. 15 Picture of the BioBadge aerosol samplers

Fig. 16 Sampling efficiency of BioBadge aerosol samplers

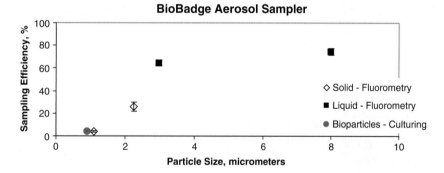

virtual impactor aerosol concentrator that concentrates particles that are in air. It is designed to sample the air at a flow rate of 800 L/min. A picture of the XMX/2A sampler is shown in Fig. 17. On top of the inlet, there is a rain cap to prevent rain water from entering the sampler, and a wire mesh in the inlet to prevent large dust particles and bugs from entering the sampler. The sampler's size is approximately 8.5″ by 18″ by 18″ and the power usage of the system is 780 W (115.6 V and 7 Amp). An external pump pulled the concentrated air flow from the third stage through a 47-mm filter at 1 L/min in tests where a sample needed to be collected. An APS (1 L/min) was connected to the concentrator's output in tests where the APS was used as the analyzer.

Sampling Efficiency was determined using polydisperse aluminum oxide particles, 5 μm oleic acid particles, and 3 and 5.5 μm cluster bioparticles. The polydisperse aluminum oxide aerosol was generated from dry powder using the sonic nozzle and analyzed by Coulter counting. The fluorescent oleic acid particles were generated using the VOAG and analyzed by fluorometry. The 3 and 5.5 μm bioparticles were generated using the IJAG and analyzed by APS counting. The Sampling Efficiency results are shown in Fig. 18.

5.3 Smart Air Sampler System, SASS 2300 Plus (SASS)

The SASS 2300 is a wetted wall cyclone sampler (Research International, Monroe, WA) and is shown in

Fig. 17 Picture of XMX/2A aerosol sampler

Fig. 18 Sampling Efficiency of XMX/2A aerosol sampler

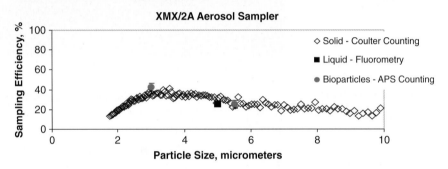

Fig. 19. The manufacturer states that distilled water is typically used in the system and that no additives or surfactants are required for maximum efficiency. The air sampler is microcontroller based and can function as a stand-alone unit or linked to other sampling, detection, or communication systems. This system continuously recycles liquid to concentrate the sample and adds makeup liquid as it evaporates. The designed air flow rate of the sampler is 302 L/min and sample volume is 5 mL. The SASS is light-weight (10 lb with the battery), and easily portable with a built-in handle on top. It sits stably on a flat surface and is relatively small in size, measuring approximately 13″ high with a slightly oval 13″ × 7″ diameter body. The power requirement of the unit is approximately 19.3 W (119 V with 0.3 Amp).

The Sampling Efficiency of the SASS aerosol sampler was determined using 1, 3, and 5 μm PSL microspheres, 5 and 8 μm fluorescent oleic acid particles, and 5 μm cluster bioparticles. The PSL microspheres were aerosolized using a Collision nebulizer and ana-

Fig. 19 Picture of smart air sampler system, SASS 2300 plus

Fig. 20 Collection and Sampling efficiencies of a SASS aerosol sampler (see text)

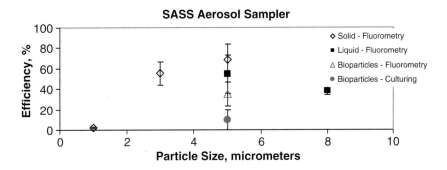

lyzed by fluorometry. The oleic acid particles were generated using the VOAG and analyzed by fluorometry. The fluorescent tagged cluster bioparticles were aerosolized from alcohol suspension of sodium fluorescein and BG spores using the Sono-Tek ultrasonic atomizing nozzle and analyzed by culturing of the samples as well as fluorometry. The Sampling Efficiency results are shown in Fig. 20. The results illustrate a difference of the fluorescence of the sample vs. directly culturing of the bioparticles. The fluorescein is released from the bioparticles when they are taken up in the cyclone and appears in the sample more quantitatively than bioparticles themselves can be extracted from the cyclone due to wall losses (i.e., bioparticles that attach to the cyclone and tubing walls and hence do not appear in the extracted collection liquid at the end of the sampling period nevertheless release their fluorescein into the collection liquid). In this case, the fluorescence results indicate the collection efficiency of the device and the culturing results indicate its Sampling Efficiency.

6 Discussion

Many aerosol generation and analysis methods are used at ECBC. Solid particles may bounce when they hit a surface and reach the collection site; on the other hand, the liquid particles are removed from the air when they hit surfaces and may not reach the collection site. After an initial quick test with polydisperse aluminum oxide particles, if the sampler has good Sampling Efficiency, then additional detailed Sampling Efficiency tests using monodisperse fluorescent oleic acid particles and monodisperse fluorescent PSL particles are conducted with fluorometer analysis. Sampling Efficiency tests using monodisperse particles are conducted one particle size at a time, which is a

time consuming procedure when the $70\,m^3$ chamber needs to be filled and evacuated after each test.

An IJAG can be used to test a sampler with monodisperse solid, liquid, and biological particles by delivering the test aerosol to the inlet of the sampler and separately to the reference unit. Tests with many particle sizes can be conducted in one day because the chamber does not need to be filled and evacuated with aerosols for each test. However, the Sampling Efficiency of particles generated by the IJAG at the inlet needs to be compared to the Sampling Efficiency of particles generated by other methods into a chamber because the IJAG test usually requires removing a rain cap from the inlet and this can modify the inlet flow to reduce collection on walls in the inlet. The use of polydisperse solid aerosols with a Coulter Multisizer analysis gives information of all particle sizes at once. Therefore, it is used as a quick test to characterize samplers over a range of particle sizes and identify the size range of interest for more detailed, accurate, and time-consuming study using monodisperse particles.

Membrane filters must be used as a reference filter medium to collect the solid particles for Coulter Multisizer and fluorometer analyses to improve sample recovery. However, the air flow rate through the membrane filters cannot be very high compared to the rate through glass fiber filters. Therefore, glass fiber filters are preferred for other types of particles.

The use of PSL microspheres is a convenient and accurate method for monodisperse, small particle (0.5–3 μm) tests. They are easy to aerosolize but are expensive. The PSL beads have a limited amount of fluorescence; therefore, very small amounts cannot be detected by the fluorometer. Also, the particle count has to be significantly above the background noise for Coulter Multisizer analysis. An IJAG can be used to cluster small (0.5–3 μm) PSL beads to generate larger PSL particles.

The use of fluorescent oleic acid with fluorometric analysis is an accurate method for 3–20 μm diameter particle tests. It is difficult to generate particles smaller than 3 μm using the VOAG. Correct excitation and emission filters should be installed in the fluorometer for the detection of sodium fluorescein. Samples have to be pH corrected for the maximum fluorescence of sodium fluorescein. All the samples have to be at the same temperature because temperature affects the amount of fluorescence (Kesavan et al., 2001).

Much care must be taken in the use of bioaerosols in sampler characterization tests. Many aerosol generation and sampling methods kill vegetative bacteria and some spores. Another disadvantage is that some organisms start growing once they are mixed with the right nutrients; therefore, care must be taken to avoid contact with nutrient material during sampling. Also, samples have to be refrigerated to prevent the growth of organisms if the analysis is planned for a later time.

Some bioparticles are hydrophobic and others are hydrophilic. Bioparticles may also be sticky. We have observed bioparticles sticking to walls and tubing while inert particles get washed off completely (Fig. 20). Researchers need to use a test method to verify that inert particle results will represent bioaerosol test results. Sampling Efficiency tests should be conducted with bioparticles if it is suspected that they can stick to walls.

Tests with no aerosols are conducted to measure background signal of the samplers as well as reference filters. In liquid based aerosol samplers, washes of the aerosol collection surfaces are conducted before each test to confirm that the samplers are free of test materials. A wash usually involves cycling the sampler with a fresh amount of it's own sampling liquid and the recommended procedure by the manufacturer is used. After the first sampling test, up to four washes are conducted to remove test material from the samplers, and to determine the number of washes required to remove all test material from the sampler after each test. Based on these results, the samplers are washed in between tests to remove all test materials from the samplers.

7 Summary

We have presented a set of definitions for various efficiencies that can be measured experimentally, and recommended adoption of Sampling Efficiency as an overall performance measure rather than the often ambiguous collection efficiency frequently found in the literature. Test aerosols, generation methods, and analysis techniques that are used at ECBC to characterize samplers and to determine particle losses have been described in this chapter. Samplers are routinely characterized in chambers at ECBC at calm air conditions using the following methods: (1) monodisperse fluorescent/non-fluorescent PSL microspheres with fluorometric analysis or Coulter Multisizer analysis, (2) polydisperse solid aluminum oxide particles with APS analysis or Coulter Multisizer analysis, (3) fluorescent oleic acid particles with fluorometric analysis, and (4) bioparticles with Coulter Multisizer, culturing, ELISA, PCR, and APS analyses. The description of three aerosol samplers, example sampling efficiencies, and the test methods used are presented in this chapter.

We routinely characterize samplers with all three aerosol types (solid, liquid, and biological). Samplers are characterized with 1–10 μm particles to determine the Sampling Efficiency and are characterized with >10 μm particles to determine the sampler's rejection efficiency of larger particles.

Solid particles can bounce when they impact onto internal surfaces, or they can be re-entrained into the airflow after deposition; however, liquid particles are permanently captured upon impact. Biological particles (microorganisms and biological substances such as proteins) fall in between solid and liquid particles with respect to their characteristics of bounce. Bioparticles may also be tacky and stick to tubing and walls which can seriously affect recovery by elution of the sample from the collector component as well as affecting the aerosol sampler's overall performance. Therefore, the test methodology should be carefully selected to answer the research question. Additionally, aerosol can either fill a chamber or be delivered to the inlet of the sampler under test and reference filter. Using an IJAG to deliver the test aerosol to the inlet of the sampler and reference filter will allow many tests to be conducted in one day; however, any disadvantages in using this method need to be fully explored before this method is accepted as a gold standard.

Acknowledgments The Department of Homeland Security sponsored the production of this material under an Interagency Agreement with the Edgewood Chemical Biological Center.

References

Aizenberg V, Grinshpun SA, Willeke K, Smith J and Baron PA (2000a) Performance characteristics of the button personal inhalable aerosol sampler. *Am. Ind. Hyg. Assoc. J.* 61, 398–404.

Aizenberg V, Reponen T, Grinshpun SA and Willeke K (2000b) Performance of Air-O-Cell, Burkard, and button samplers for total enumeration of airborne spores. *Am. Ind. Hyg. Assoc. J.* 61, 855–864.

Bottiger J and DeLuca P (1999) Low concentration aerosol generator. United States Patent Number 5, 918, 254. USA.

Gao P, Dillon HK and Farthing WE (1997) Development and evaluation of an inhalable bioaerosol manifold sampler. *Am. Ind. Hyg. Assoc. J.* 58, 196–206.

Hering S (1995). Impactors, Cyclones, and Other Inertial and Gravitational Collectors. In, Beverly C and Susanne H (eds.), *Air Sampling Instruments for Evaluation of Atmospheric Contaminants* 279–322. ACGIH, Cincinnati, Ohio.

Hinds WC (1999) *Aerosol Technololgy: Properties, Behavior, and Measurement of Aerosol Particles. 2nd Ed.* John Wiley & Sons, New York.

Jensen PA, Todd WF, Davis GN and Scarpino PV (1992) Evaluation of eight bioaerosol samplers challenged with aerosols of free bacteria. *Am. Ind. Hyg. Assoc. J.* 53, 660–667.

John W, Kreisberg NM (1999) Calibration and testing of samplers with dry, polydisperse latex. *Aerosol Sci. Technol.* 31, 221–225.

Kesavan J and Doherty R (2001) Comparison of Sampler Collection Efficiency Measurements using a Polydisperse Solid Aerosol and a Monodisperse Liquid Aerosol. ECBC-TR-137. U.S. Army, Edgewood Chemical Biological Center, Edgewood, Maryland.

Kesavan J, Doherty R, Wise D and McFarland AR (2001) Factors that Affect Fluorescein Analysis. ECBC-TR-208. U.S. Army, Edgewood Chemical Biological Center, Edgewood, MD.

Kesavan J, Carlile D, Sutton T, Hottell KA, and Doherty RW (2002) Characteristics and Sampling Efficiency of PHT-LAAS Air Sampler. ECBC-TR-267. U.S. Army, Edgewood Chemical Biological Center, Edgewood, Maryland.

Kesavan J and Hottell KA. (2005) Characteristics and Sampling Efficiencies of BioBadge Aerosol Samplers. ECBC-TN-024.

U.S. Army, Edgewood Chemical Biological Center, Edgewood, Maryland.

Li CS, Hao ML, Lin WH, Chang CW and Wang CS (1999) Evaluation of microbial samplers for bacterial microorganisms. *Aerosol Sci. Technol.* 30, 100–108.

Li SN, Lundgren DA and Rovell-Rixx D (2000) Evaluation of six inhalable aerosol samplers. *Am. Ind. Hyg. Assoc. J.* 61, 506–516.

Lin X, Reponen T, Willeke K, and Wang Z (2000) Survival of airborne microorganisms during swirling aerosol collection. *Aerosol Sci. Technol.* 32, 184–196.

Lippmann M (1995). Filters and Filter Holders. In Beverly C and Susanne H (eds.), *Air Sampling Instruments for Evaluation of Atmospheric Contaminants* 247–278. ACGIH, Cincinnati, Ohio.

Maynard AD (1999) Measurement of aerosol penetration through six personal thoracic samplers under calm air conditions. *J. Aerosol Sci.* 30, 1227–1242.

Maynard AD, Kenny LC and Baldwin PEJ (1999) Development of a system to rapidly measure sampler penetration up to 20 um aerodynamic diameters in calm air, using the Aerodynamic Particle Sizer. *J. Aerosol Sci.* 30, 1215–1226.

McFarland AR, Bethel EL, Ortiz CA and Stanke JG (1991) A CAM sampler for collection and assessing α-emitting aerosol particles. *Health Phys.* 61, 97–103.

Olan-Figueroa E, McFarland AR and Ortiz CA (1982) Flattening coefficients for DOP and oleic acid droplets deposited on treated glass slides. *Am. Ind. Hyg. Assoc. J.* 43, 395–399.

Rule A, Kesavan J, Schwab K and Buckley Timothy (2007) Application of flow cytometry for the assessment of preservation and recovery efficiency of bioaerosol samplers spiked with Pantoea agglomerans. *Environ. Sci. Technol.* Apr., 41(7), 2467–2472.

Swift D and Lippmann M (1995) Electrostatic and Thermal Precipitators. In Beverly C and Susanne H (eds.), *Air Sampling Instruments for Evaluation of Atmospheric Contaminants*, 323–336. ACGIH, Cincinnati, Ohio.

Willeke K, Lin X and Grinshpun SA (1998) Improved aerosol collection by combined impaction and centrifugal motion. *Aerosol Sci. Technol.* 28, 439–456.

Witschger O, Willeke K, Grinshpun SA, Aizenberg, V, Smith J and Baron PA (1998) Simplified method for testing personal inhalable aerosol samplers. *J. Aerosol Sci.* 29, 855–874.

Smog Chamber Measurements

Seung-Bok Lee, Gwi-Nam Bae and Kil-Choo Moon

Abstract Photochemical smog still remains an issue in urban areas. Various smog chambers have been used to examine atmospheric processes of the formation of ozone and secondary organic aerosols. Prior to conduct smog chamber experiments, spectrum and intensity of light sources and chamber wall effects need to be characterized. Experimental techniques such as light intensity control, temperature control, comparison of twin chambers are also required to obtain more reliable and useful data from smog chamber experiments. Smog chamber experiments can be classified into indoor and outdoor chamber studies or VOCs–NO$_x$–air mixture and ambient air experiments. Some typical and important investigations from previous smog chamber experiments are introduced here. Finally, applications of smog chamber experiments are demonstrated for diesel exhaust and indoor air chemistry.

Keywords Ozone · Photochemical smog · Secondary organic aerosol · Smog chamber · Visibility

1 Introduction

1.1 Smog Phenomena

The term 'smog' is derived from a combination of smoke and fog. Extensive air contamination by aerosols is also referred to as smog, and the term is sometimes used loosely to describe any air contamination (Seinfeld and Pandis 1998). Visibility impairment is also used instead of smog phenomena because it is easily detected by the eyes. 'London-type' sulfurous smog has been known for at least eight centuries (Finlayson and Pitts 1976). The famous London smog episode that occurred in 5~9 December 1952 was the first typical case of a severe smog episode that caused the death of many people. As for London smog, the primary pollutants emitted directly from the combustion of coal in winter were considered to be the major reason for the smoke in thick fog in the morning. The rapid oxidation reaction of sulfur dioxide gas to sulfate aerosols inside an aqueous phase may be another reason for the London smog episode. The total death toll related to the London smog episode was estimated to be 4,000~12,000.

Although efforts had been made to reduce primary pollutants, such as sulfur dioxide, since the London smog disaster, the visibility impairment that appears as fog occurs frequently in metropolitan areas. The cause of the smog in Los Angeles, USA is different to that of London smog. The secondary oxidation products of organic gases and nitrogen oxides in ambient air under solar irradiation are the major reason for the decrease in visibility of Los Angeles. Visibility impairment of this smog is known as photochemical smog, even though it is neither smoke nor fog. Smog occurs in many mega cities such as Mexico City, Bombay, Singapore, Tokyo, and Seoul, which is known as a representative air pollution phenomenon. The visibility of Seoul, Korea is much lower than in the rest of Korea, as shown in Fig. 1 (Ghim et al. 2005).

There are many difficulties in examining the major reasons for smog phenomena by monitoring only the concentrations of reactants and products from complex photochemical chemistry because many factors, such

S.-B. Lee (✉)
Korea Institute of Science and Technology, 39-1 Hawolgok-dong, Seongbuk-gu, Seoul 136-791, Korea,
e-mail: sblee2@kist.re.kr

Y.J. Kim et al. (eds.), *Atmospheric and Biological Environmental Monitoring*,
DOI 10.1007/978-1-4020-9674-7_8, ⓒ Springer Science+Business Media B.V. 2009

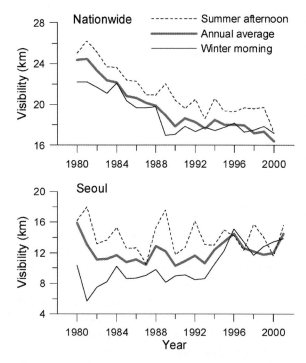

Fig. 1 Visibility trends nationwide and in Seoul, Korea (Ghim et al. 2005)

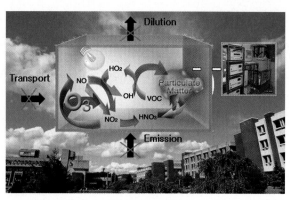

Fig. 2 Concept of the smog chamber

as meteorological conditions and pollutants emission rates, cannot be controlled artificially.

In order to determine effect of a single parameter, the other factors and conditions should not be changed during the experiments. Considerable information, such as the reaction rate of a certain reaction, is needed to simulate and predict the chemical changes in the real atmosphere using complex chemical reaction mechanisms. A controlled and simulated reaction chamber, so called 'smog chamber' is a powerful tool for the above research needs. Experiments using a controlled smog chamber for understanding atmospheric phenomena can be called 'smog chamber experiments'.

1.2 Concept of a Smog Chamber

A controlled system is needed to examine the nature of the atmosphere. A smog chamber is isolated using a reaction vessel wall. This system can be considered a perfect smog chamber for examining the nature of the atmosphere if all the conditions of the system and its surroundings can be controlled artificially. Although

there are many limitations in simulating a real atmosphere, smog chambers can be used in a wide variety of research topics on the atmosphere. There are four major components of a smog chamber, a reaction bag, precursors, light sources, and monitoring instruments, as shown in Fig. 2. Once the initial concentrations of the precursors, such as volatile organic compounds and nitrogen oxides are determined and UV irradiation has started, the additional emission of precursors and transport of different air masses are prohibited in order to minimize the independent variables. However, for a rigid and non-collapsible reaction bag, makeup air should be supplied to the reaction bag during the experiments to replace the air sampled by the monitoring instruments.

1.3 World-Wide Smog Chambers

Historically, smog chamber experiments began in the 1950s (Finlayson and Pitts 1976). The basic ozone chemistry and formation of aerosols from gaseous air pollutants related to visibility impairment phenomena were the main topics of smog chamber experiments in the initial periods (Prager et al. 1960; Stephens et al. 1956). Outdoor smog chambers were introduced to overcome the limitation of artificial light sources (Roberts and Friedlander 1976). However, outdoor smog chambers are dependent on the weather conditions, such as rain, cloudiness and variations in light intensity with the seasons.

There are some research groups, such as University of California, Riverside (UCR), California Institute of Technology (Caltech), and University of

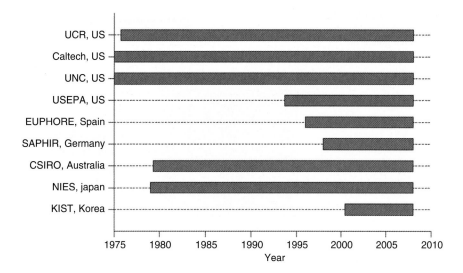

Fig. 3 Research periods of the smog chamber measurements conducted by each research group

North Carolina (UNC), USA, that have led this research field for more than 35 years. Figure 3 summarizes the research periods in smog chamber measurements conducted by many research groups. The two largest outdoor smog chambers in the world, whose volume range from 200 to 300-m³, were established for collaborating research in the EU from the late of 1990s, as shown in Fig. 4. The size of the smog chambers depends on the research topics. In order to minimize rate of material loss to the chamber walls, Heisler and Friedlander (1977) at the Caltech used a large vessel, 134-m³ in size, to make precise measurements of the growth rate of aerosols due to gas-to-particle conversion over a 1 h period. For the comparison experiments, some smog chambers were operated with a dual mode condition or twin smog chambers were used at the same time.

The smog chambers can be classified according to light sources, as shown in Fig. 5. Backlights and/or arc lamps have been used for indoor chambers, such as the new UCR-EPA chamber (so called 'next generation chamber'), new CSIRO (Commonwealth Scientific and Industrial Research Organization) chambers, and KIST (Korea Institute of Science and Technology) chambers, and natural sunlight has been used for outdoor chambers, such as the old Caltech chamber, UNC chambers, and the European largest chambers.

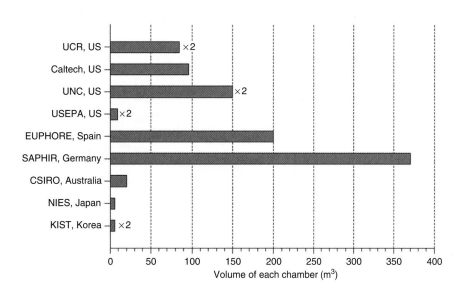

Fig. 4 Volume of the smog chambers used by each research group

Fig. 5 Light sources of the smog chamber facilities by the research groups

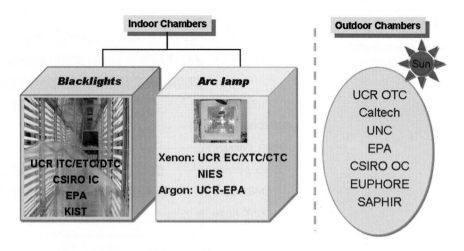

A large number of smog chamber experiments have been carried out since the 1960s. Major research topics with the time are roughly summarized in Fig. 6. The research topics changed as follows: from the gas phase to particulate phase for measuring the target compounds; ambient air to synthetic mixtures in purified air for sample mixtures; high concentration to low concentration for the level of precursor gases; and slow response measurements of limited compounds using traditional instruments to rapid response measurements of various compounds using newly developed instruments.

The following gives a short description with photographs or schematic diagrams of some smog chambers, in which smog chamber experiments were carried out for a long time or remarked results are expected.

The UCR-EPA chamber was established at Bourns College of Engineering- Center for Environmental Research and Technology (CE-CERT), University of California, Riverside. As shown in Fig. 7, its chamber facility allows for experiments at very low reactant concentrations because the enclosure in which the Teflon reactors reside is flushed continually with purified air, resulting in low background contamination, which allows high quality data at low pollutant levels. Some information is available from http://www.cert.ucr.edu/about/apl.asp, or http://pah.cert.ucr.edu/~carter/epacham/.

The University of North Carolina has four types of outdoor smog chambers; one 300-m^3 dual gas-phase chamber, two dual 25-m^3 aerosol chambers and one 190-m^3 aerosol chamber, as well as a new 120-m^3 rooftop chamber for gas/aerosol-phase chemistry and in vivo/in vitro exposure experiments (See Fig. 8). Some information is available from http://airchem.sph.unc.edu/Research/Facilities/.

The European photoreactor, EUPHORE chamber, is one of the largest smog chambers (200-m^3) for

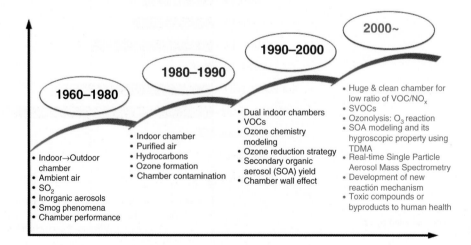

Fig. 6 Trend of major research topics using smog chambers

1960–1980
• Indoor→Outdoor chamber
• Ambient air
• SO$_2$
• Inorganic aerosols
• Smog phenomena
• Chamber performance

1980–1990
• Indoor chamber
• Purified air
• Hydrocarbons
• Ozone formation
• Chamber contamination

1990–2000
• Dual indoor chambers
• VOCs
• Ozone chemistry modeling
• Ozone reduction strategy
• Secondary organic aerosol (SOA) yield
• Chamber wall effect

2000~
• Huge & clean chamber for low ratio of VOC/NO$_x$
• SVOCs
• Ozonolysis: O$_3$ reaction
• SOA modeling and its hygroscopic property using TDMA
• Real-time Single Particle Aerosol Mass Spectrometry
• Development of new reaction mechanism
• Toxic compounds or byproducts to human health

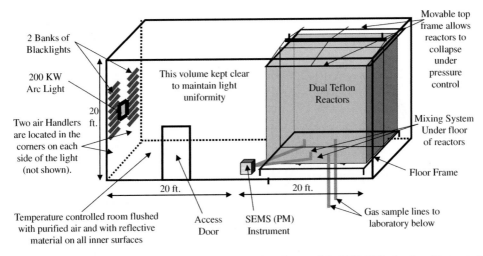

Fig. 7 Schematic diagram of the environmental chamber reactors and enclosure of the UCR-EPA chamber (Carter et al. 2005)

examining a variety of chemical processes under atmospheric conditions. It was installed at CEAM (Centro de Estudio Ambientales del Mediterraneo), Valencia, Spain. A white mirror system for FTIR was located inside the chamber and diesel emissions could be generated on site and introduced to the EUPHORE chamber, as shown in Fig. 9. Some information is available from http://www.ceam.es/html/index_i.htm. There is an on-line centralized database, EUROCHAMP, on environmental simulation chamber studies for making the results obtained in the different simulation chambers more available to the scientific community. This EUROCHAMP can be directly accessed at http://www.eurochamp.org.

The SPHIRE chamber is the largest outdoor chambers (370-m^3) made from double-walled Teflon casing with a high-purity air flow, as shown in Fig. 10. SPHIRE stands for 'Simulation of Atmospheric Photochemistry In a large Reaction Chamber', and is managed by the Institute of Chemistry and Dynamics of the Geosphere, Forschungszentrum Jülich. There is a laser absorption spectroscope for OH measurements. The major research topics are the verification and improvement of tropospheric chemistry models and the development of emission reduction strategies. Some information is available from http://www.fz-juelich.de/icg/icg-2/index.php?index=445.

A new indoor environmental camber (18-m^3) has been developed by the CSIRO Division of Energy Technology to replace the decommissioned, dual outdoor chambers (20-m^3) which were located at the Mineral Research Laboratories (MRL), North Ryde, Sydney, Australia. The new indoor chamber was originally located at MRL but was re-located to the Lucas Heights Science and Technology Centre, Lucas Heights, Sydney Australia, in 2004. The major research areas are the development of models for ozone and secondary organic aerosol formation, the fate of toxic compounds emitted from industrial, automotive and biogenic sources, and the impact of ethanol as a fuel additive on air quality (Fig. 11). The chamber is described in Angove et al. (2000) and in more detail in Hynes et al. (2005). Some information is also available from http://www.det.csiro.au/science/e_e/e_e_topics. htm#smog or http://www.det.csiro.au/PDF%20files/CET_Div_Brochure.pdf.

2 Smog Chamber Experiments

2.1 Outline

Smog chamber experiments can determine the correlation between primary pollutants as precursors and secondary products, such as ozone and secondary organic aerosols. This correlation is also referred to as physical and chemical mechanisms, as shown in Fig. 12 (McMurry et al. 2004). Therefore, smog chamber experiments have significant advantages over other field studies that examined the contribution of aerosol chemical components to visibility for deriving empirical correlations between those mass concentrations and visibility. The amount of ozone and secondary aerosols formed during the daytime can be estimated using the results of smog chamber measurements, but one cannot predict these using results of a field study.

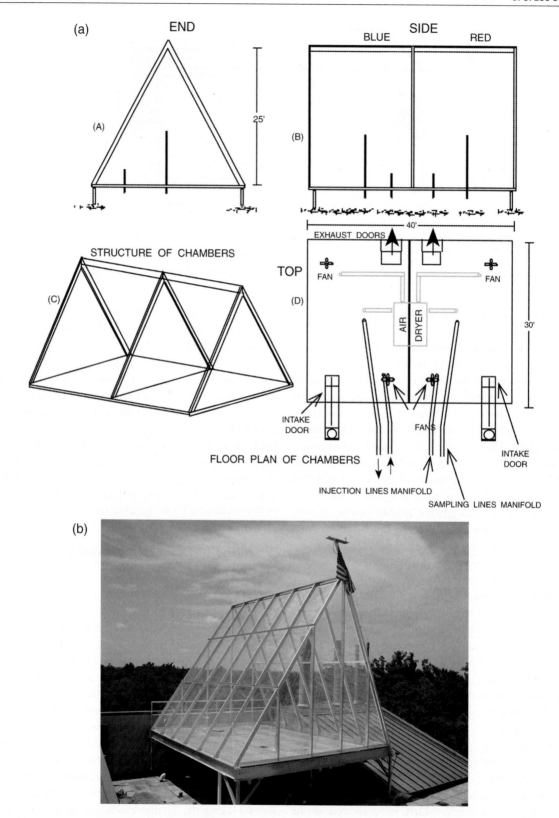

Fig. 8 Schematic diagram and photographs of the old UNC chambers (http://airchem.sph.unc.edu/Research/Facilities/). (**a**) 300-m^3 dual gas-phase chamber. (**b**) 120-m^3 rooftop chamber

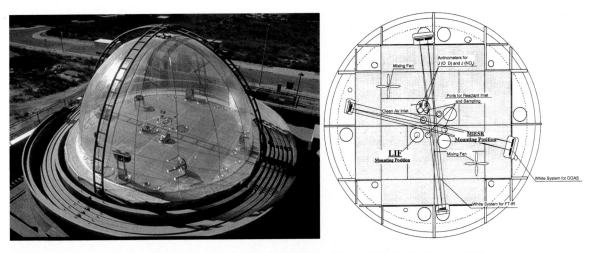

Fig. 9 Photographs and schematic diagram of the EUPHORE chamber (Wiesen 2000; Zielinska et al. 2007)

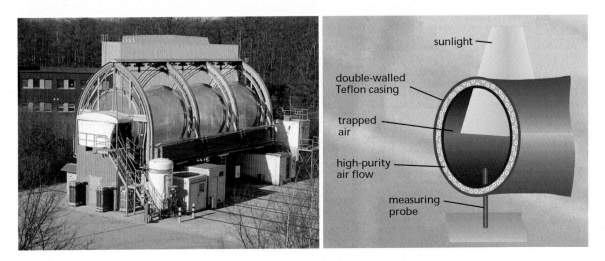

Fig. 10 Photographs and schematic diagram of the SAPHIR chamber (http://www.fz-juelich.de/icg/icg-2/index.php?index=445, Reproduced with the permission of the Institute of Chemistry and Dynamics of the Geosphere)

As mentioned earlier, smog chamber facilities have many parts, such as a reaction bag, precursors and pure air supply systems, light sources, and monitoring instruments. Smog chambers can be classified into two types according to the light sources used. One is an indoor smog chamber using artificial light inside a certain enclosure, and the other is an outdoor smog chamber installed outside of buildings and exposed to natural sunlight. There are several types of artificial light, such as blacklight, Xenon arc lamps, and Argon arc lamps. Some types of artificial lamps are combined to simulate the spectrum of natural sunlight.

Real ambient air was used as a sample to characterize smog phenomena at certain areas, such as Pasadena, Riverside, Detroit, LA, Claremont, Mexico City, Guadalajara, Tokyo, Kawasaki, and Seoul, which is known as captive-air experiments normally carried out in outdoor chambers except for Seoul (Heisler and Friedlander 1977; Jaimes et al. 2003, 2005; Kelly 1987; Kelly and Gunst 1990; Moon et al. 2004b, 2006; Pitts et al. 1977; Roberts and Friedlander 1976; Shibuya et al. 1981). Secondary aerosol formation was investigated with ozone formation by these captive-air experiments in the 1970s, and then the change in ozone formation due to added species, such as hydrocarbon, NO_x, SO_2, diethylhydroxylamine, and LPG, became the main target since the 1980. For Seoul, the formation of ozone and secondary aerosols from

Fig. 11 Photographs and
schematic diagram of a new
indoor chamber of CSIRO
(Angove et al. 2000; http://
www.det.csiro.au/science/e_
e/e_e_topics.htm#smog,
Reproduced with the
permission of the CSIRO)

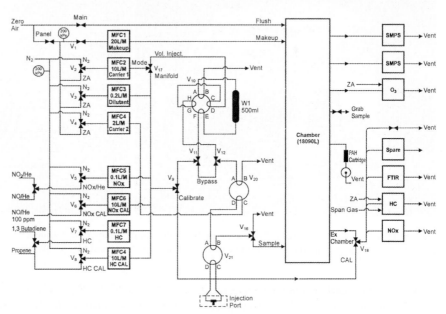

photochemical reactions of real ambient air itself in
indoor twin smog chambers has recently been investigated. There are some disadvantages with these
captive-air experiments using real ambient air in that
characterization of the initial air quality and reproducibility are difficult because ambient air including
many chemical compounds are not reproducible. Special experimental and analytical tools are needed to

eliminate the effect of a difference in ambient air quality, as described later.

A synthetic mixture in purified air is a better sample than real ambient air when using the results of
smog chamber experiments to examine chemical and
physical mechanisms because the experiments can
be repeated and are reproducible. Before middle of
the 1970s, hydrocarbon and NO_x were only used as

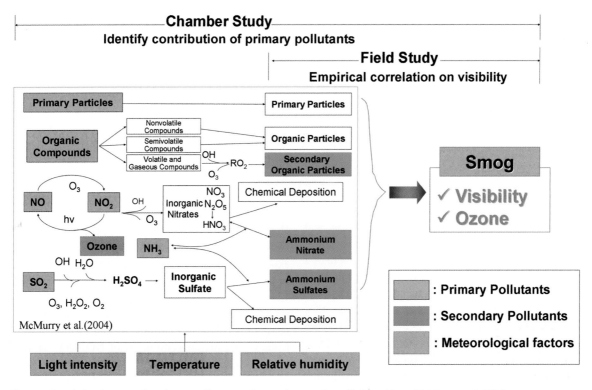

Fig. 12 Correlation between the primary pollutants and secondary products (Adapted from McMurry et al. 2004)

synthetic precursor gases for photochemical reactions. SO_x was not considered to be a precursor because sulfurous and photochemical air pollution were treated separately, partly because the conditions favoring their formation differ markedly.

2.2 Performance of Smog Chamber

The performance of a smog chamber facility is generally evaluated before carrying out the main experiments. The major topics on a performance evaluation of a smog chamber facility are artificial light sources, zero air supply system, sampling systems for real ambient air, and chamber wall effects. In addition, spatial distribution of temperature inside the enclosure when the artificial light is turned on can become an important issue.

2.2.1 Light Sources and Intensity

NO_2 is photolyzed to NO by UV light. The photolysis rate of NO_2, k_1, is used as an index of the light inten-

sity. There are two methods for measuring k_1 of light sources. One is the 'quartz tube actinometry method' based on the calculation equation reported by Zafonte et al. (1977) (Carter et al. 1995). k_1 is calculated with NO, NO_2, and NO_x concentration data under both irradiation and dark conditions using Equation (1) for a long tubular reactor, as shown in Fig. 13. The NO_2 concentration from a few hundred ppb to a few ppm can be prepared from a NO_2 test gas in a N_2 base by dynamic dilution with ultra-high purity N_2 gas (99.9999%). The flow rate of NO_2 gas passing through the quartz

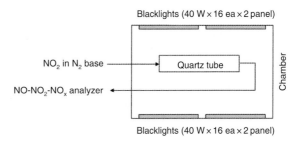

Fig. 13 Example of a schematic diagram for the k_1 measurement

tube can be adjusted to the sampling flow rate of a NO_x analyzer. Excess gas should be vented before the NO_x analyzer. The NO_x concentration should be maintained within $\pm 3\%$ because the NO_x concentration is unaffected by interconversion between NO and NO_2

(Carter et al. 1995). However, if the efficiency of the NO_x converter inside a NO_x analyzer is not correct, there will be a significant difference between the NO_x concentrations under irradiation and dark conditions.

$$k_1 = \frac{[NO]^{light} - [NO]^{dark}}{[NO_2]^{light} + \frac{1}{2}([NO]^{light} - [NO]^{dark})} \times \frac{F}{V} \times \frac{1}{\phi} \tag{1}$$

where, F, V, and ϕ are the flow rate of NO_2, the volume of the quartz tube exposed to the blacklight, and the effective quantum yield for the production of nitric oxide during photolysis, respectively. An example of a quartz tube is 1,000 mm in length and 25 mm in outer diameter with 6 mm in outer diameter extensions at each end. A mass flow controller controls the NO_2 flow rate at a concentration of approximately 0.8 ppm to 3 L/min. The value of ϕ was changed from 1.75 to 1.66 based on computer model simulations of a large number of experiments (Carter et al. 1995). Note that

instrument span errors that affect the measured NO and NO_x equally would not affect the calculated k_1.

Another method for measuring k_1 is the 'chamber experiment' used by the CSIRO (Angove et al. 2000). In this case, the O_3 concentration should be measured with NO and NO_2 concentrations. The reaction constant, k_3, of a NO titration by O_3 as a function of temperature is calculated initially, k_1 can then be derived from k_3 and the concentrations of NO, NO_2, and O_3 can be determined as follows:

$$NO_2 + h\nu \xrightarrow{k_1} NO + \overset{\bullet}{O}$$
$$\overset{\bullet}{O} + O_2 \xrightarrow{k_2} O_3$$
$$NO + O_3 \xrightarrow{k_3} NO_2 + O_2$$
$$\frac{d[NO_2]}{dt} = -k_3[NO][O_3] = -k_1[NO_2]$$
$$\therefore k_1 = k_3 \frac{[NO][O_3]}{[NO_2]} \; at \; steady \; state \tag{2}$$

where k_3 is calculated from the Arrhenius equation as follows:

$$k_3(T) = A \exp \left\{ \left(-\frac{E}{R} \right) \left(\frac{1}{T} \right) \right\} \tag{3}$$

where A and E/R are 2.0×10^{-12} and 1,400, respectively. For example, when the concentrations of NO, NO_2, and O_3 at 300 K are 72, 178, and 73 ppb, respectively, the calculated k_3 and k_1 values are 1.9×10^{-14} cm^3 molecule^{-1} s^{-1} and 0.83- min^{-1}, respectively.

A blacklight can be backed by aluminum-coated plastic reflectors (Carter et al. 1995). The other surfaces are covered with polished aluminum panels to ensure a uniform light intensity inside the chamber and to conserve light energy. A rapid decrease in light intensity with time was observed with the new blacklights (Carter et al. 1995). Because the intensity of the blacklights can be stabilized after the lights are aged, it is recommended that researchers turn the lights on for a period of time prior to use to allow aging. Figure 14 shows the change in k_1 with the number of blacklights for some indoor chambers. k_1 is linearly proportional to the number of blacklights turned on.

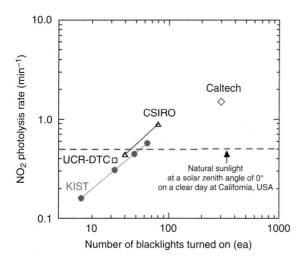

Fig. 14 Change in k_1 with the number of blacklights for indoor chambers

The k_1 of natural sunlight in the summer daytime of Seoul is similar to that of California (Cocker III et al. 2001; Gery and Corouse 2002)

For arc lamps, the spatial uniformity should be examined because only one lamp at a fixed position is used. The spectra of xenon arc lights are expected to change gradually with aging, which is unlike black-lights. The relative spectral distributions of the light sources can be measured using a spectrometer, such as a LiCor Li-1800 spectrometer.

More than two light sources are used together to simulate the spectral distribution of the lamps to be similar to natural light, as in the UCR-EPA chamber whose light spectrum is shown in Fig. 15 (Carter et al. 2005).

2.2.2 Chamber Wall Effects

According to Dodge (2000), the greatest uncertainty of approximately 50% for the smog chamber data is caused by chamber wall effects including film contamination. The chamber walls can serve as both the sources of gaseous species including radicals and the sinks of gaseous species and particulate matter. Heterogeneous processes are also related to the surface of the chambers.

Wall Loss

Gases and particles in the atmosphere can be removed by various surfaces, such as the leaves, rivers, and buildings. The surface area to volume ratio of a real environment is much lower than that of smog chambers. Therefore, the chamber walls may produce unrealistic data. In order to adapt the experimental results of the smog chambers to a real atmosphere, the wall loss rate of the species should be examined and corrected. The accurate wall loss rates for gaseous and particulate matter in smog chambers are important for validating computer kinetic models, deriving the carbon, sulfur, and nitrogen mass balances in chemically reactive mixtures, and determining the organic aerosol formation rates from precursor hydrocarbons (Grosjean 1985). Some studies have carried out theoretical predictions of particle deposition on the chamber walls (Crump and Seinfeld 1981; McMurry and Rader 1985). Wall loss has also been studied experimentally using monodisperse latex spheres, ambient aerosols or aerosols generated by photochemical reactions (Behnke et al. 1988; McMurry and Grosjean 1985).

Crump and Seinfeld (1981) examined the wall deposition rate of particles in a vessel of an arbitrary shape, as shown by Equation (4). This theory accounted for aerosol transport through convection, Brownian diffusion and gravitational sedimentation.

$$\frac{\partial N(D_p, t)}{\partial t} = -\beta(D_p)N(D_p, t) \qquad (4)$$

$$N(D_p, t) = N(D_p, t_0) \exp\{-\beta(D_p)t\} \quad (5)$$

$$\log\{N(D_p, t)\} = \log\{N(D_p, t_0)\} - \beta(D_p)t \quad (6)$$

where $N(D_p, t)$ is the number concentration of aerosols with a diameter D_p at time t, and fractional loss rate, $\beta(D_p)$ is the wall deposition or wall loss rate. McMurry and Rader (1985) extended this theory to include electrostatic deposition observed in Teflon bags. Because Equation (4) is converted to Equation (5), the wall loss rate of aerosols can be derived from a regression curve of a plot of the measured particle number concentration versus time, as shown in Fig. 16. For a 2.5-m^3 Teflon film bag, the wall loss rates of particles ranging from 18~31 and 57~98 nm in diameter are 1.6×10^{-4} and 6.5×10^{-5} sec^{-1}, respectively, as shown in Fig. 16 (Lee et al. 2004a).

Figure 17 shows the measured wall loss rates for large bags. The measured wall loss rates show a large difference for particles > 100 nm compared with the theoretical values suggested by Crump and Seinfeld (1981) due to an electrostatic effect. However, they are

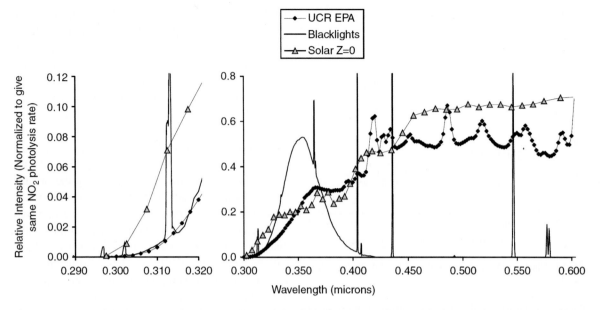

Fig. 15 Spectrum of the argon arc light source used in the UCR-EPA chamber blacklight and representative solar spectra, with the relative intensities normalized to give the same NO_2 photolysis rate (Carter et al. 2005)

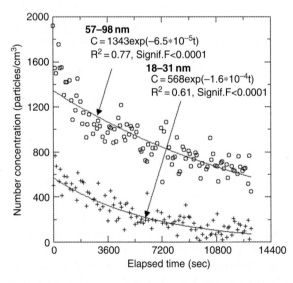

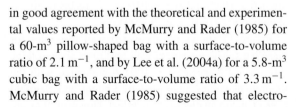

Fig. 16 Example of deriving the wall loss rates according to the particle size for a 2.5-m^3 Teflon bag filled with ambient air (Lee et al. 2004a)

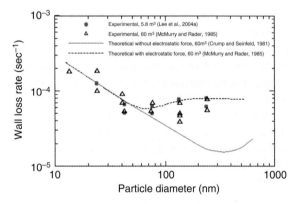

Fig. 17 Wall loss rate of aerosols as a function of the particle size in Teflon bags filled with ambient air (Lee et al. 2004a)

in good agreement with the theoretical and experimental values reported by McMurry and Rader (1985) for a 60-m^3 pillow-shaped bag with a surface-to-volume ratio of 2.1 m^{-1}, and by Lee et al. (2004a) for a 5.8-m^3 cubic bag with a surface-to-volume ratio of 3.3 m^{-1}. McMurry and Rader (1985) suggested that electro-

static forces affect the overall deposition rates for particles ranging in diameter from 50 to 1,000 nm. The wall loss rate varies with time because the charge distribution is time-dependent due to collisions with ions existing in ambient air (McMurry and Rader 1985).

For a specific particle size bin, the corrected particle number concentration, N_c, at a given time is the measured value, N_m, at a given time plus the sum of the decrease in the particle concentration due to wall loss from time = 0 (i.e., t_0) to a given time, as shown in Equation (7).

$$N_c(D_p, t_i) = N_m(D_p, t_i) + \sum_{j=0}^{j=i-1} N_m(D_p, t_j) \times [1 - \exp\{-\beta(D_p)(t_{j+1} - t_j)\}], \ i \geq 1 \qquad (7)$$

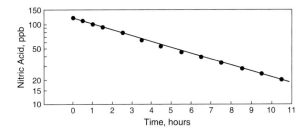

Fig. 18 Example of the loss rate measurement of gas species: nitric acid in the 4-m³ Teflon chamber under dark condition (Reprinted with permission from Grosjean 1985. Copyright 1985 American Chemical Society.)

The above methodology for determining the aerosol wall loss rates was also extended to loss rates of gaseous species. When Equation (6) was used instead of Equation (5), the wall loss rate of gas species can be derived from a regression line in a plot of the measured gas concentration as a function of time, as illustrated in Fig. 18 (Grosjean 1985). Table 1 summarizes the reported wall loss rates of aerosols and gases under dark conditions.

Wall Contamination

Extensive studies of the wall contamination in smog chambers have been carried out since the 1970s. Newly constructed smog chambers should first be 'baked' in order to minimize the level of organic impurities desorbed from the Teflon film (Kelly 1982). The chambers used should be also cleaned by purging with purified air under irradiation after each experiment because reactants and products from one run can contaminate another (Grosjean 1985). The pollutants out-gassed from contaminated chamber walls include chemical compounds that exist in the Teflon film itself, precursors and products absorbed on the chamber walls during photochemical reactions, and outdoor pollutants that can penetrate the Teflon film and enter the bags (Bufalini et al. 1977; Dodge 2000; Kelly 1982). Therefore, the 'next generation smog chamber' of UCR was installed inside a clean enclosure supplied with purified air in order to minimize contamination from outdoor pollutants (Carter et al. 2005; Carter and Fitz 2003). The 'next generation smog chamber' of UCR is now called the UCR-EPA chamber.

The following gives a brief review of studies on the out-gassing of pollutants from the chamber walls. Bufalini et al. (1977) reported that HCHO, HONO$_2$, HONO, NO$_x$ containing species, organic fragments were out-gassed from the Teflon chamber walls by experiments using purified air and through simulations. Lonneman et al. (1981) showed that heat treatment or

Table 1 Wall loss rates of gases and aerosols in Teflon chambers

Compounds	Chamber size (m³)	Surface to volume ratio (m⁻¹)	Wall loss rate (×10⁻⁶ sec⁻¹)	References
O$_3$	4	3.8	5.6	Grosjean (1985)
	80	1.4	2.2	Grosjean (1985)
	2.5	4.4	20~40	Bae et al. (2003)
NO	4~80	1.4~3.8	0.7	Grosjean (1985)
NO$_2$	4~80	1.4~3.8	−2.7	Grosjean (1985)
Toluene	4~80	1.4~3.8	< 4.2	Grosjean (1985)
Nitric acid	4	3.8	25	Grosjean (1985)
	60	2.1	58~140	McMurry and Rader (1985)
Aerosols	28	1.0	25~50	Cocker III et al. (2001)
(20~300 nm)	5.8	3.3	50~130	Lee et al. (2004a)
	2	5	30~650	Hurley et al. (2001)

some subsequent photochemical reaction experiments of an ozone formation can reduce the amount of out-gassed pollutants. Kelly et al. (1985) monitored the gas-phase organic pollutants of 0.06 ppmC two days after injecting purified air into a newly constructed Teflon chamber. Dodge (2000) reviewed the wall effects on the chemical mechanisms for how gas-phase species such as O_3, HNO_3, N_2O_5, H_2O_2, and HCHO adhered or adsorbed on the chamber walls might out-gas to the inside of the chamber during the experiments. Researchers have simulated the out-gassing of OH radicals as an emission from the wall and subsequent photolysis of HONO and HCHO (Dodge 2000).

Izumi and Fukuyama (1990) examined aerosol formation of purified air to which NO_x (about 10 ppb) and water vapor (relative humidity of 50%) was added in order to examine the background aerosol formation probably due to wall contamination or impurities of purified air. Because the new aerosol formation was approximately 3~4 $\mu m^3/cm^3$ after 200-min irradiation for a background experiment, they could not reduce the hydrocarbon level to as low as 0.01 ppm and had to increase the concentration to 1 ppm. The background photochemical reactivity could be enhanced without injecting NO_x to purified air. Therefore, the base photochemical experiment using purified air only might be a more sensitive case for examining the background formation potential of ozone and aerosols due to wall contamination.

A flushing procedure was carried out before each experiment to minimize the level of wall contamination and its effect on the experimental results (Carter et al. 1995; Cocker III et al. 2001; Hurley et al. 2001; Odum et al. 1996). For example, Hurley et al. (2001) flushed a 2-m^3 smog chamber by passing purified air at a flow rate of 15 L/min for 40 h under irradiation (baking flush) and then for 6 h under dark conditions (non-baking flush). The ozone, NO_x, hydrocarbons, and particle concentrations should be as low as possible before the start of each experiment. The total volume of purified air that used for flushing the bag was approximately 21 times the chamber volume.

2.3 Measuring Instruments

The concentrations of both reactants and products in the gas and aerosol phases are measured during the smog chamber experiments. The total sampling flow rate of the measuring instruments should be less than a few liters per minute in order to prevent rapid collapse of the reaction bag unless additional dilution air is supplied. If the volume of the reaction bag is a few m^3,

Table 2 List of the analytical instruments commonly used in smog chamber experiments (Carter and Fitz 2003)

Species	Instrument	Approx. sensitivity (ppb) or comment
O_3	UV photometric analyzer	1~2
NO–NO_2–NO_x	Chemiluminescence analyzer	~1
		Note inference of NO_2 by NO_z
SO_2	Pulsed fluorescent analyzer	~1
CO	Gas filter correlation analyzer	40~50
Volatile organic compounds (VOCs)	GC-FID	1~10 ppbC
		Every 5~30- min measurement
		C_6F_6 as an internal standard
Aldehydes	HPLC	0.04 μg/cartridge
Identification of organics	GC-MS	–
Aerosol size distribution	Scanning electrical mobility spectrometer, scanning mobility particle sizer	Every one or several minute measurement 20~300 nm in mobility diameter
Aerosol response to changes in RH or temperature	Tandem differential mobility analyzer	–
Aerosol total number concentration	Condensation particle counter	Single particle larger than 3 nm in diameter Every several second measurement
Aerosol density	Gravimetric method	Particle collection on Teflon filters
Aerosol morphology	Electrostatic precipitation	SEM or TEM analysis

discrete measurements are recommended for several-hour experiments. Table 2 lists the analytical instruments used in smog chamber experiments. In addition, instruments such as portable temperature and humidity sensors equipped with a data logger and gas calibrators are required.

2.3.1 Monitoring

Gas-Phase Pollutants

Continuous gas analyzers and a GC are calibrated for each experiment or every week (Cocker III et al. 2001; Wang et al. 1992). The NO_2 concentration determined using conventional chemiluminescence $NO–NO_2–NO_x$ analyzer includes some NO_z (i.e. NO_y except NO_x) species such as HNO_3 and peroxyacetyl nitrate (PAN) due to interference from the analyzer (Winer et al. 1974). For this reason, a luminol gas chromatograph and a tunable diode laser absorption spectrometer (TDLAS) were developed to monitor the NO_2 concentration accurately (Carter and Fitz 2003). GC-FID can be used to measure the light or heavy hydrocarbons by changing columns for loop or trap sampling. A hexafluorobenzene (C_6F_6) tracer can provide an internal standard for the GC measurements and serve as an indicator of air leaks in the chamber (Cocker III et al. 2001). A GC-ECD is used to determine the concentration of PAN and other species for which ECD is more sensitive than FID.

Particulate-Phase Pollutants

The particle size distribution, number concentration, and hygroscopic nature of the chamber aerosol were measured using a scanning electrical mobility spectrometer (SEMS) or a scanning mobility particle sizer (SMPS), a condensation particle counter (CPC), and a tandem differential mobility analyzer (TDMA), respectively. SEMS and SMPS consist of a differential mobility analyzer (DMA) and a CPC. Normally, a DMA operates with sheath flow rate 10 times higher than the sample inlet flow rate and the classified aerosol outlet flow rate. The measured particle size may be slightly different if the relative humidity of the sheath air is different from that of the sample air. When humid air is used as sheath air, the aerosol volume is approximately 10~15% higher than when dry air is used as sheath air (Hurley et al. 2001; Izumi and Fukuyama 1990). This difference in aerosol volume with the relative humidity depends on the hygroscopic nature of the secondary organic aerosols. TDMA is useful for examining the effects of humidity (hygroscopicity), temperature (volatility), or other environmental changes on the aerosol size distribution. TDMA consists of a classifying DMA that selects particles of a single size range, a small environment chamber exposing the particles to humidity, heat or other conditions of interest, and a scanning DMA that measures the size distribution of the particles passing through the chamber.

The measured aerosol data should be corrected for any diffusion loss at the inside of the sampling tube and instrument, particle charging efficiency, and deposition loss on chamber walls. Unlike gas-phase measurement, switching aerosol sampling inlets to each bag of a dual mode chamber with a solenoid valve is not recommended because of the severe particle loss and modification while passing through the valve channel.

The particle size distribution data measured by SMPS can be converted into a surface area, volume, and mass concentration, assuming that the aerosols are spherical with a uniform density. The mean density of secondary organic aerosols is normally assumed to be $1.0\,g/cm^3$ (Bowman et al. 1997; Odum et al. 1996; Wang et al. 1992). The volume concentrations of aerosols are preferred in a presentation when aerosol mass concentration data are not required, unlike the calculation of the aerosol mass yield of each volatile organic compound (VOC).

2.3.2 Sampling and Analysis of Aerosols

Gas and particulate-phase chemical compounds produced from photochemical reactions in a smog chamber are identified mainly by GC-MS along with infrared spectroscopy (Forstner et al. 1997; Jang and Kamens 2001). Advanced analysis techniques, such as laser desorption ionization-mass spectrometry (LDI-MS) and aerosol mass spectrometry (AMS), can allow detection of the polymerization of secondary organic aerosols occurring within aerosol particles. The molecular mass of the polymers can increase up to 1,000 daltons with an aging time > 20 h, which can result in a lower volatility of this secondary organic aerosol and

Fig. 19 SEM image of
secondary organic aerosols
after 5-hr irradiation of a
toluene–NO_x–air mixture
without seed particles
(Lee et al. 2005b)

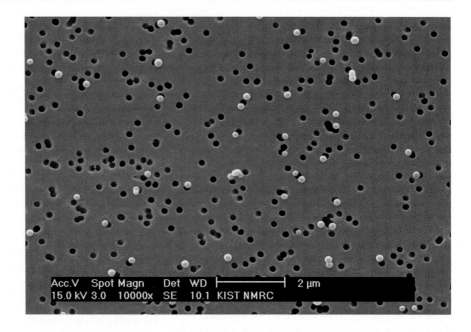

a higher aerosol yield than that predicted by the model using the vapor pressures of individual organic species (Kalberer et al. 2004). This phenomenon is so-called oligomerization.

Figure 19 shows SEM images of secondary organic aerosols produced during the photochemical reactions of a toluene–NO_x–air mixture without initial seed particles (Lee et al. 2005b). The aerosol sample was collected on a polycarbonate membrane filter with a pore size of 200 nm. Before SEM analysis, the sample was coated with gold using a sputtering coater to prevent surface charging. Most of the aerosols are single spheres, which supports the assumption that secondary organic particles are spherical when estimating the particle mass concentration from the SMPS data.

In the case of experiments with soot particles, as initial seeds, the condensation of secondary organic aerosols led to compaction of the original fractal aggregates, as shown in Fig. 20.

A commercial TEM grid can be used for TEM analysis to collect aerosols using an electrostatic precipitator type sampler with an electric field between several kV nodes or a cascade impactor.

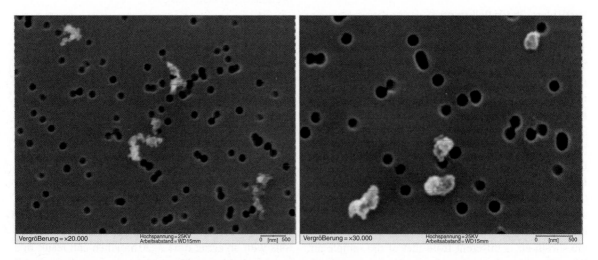

Fig. 20 SEM images of soot before (*left*) and after coating (*right*) with α-pinene ozonolysis products (Saathoff et al. 2003)

However, possible distortion of the aerosol morphology during sampling and analysis should be considered. Aggregated particles can form during sampling when the sampling time is too long for high particle concentrations. A TEM image of the secondary organic aerosols showed signs of melting or evaporation from the outer layer of organic or inorganic aerosols, suggesting their low volatility (Wentzel et al. 2003).

2.4 Experimental Techniques

As for an ideal smog chamber, all the parameters should be controlled according to the researchers' requirements. However, for the environment in smog chamber experiments, there are many limitations in controlling all the parameters independently and identically to real atmospheric conditions. Therefore, experimental techniques are required.

2.4.1 Light Intensity Control

There are some dual-mode smog chambers that can examine the effect of a single parameter, such as a single precursor concentration, through comparison experiments. A separation tool was used to divide one large bag into two independent bags. However, the light intensity of each reaction bag cannot be controlled independently if there is a single enclosure and light source. On the other hand, if each reaction bag is installed in a separated enclosure that has a single light source, the light intensity of the two chambers could be different from each other. These smog chambers are called twin smog chambers. In the case of blacklights, one power switch can be used to turn on one or two blacklights. In order to reduce the light intensity to the half of the normal experimental condition, the number of power switches turned on can be set to 50%. It should be considered that the temperature can change with light intensity.

2.4.2 Temperature Control

The temperature inside the smog chamber increases immediately after the light source is switched on, and

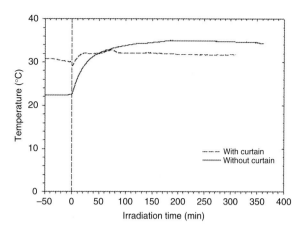

Fig. 21 Temperature change in the smog chambers with or without a UV blocking curtain (Ju 2006)

thereafter reaches a steady state value. A curtain blocking UV light can be used to minimize the change in temperature at the initial stages of the experiments as a result of turning on the light source. After the temperature reaches a steady state value, the curtain can be removed when irradiation begins. The temperature change using a UV blocking curtain is less than 3 °C during irradiation, and within 1 °C after 10-min irradiation, as shown in Fig. 21 (Ju 2006).

2.4.3 Comparison of Twin Chambers

The photochemical reactions of real ambient air might be strongly dependent on the initial urban air quality. Because the ambient air quality changes with time and is not reproducible, two experiments should be carried out simultaneously to determine the difference in the experimental results due to differences in a single parameter for examining the effect of a parameter on smog chamber experiments using real ambient air. A dual-mode chamber can be used for this purpose. For studies on the effect of light intensity, twin smog chambers are needed because the light intensity of each reaction bag should be controlled independently. A comparison of the data from the twin smog chambers with the same initial urban air quality can exclude the effect of the initial urban air quality.

Initial Particle Control

There are two pathways for the formation of secondary organic aerosols in a smog chamber. One is the

Fig. 22 A schematic diagram
of twin smog chambers
equipped with filters as
control devices

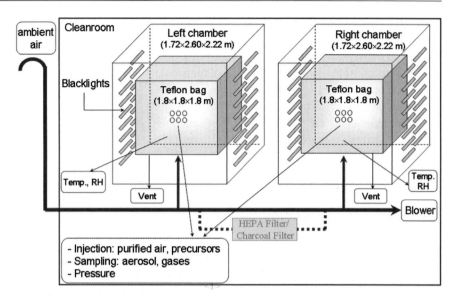

homogeneous nucleation of new particles from a
supersaturation of semi-volatile organics derived from
the gas-phase oxidation products, and the other is heterogeneous condensation of condensable vapors onto
the surface of pre-existing aerosols (Bowman et al.
1997). However, the organic nucleation process in the
atmosphere is a rather controversial issue (Kanakidou
et al. 2005). Seed particles such as ammonium sulfate
have been injected into the chamber prior to the experiments to suppress the nucleation burst of new particles
because aerosol instruments in older models could not
detect the nucleated particles or follow the extremely
rapid process (Cocker et al. 1999). Studies on the effect
of ambient aerosols itself as seed particles, such as
ammonium sulfate and diesel exhaust particles, to the
photochemical reactions or secondary aerosol formation have been carried out (Edney et al. 2005; Geiger
et al. 2002; Oh and Andino 2000, 2001, 2002; Wang
et al. 1992). A stainless steel constant-rate atomizer
is used to generate seed aerosols from inorganic solutions in ultra pure water. The seed aerosols generated
should be passed through a heated zone and diffusion
dryer to remove the water content and be mixed with
air that has been passed through a neutralizer to eliminate the charge produced during atomization (Cocker
III et al. 2001).

In order to control the ambient aerosol number concentration for captive-air experiments using ambient
air, a high efficiency particulate air (HEPA) filter can
be installed at the middle of the sampling ducts for
twin smog chambers, as shown in Fig. 22. The injection time during which ambient air pass through the
HEPA filter could be adjusted according to the target
number concentration of the ambient aerosols.

Initial Gas Control

In order to examine the effect of initial concentration of specific precursor gas, an additional injection can be made at the injection ports after fully
introducing the ambient air sample to the twin smog
chambers. As for the injection of liquid VOCs, a
hot heating zone is useful for evaporating the VOCs
in carrier air (Angove et al. 2000). Carter and Fitz
(2003) introduced a new evaporation device without
heating the VOCs liquid to the next generation smog
chamber of UCR. The device allows the liquid to
be deposited on a fabric held in an air circulation
system. Hence, the liquid is evaporated by forcing
large quantities of air past it, which appears to be
good means of preventing the thermal transformations
of VOCs.

Appropriate filters can be installed instead of the
HEPA filter, as shown in Fig. 22, to remove target
gas-phase compounds of the right chamber. For example, a charcoal filter is used to control the VOCs
concentration.

2.5 Typical Smog Chamber Experiments

2.5.1 Indoor Chamber Studies

VOCs–NO$_x$–Air Mixture Experiments

Figure 23 shows typical results of a concentration change of gases and aerosols as a function of the irradiation time during photochemical reactions of VOCs–NO$_x$–air mixture in the 6-m^3 KIST chamber (Lee et al. 2005a). Synthetic mixtures with NO$_x$ and VOCs concentrations higher than atmospheric concentrations have been used in smog chamber experiments to determine dependence of the formation of ozone and secondary aerosols on a single parameter, such as the initial precursor concentration, and to identify the chemical mechanisms (Dodge 2000; Song et al. 2005). According to Equation (2), NO begins to convert to NO$_2$ immediately after the artificial light source is turned on. NO$_2$ is photolyzed into NO under irradiation, and NO reacts with O$_3$ to return to NO$_2$. Therefore, there is no net change in ozone for NO$_x$ alone without VOCs. The chemical process of ozone formation occurs when NO is converted to NO$_2$ by a reaction, not with O$_3$, but with other oxidants, such as organic peroxy radicals (RO$_2$) (Equation (9)), which are formed through reaction sequences initiated by reactions between the VOCs and hydroxyl radicals

(Equation (8)):

$$VOCs + OH(+O_2) \rightarrow RO_2 + H_2O \qquad (8)$$

$$RO_2 + NO(+O_2) \rightarrow R'CHO + HO_2 + NO_2 \quad (9)$$

where R$'$CHO represents the intermediate organic species.

For this experiment of toluene–NO$_x$–air mixture, when the concentration of NO$_2$ exceeds that of NO after approximately 140-min irradiation, toluene and ozone concentrations begin to be consumed rapidly and increased significantly, respectively. The NO$_2$ concentration reaches its maximum value and NO is almost exhausted at that time. There was no significant decrease in NO$_2$ concentration because the NO$_2$ concentration measured by the chemiluminescence NO–NO$_2$–NO$_x$ analyzer includes the interference effect from NO$_z$ species, as mentioned in Section 2.3.1. This may be due to the significant NO$_2$ consumption, which reduces the slope of ozone concentration to reach a maximum value after approximately 400-min irradiation. Due to the indoor chamber size (6 m^3), the experimental duration shown in Fig. 23 could be up to 8 h under the same irradiation intensity.

As shown in Fig. 24, the final wall-loss-corrected aerosol mass concentrations of the four m-xylene–NO$_x$–air experiments with runs of 104A, 129A, 142A, and 142B ranged from 21.0 to 21.7 μg/m^3 with a standard deviation of 0.249 μg/m^3 when the initial concentrations of m-xylene and NO$_x$, and their ratio

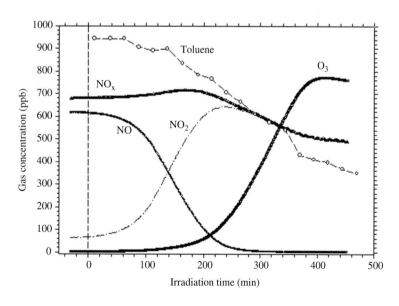

Fig. 23 Observed concentrations of gases as a function of the irradiation time during a photooxidation experiment on a toluene–NO$_x$–air mixture. A09R10 data of Fig. 5 in Lee et al. (2005a)

Fig. 24 SOA mass concentrations as a function of the irradiation time for four photooxidation experiments on a m-xylene–NO_x–air mixture with the same initial conditions to test the side-to-side reproducibility between experiments 142A and 142B and the run-to-run reproducibility among experiments 104A, 129A, and 142A. The detailed conditions of the four experiments can be found in Table 1 in the reference (Reprinted with permission from Figure 2 of Song et al. 2005. Copyright 2005 American Chemical Society)

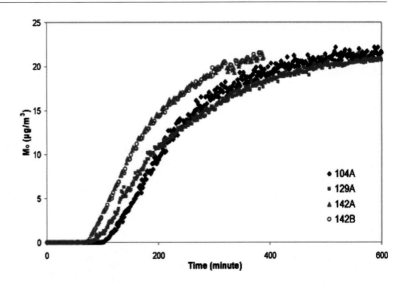

ranged from 341∼366 µg/m³, 46.6∼66.5 ppb, and 9.6∼14.7 ppbC/ppb, respectively (Song et al. 2005). The reproducibility in the SOA mass concentration is believed to be the best for smog chamber measurements thus far. However, it is possible that the organic mass formed at the same irradiation time can change significantly with each experiment. In other words, each experiment might require a different irradiation time to reach the equilibrium SOA mass concentration.

Propene (C_3H_6) is widely added as a photochemical initiator in smog chamber experiments to increase the OH levels (Cocker III et al. 2001; Jang and Kamens 2001; Odum et al. 1996, 1997; Wang et al. 1992). Although it is commonly accepted that the aerosol formation potential of alkenes with fewer than six carbon atoms is zero, recent research has suggested that

propene has slight aerosol formation potential in the propene–NO_x–air system and reduces the OH level in the m-xylene–NO_x–air system and consequently reduces its SOA yield (Song et al. 2007). Therefore, caution should be taken when comparing the aerosol yield of experiments with added propene to that of other studies.

Figure 25 shows the typical time variation of total number (N_p), surface area (S_p), volume (V_p) concentrations of the aerosols, and geometric mean diameter (D_p) based on (N_p). The data was obtained from the photooxidation experiment on a toluene–NO_x–air mixture (Izumi and Fukuyama 1990). After 25-min irradiation, new ultrafine particles began to nucleate up to total number concentration of 60, 000 particles/cm³, and then decreased slowly due mainly to particle wall loss. Although the total number concentration

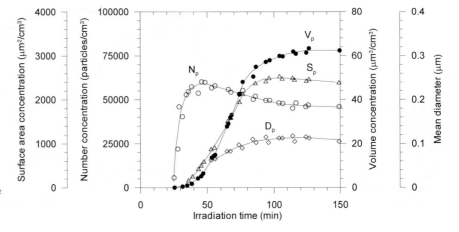

Fig. 25 Change in particle concentrations during photooxidation experiments on a toluene–NO_x–air mixture (Adapted from Izumi and Fukuyama 1990)

of particles decreased, the total volume concentration increased up to $60\,\mu m^3/cm^3$ (i.e. $60\,\mu g/m^3$ for unit-density aerosols) due to the condensation of condensable vapors onto the surface of existing particles. At that time, the grown aerosol size was approximately $0.1\,\mu m$. The change in aerosol concentration according to surface area followed a similar trend to that of volume.

Efforts to represent the SOA formation potential of each VOC precursor in ambient models have been made using a variety of approaches. Grosjean and Seinfeld (1989) used the 'fractional aerosol coefficient (FAC)' in terms of the molar, volume, mass, or carbon concentration, which was defined as Equation (10). The FAC was also called the gross gas-to-particle conversion (Wang et al. 1992).

$$FAC = \frac{\text{aeosol formed from VOCs } (\mu m^3/cm^3, \, \mu g/m^3, \, \text{or } \mu gC/m^3)}{\text{initial concentration of VOCs (mol, ppm, } \mu g/m^3, \, \text{or } \mu gC/m^3)} \qquad (10)$$

However, this concept of aerosol yield not only depends on the experimental duration but is also difficult to use as an input in air quality models where the 'initial concentration' cannot be identified due to continuous emission. For this reason, a new concept of the aerosol yield using the amount of reacted VOCs instead of the initial concentration of VOCs in Equation (10) was introduced in 1990 and 1991 (Izumi and Fukuyama 1990; Pandis et al. 1991). It is called the aerosol carbon yield if the unit of $\mu gC/m^3$ is used in Equation (11). Currently, most researchers express the aerosol formation potential as a dimensionless aerosol yield in terms of the mass concentrations suggested by Wang et al. (1992).

$$\text{Aerosol yield} = \frac{\text{aeosol formed from VOCs } (\mu m^3/cm^3, \, \mu g/m^3, \, \text{or } \mu gC/m^3)}{\text{amount of reacted VOCs (ppm, } \mu g/m^3, \, \text{or } \mu gC/m^3)} \qquad (11)$$

The new concept of aerosol yield includes two values whose variation can be plotted as a function of the irradiation time, as shown for the toluene in Fig. 26 (a) from Hurley et al. (2001). In Fig. 26, the number of data points for Odum et al. (1997) and Gery et al. (1985) means the number of experiments. The concentration of toluene and NO_x ranged from $0.7\sim24.1$ ppm and $0.045\sim13$ ppm, respectively, and the toluene/NO_x ratios were $1.4\sim124.4$ for all experiments shown in Fig. 26. The solid symbols denote the experiments called Group A below, where excess NO was added to suppress the formation of ozone during irradiation. There was also a considerable difference in the light source, light intensity, temperature, relative humidity, experimental scheme, etc. Hurley et al. (2001) reported that the toluene threshold (i.e. latent consumption necessary for starting of aerosol formation, which implies that the aerosol was produced via secondary reactions of the primary gaseous products) is proportional to the initial toluene concentration, and there was a regime of a constant aerosol yield of about 10% when this

toluene threshold is disregarded, as reported by Izumi and Fukuyama (1990). Therefore, they suggested that the aerosol yield be calculated using the toluene that reacted after the threshold instead of the total toluene consumed in Equation (11). They explained that the higher aerosol yield in the presence of ozone (Group B) than in the absence of ozone (Group A) was due to the lower vapor pressure of the secondary products formed by the reaction with ozone. This is because the levels of the aerosol mass formed were classified into two groups according to the existence of ozone during photooxidation when the toluene threshold was excluded, and because the simple model based on the saturation vapor pressures of the secondary products could fit the experimental data.

Originally, the aerosol yield was calculated using the final aerosol mass concentration and the final reacted VOCs mass concentration when aerosol formation was terminated. The aerosol yields of four experiments reported by Odum et al. (1997) increased with increasing aerosol mass concentrations. Because

Fig. 26 Plot of the aerosol mass formed as a function of the reacted organic gas concentration considering the threshold for aerosol formation (Reprinted with permission from Hurley et al. 2001. Copyright 2001 American Chemical Society). (a) With the total toluene consumed (b) With the toluene consumed excluding a threshold

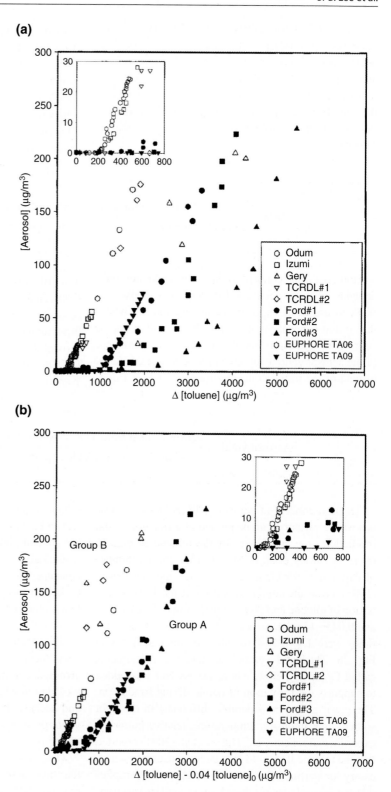

the aerosol yields for an individual VOC showed wide variations both between and within laboratories, Pankow (1994a, b) suggested a new gas/particle absorptive partitioning model, where even the products whose gas-phase concentrations are below their saturation concentrations will partition a portion of their mass into a condensed organic phase. Odum et al. (1996) reported that the aerosol yield for an individual VOC is not uniquely valued but is instead a function of the available absorbing organic aerosol concentration using the following equations.

$$Y = \frac{\Delta M_o}{\Delta VOC} = \frac{\sum F_{om,i}}{\Delta VOC} = \sum \left(\frac{\alpha_i F_{om,i}}{\alpha_i \Delta VOC} \right) = \sum \left(\frac{\alpha_i F_{om,i}}{C_i} \right) = \sum \left(\frac{\alpha_i F_{om,i}}{A_i + F_{om,i}} \right)$$

$$= \sum \left(\frac{\alpha_i F_{om,i}/A_i}{1 + F_{om,i}/A_i} \right) = \sum \left(\frac{\alpha_i K_{om,i} M_o}{1 + K_{om,i} M_o} \right) = M_o \sum \left(\frac{\alpha_i K_{om,i}}{1 + K_{om,i} M_o} \right) \qquad (12)$$

where, M_o, α_i, C_i, A_i, $F_{om,i}$, $K_{om,i}$ are the total aerosol organic mass concentration, mass-based stoichiometric fraction of product i formed from the parent VOC, total concentration of product i, gas phase concentration of product i, aerosol phase concentration of product i, partitioning coefficient for product i defined in terms of the organic mass concentration as $K_{om,i} = (F_{om,i}/M_o)/A_i$ (Odum et al. 1996), respectively. This constant equilibrium partitioning coefficient suggests that a fraction of a compound's total mass residing in the particulate phase will increase with increasing organic mass concentration, as shown in Fig. 27.

Although there are many products in the particle phase from a given VOC, a two products model is widely used to express characteristics of aerosol yield for a given VOC because two products are the minimum number to fit the behavior of the system. The values of α_1, α_2, $K_{om,1}$, $K_{om,2}$ for the curve fit for the high-yield aromatics data (toluene, ethyltoluenes, ethylbenzene, and n-propylbenzene) reported by Odum et al. (1997) were 0.071, 0.138, 0.053, 0.0019, respectively. These fitted values do not apply to low-yield aromatics, which appears to be a misprint in the original reference. A higher aerosol yield was

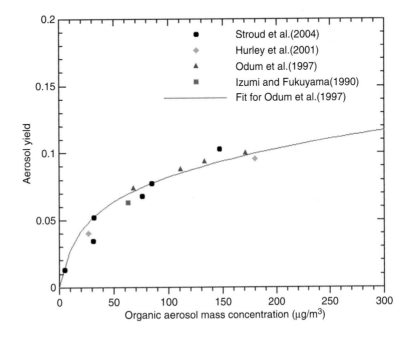

Fig. 27 Comparison of the aerosol yields for toluene. The *solid line* represents the best-fit two-product model for the high-yield aromatics data reported by Odum et al. (1997), whose fitting values of α_1, α_2, $K_{om,1}$, $K_{om,2}$ are 0.071, 0.138, 0.053, 0.0019, respectively. A particle density of $1.0\,g/cm^3$ was used to convert the volume to mass. Data reported by Stroud et al. (2004) was used as the measured values before a wall loss correction

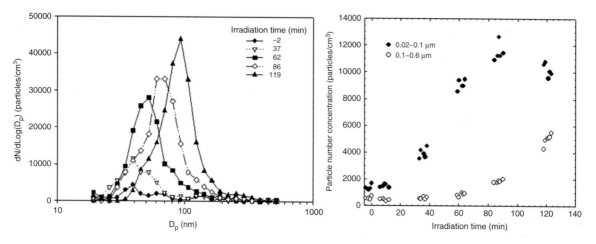

Fig. 28 Change in the size distribution (*left*) and number concentration (*right*) of particles at different irradiation times for ambient air experiments (Bae et al. 2008)

obtained at lower temperatures due to the temperature dependence of the saturation vapor pressure.

Recently, the aerosol yield for *m*-xylene was reported to be higher with lower NO_x levels than with higher NO_x levels when the amount of *m*-xylene was kept constant. However, this was not proportional to the ozone concentration, indicating that the aerosol yield is only slightly dependent on or unaffected by O_3 and NO_3 existing in the smog chamber, which is in contrast to that suggested by Hurley et al. (2001) (Song et al. 2007). A greater understanding of the characteristics of the aerosol yield is expected using more advanced measurement techniques in smog chambers with new experimental approaches. The assumed particle density should be considered when comparing the aerosol yield.

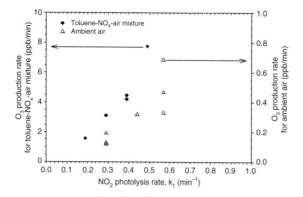

Fig. 29 Comparison of the ozone production rates for ambient air with those of a toluene–NO_x–air mixture (Bae et al. 2008)

Ambient Air Experiments

Most ambient air experiments have been carried out using outdoor chambers, except at KIST (Bae et al. 2008). Bae et al. (2008) examined the effect of light intensity on ozone formation and the aerosol number concentration during the photochemical reactions of ambient air using a 2.5–m^3 indoor smog chamber. In this work, the photolysis rate of NO_2, k_1, was used as an index of light intensity. Three light intensities were controlled by changing the number of blacklights switched on out of a total of 64 blacklights. The ozone concentration increased rapidly within 10-min after irradiation irrespective of the light intensity, and it increased linearly thereafter during 2 h irradiation. The rate of ozone production and the change in aerosol number concentration appear to be dependent on both the light intensity and quality of ambient air introduced into the reaction bag. The aerosol formation and growth processes were clearly observed when the number concentration of particles, approximately 0.1 μm in diameter, was lower than that of the particles, approximately 0.04 μm in diameter, as shown in Fig. 28.

As shown in Fig. 29, the rates of ozone production in ambient air at fixed light intensity vary with experiment. This is due to the difference in ambient air quality. Some results of twin smog chambers using ambient air were reported by Moon et al. (2004b, 2006), showing that the ozone production rates could be expressed as a power function of the light intensity and toluene/NO_x ratio through twin chamber comparison analysis that can avoid differences in ambient air quality.

2.5.2 Outdoor Chamber Studies

The outdoor chamber has the advantage of using natural sunlight, which is unlike indoor chambers. However, a direct comparison of the data for outdoor smog chamber experiments is difficult because the photolytic parameter varies with time due to different zenith angle of the sun, and day by day as a result of different weather conditions (Geiger et al. 2002). Solar radiation is monitored continuously in outdoor smog chamber experiments. The data for outdoor chambers is used mainly to develop chemical mechanisms for the formation of ozone and secondary organic aerosols, in particular, for the dual outdoor chamber of UNC (Hu and Kamens 2007; Hu et al. 2007).

VOCs–NO_x–Air Mixture Experiments

A kinetic mechanism to predict secondary organic aerosol formation from the photooxidation of toluene was developed, and compared with the experimental data from the UNC outdoor chamber, the European photoreactor (EUHPORE), and smog chambers at the California Institute of Technology (Hu and Kamens 2007; Hu et al. 2007). The experimental duration was approximately 8 h from morning to late evening, as shown in Fig. 30. The maximum ozone concentration normally was observed from 11 a.m. to 3 p.m. depending on the experimental conditions. The model developed by the UNC can simulate the decay of toluene and the NO–NO_2 conversion very well. In addition, the model fits quite well the ozone concentration profiles for high toluene concentration runs, as shown in Fig. 30. The model tends to over-predict the afternoon ozone concentrations by approximately 30% for experiment 072705S of low-concentration systems probably because of a lack of reaction pathways to remove the excess NO_2 (Hu et al. 2007). It also provides a reasonable prediction of secondary organic production under different conditions ranging from 15 to 300 $\mu g/m^3$, and the detailed speciation of simulated aerosols including oligomers contribution (Hu et al. 2007). The particle-phase reaction of oligomerization was reported to be promoted by acidic seed aerosols (Jang et al. 2003).

Figure 31 shows a comparison of the calculated SMPS data with measured gravimetric particle mass concentrations for a toluene (1 ppm)–NO_x (0.11 ppm)–

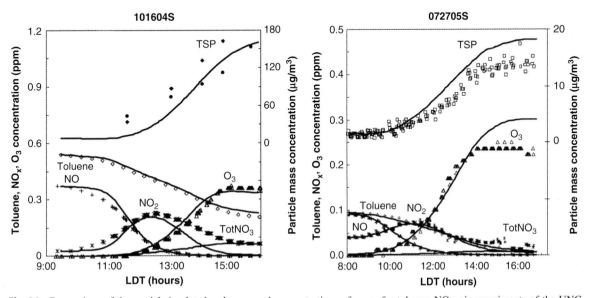

Fig. 30 Comparison of the model simulated and measured concentrations of gases for toluene–NO_x–air experiments of the UNC chamber. Figure 4 in Hu et al. (2007). The solid lines represent the model simulations, and the symbols denote the experimental data. The titles of the figures, such as 101604S and 072705S, represent experimental dates (mmddyy) and the south chamber

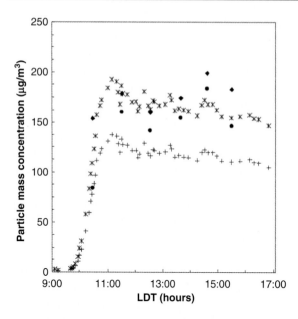

Fig. 31 Comparison of the calculated SMPS masses and gravimetric masses for a toluene–NO$_x$–air experiment. The solid squares and solid dots represent the uncorrected front filter mass concentration and filter mass concentration corrected by backup filter subtraction, respectively. The asterisks and plus represent the calculated SMPS particle mass concentrations assuming an aerosol density of 1.4 g/cm^3, and 1.0 g/cm^3, respectively (Hu et al. 2007)

air experiment (Hu et al. 2007). Under natural sunlight, new particles were generated at around 10 a.m. This figure shows very good agreement between the SMPS mass and measured filter mass if a density of 1.4 g/m^3 is applied to correct the SMPS data instead of 1.0 g/m^3.

Ambient Air Experiments

Experiments on the irradiation of real ambient air are known as captive-air irradiation experiments. Captive-air irradiation experiments are normally carried out in outdoor chambers in order to examine the real effect and/or reduction strategies of the added target gas, such as SO$_2$, liquefied petroleum gas, hydrocarbons, and NO$_x$ to ozone and/or particle formation (Heisler and Friedlander 1977; Jaimes et al. 2003, 2005; Kelly 1987; Kelly and Gunst 1990; Pitts et al. 1977; Roberts and Friedlander 1976). Kelly and Gunst (1990) used eight reaction bags, whose initial concentrations of hydrocarbons and NO$_x$ were changed by ±25 to ±50%, respectively, by adding various combinations of a mixture of hydrocarbons, clean air, or NO$_x$ to ambient air to obtain an empirical model for the maximum ozone as a function of the concentration. For example, the temporal ozone profiles for the mixtures obtained at a suburban site by Kelly and Gunst (1990) showed that all perturbations except for the increase in NO$_x$ by 22% decreased the maximum ozone concentrations from that in the control bag on that day, as shown in Fig. 32.

Figure 33 shows the change in the number, surface area, and volume concentrations of aerosols formed in an outdoor chamber filled with filtered ambient air (McMurry and Friedlander 1978). This figure is similar to that of the toluene–NO$_x$–air mixture in Fig. 25 from Izumi and Fukuyama (1990) except for the concentrations. The total number concentration of particles > 0.01 μm for filtered ambient air was approximately double that of the toluene–NO$_x$–air mixture. However, the surface area and volume concentrations were < 7%

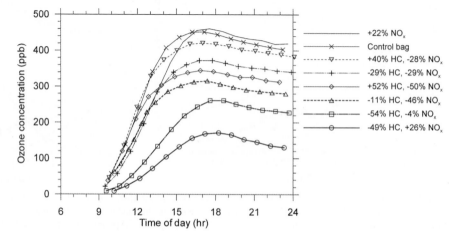

Fig. 32 Temporal O$_3$ profiles of the photochemical reactions under natural sunlight in eight chambers filled with ambient air from a suburban area in California, USA in 1987 (Adapted from Kelly and Gunst 1990)

Fig. 33 Change in the number (N_p), surface area (S_p), and volume (V_p) concentrations of aerosols formed in an outdoor chamber filled with filtered ambient air (Adapted from McMurry and Friedlander 1978)

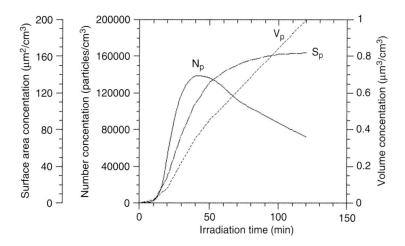

of those for the toluene–NO_x–air mixture. This difference appears to be due mainly to differences in the chemical species of the aerosols formed, i.e. between sulfate and organics.

3 Applications

3.1 Secondary Pollution of Diesel Exhaust

Automobiles are major sources of hydrocarbon emissions as the precursors of photochemical smog. Hence, smog chambers have been used for approximately fifty years to examine the effect of photochemical products of automobile exhaust on ozone and particle formation, as well as the effect of fuels and additives on human health (Jeffries et al. 1998; Kleindienst et al. 1992; Stephens and Schuck 1958; Wiesen 1999). Due to the increase in the number of diesel vehicles, particularly in Europe, the considerable emission of particulate matter and NO_x from diesel vehicles has become an important problem. The secondary pollution of diesel exhaust instead of other fueled vehicles has attracted considerable research interest (Geiger et al. 2002; Lee et al. 2004b, 2006, 2007; McDonald et al. 2004; Wiesen 2000). Figure 34 shows a schematic diagram of smog chambers for an experimental study on secondary pollution from diesel exhaust (Lee et al. 2006; Zielinska et al. 2007). When a synthetic hydrocarbon mixture of approximately 1 ppm was injected into diluted diesel exhaust to enhance its photochemical reactivity, the level of ozone formation increased

(a)

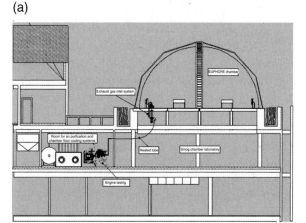

(b)

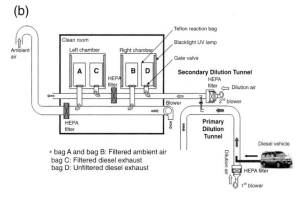

Fig. 34 Schematic diagram of the smog chambers used for experiments on diesel exhaust (a) EUPHORE outdoor chamber (Zielinska et al. 2007) (b) KIST twin indoor chambers (Lee et al. 2006)

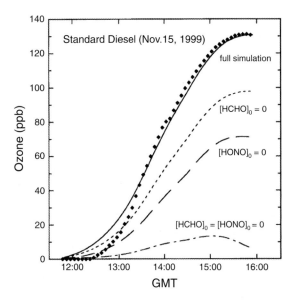

Fig. 35 Experimental (*symbol*) and simulated (*lines*) ozone profiles for the EUPHORE chamber experiment for standard diesel. The HONO and HCHO concentrations determined by the best fit to the measured data were 1.5 and 9 ppb, respectively (Geiger et al. 2002)

significantly due to the possible strong radical sources present in diesel exhaust, such as nitrous acid and formaldehyde, as shown in Fig. 35 (Geiger et al. 2002). On the other hand, when a synthetic mixture was not injected into the diluted diesel exhaust, the formation of ozone and aerosol was reduced due to the higher NO_x concentration than that of ambient air (Lee et al. 2006). The enhanced partitioning rate of secondary vapor products formed from photochemical reactions of a diesel exhaust-α-pinene mixture was explained by acid catalytic reactions probably due to sulfate in diesel exhaust aerosols (Lee et al. 2004b).

3.2 Secondary Pollution from Construction Materials and Air Fresheners

Smog chambers can be used as reaction vessels not only for photochemical reactions of outdoor environment but also as ozone reactions of indoor environment. In indoor environments, significant amount of VOCs can be emitted from a variety of areas, such as construction materials of buildings,

air fresheners, and cleaning products. Moreover, the VOCs can react with indoor ozone to form ultrafine particles and hazardous products, such as formaldehyde (Morrison and Nazaroff 2002; Nazaroff and Weschler 2004). Figure 36 shows a schematic diagram of the ozone reaction chamber used to examine the secondary pollution from natural paint (Moon et al. 2004a).

The total number concentration of ultrafine particles produced by a reaction between ozone and terpenes emitted from natural paints was reported to increase with increasing ozone concentration, as shown in Fig. 37 (Lamorena et al. 2007).

3.3 Other Applications

These smog chamber experiments can be used in the following areas: (i) to assess environmental processes, such as long-ranged transport; (ii) to suggest a strategy for reducing atmospheric pollution; (iii) to measure the outgas from building materials; (iv) to study the mechanism of nanoparticle formation; (v) to analyze and control characteristics of nanoparticles; (vi) to apply the manufacturing process of materials made from nanoparticles; and (vii) to apply gas to particle conversion in a cleanroom in the semiconductor industry.

4 Summary

Visibility impairments caused by photochemical smog phenomena still remains a problem both in many metropolitan areas throughout the world as well as in national parks with good views. The formation of ozone and secondary aerosols, which is the representative phenomenon of photochemical reactions in atmosphere, can reduce the ambient air quality during smog episodes. Knowledge about this transformation of primary pollutants to secondary pollutants is needed in order to develop effective and efficient environmental policies for improving the ambient air quality. For this goal, there are some research strategies, such as field monitoring, numerical modeling and smog chamber experiments.

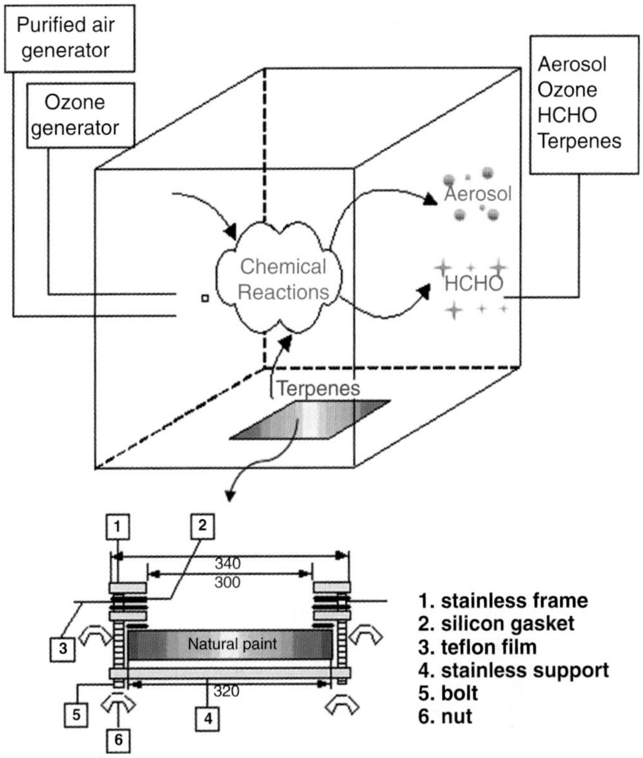

Fig. 36 Schematic diagram of the ozone reaction chamber for experiments on natural paint (Moon et al. 2004a)

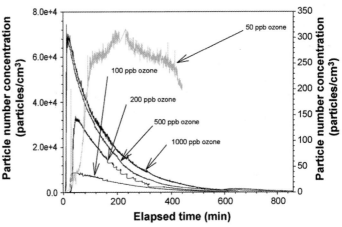

Fig. 37 Change in the particle number concentrations during the ozone-initiated oxidations at different ozone concentrations (50, 100, 200, 500, and 1,000 ppb) of paint (10 mL) (Lamorena et al. 2007). The right y-axis shows the data for 50 ppb ozone only

Smog chamber experiments are very powerful and useful tools for examining atmospheric processes and developing chemical mechanisms because controlled atmospheric conditions can be simulated experimentally. For this reason, smog chamber experiments have been carried out throughout the world for more than fifty years. These experiments can also determine the best reduction strategies through a variety of approaches. Smog chamber components, such as light sources, injection/sampling systems, and instru-

ments, are being developed to simulate more realistic atmospheric conditions. Recently, large and clean smog chambers are being constructed to minimize the experimental uncertainty.

In order to obtain more reliable and useful data from smog chamber experiments, several points in smog chamber measurements need to be considered carefully. First, the light intensity and spectrum of the light source need to be characterized and maintained during the same experimental set. The photolysis rate of NO_2

can be used as an index of light intensity. Second, the wall effect of the reaction bags needs to be considered. The wall effect includes the loss of gases and particles to the chamber walls and the outgassing of reactive species from the walls. The amount of particles lost on the wall should be corrected for when calculating the aerosol yield or the aerosol mass formed from given VOC precursors per unit amount of reacted VOCs. A wall loss correction is more important for particles in a Teflon bag due to the electrostatic effect. Third, the measuring principles of various instruments used in smog chamber experiments need to be understood to determine the measurement methods appropriate to the experimental purpose. The specific interference of some detectors should be also taken into account when comparing data, for example, the NO_2 data measured using a chemiluminescence $NO–NO_2–NO_x$ analyzer.

This book discussed some experimental techniques, such as light intensity control, temperature control, and twin chambers experiments using real ambient air. The typical outputs for ozone formation and the nucleation and growth of secondary aerosols obtained from smog chamber experiments using synthetic $VOCs–NO_x–air$ mixtures or real ambient air were summarized from the literature. Some applications of smog chamber facilities were introduced to expand the smog chamber techniques to many research fields.

In order to fully utilize the infinite capacity of smog chamber facilities, new advanced experimental techniques and analytical instruments need to be developed both with innovative ideas for breakthroughs and with improvements based on an improved understanding of the physical and chemical transformation of gases and aerosols in the atmosphere.

References

Angove DE, Halliburton BW, Nelson PF (2000) Development of a new indoor environmental chamber at MRL, North Ryde, 15th International Clean Air and Environment Conference, 270–274, Reproduced with the permission of the CSIRO

Bae GN, Kim MC, Lee SB, Song KB, Jin HC, Moon KC (2003) Design and performance evaluation of the KIST indoor smog chamber (in Korean), J Korean Soc Atmos Environ, 19(4), 437–449

Bae GN, Park JY, Kim MC, Lee SB, Moon KC, Kim YP (2008) Effect of light intensity on the ozone formation and the aerosol number concentration of ambient air in Seoul (in Korean), Part Aerosol Res, 4(1), 9–20, Reprinted with permission of KAPAR

Behnke W, Holländer W, Koch W, Nolting F, Zetzsch C (1988) A smog chamber for studies of the photochemical degradation of chemicals in the presence of aerosols, Atmos Environ, 22(6), 1113–1120

Bowman FM, Odum JR, Seinfeld JH (1997) Mathematical model for gas-particle partitioning of secondary organic aerosols, Atmos Environ, 31(23), 3921–3931

Bufalini JJ, Theodore AW, Marijon MB (1977) Contamination effects on ozone formation in smog chambers, Environ Sci Technol, 11, 1181–1185

Carter WPL, Luo D, Malkina IL, Pierce JA (1995) Environmental chamber studies of atmospheric reactivities of volatile organic compounds – Effects of varying chamber and light sources, Final report to National Renewable Energy Laboratory, Contract XZ-2-12075, Coordinating Research Council, Inc., Project M-9, California Air Resources Board, Contract A032–0692, South Coast Air Quality Management District, Contract C91323, University of California, Riverside

Carter WP, Fitz DR (2003) A smog simulation chamber for determining atmospheric reactivity and evaluating measurement techniques, Air and Waste Management Association's 2003 Annual Conference and Exhibition Proceedings, paper #111

Carter WPL, Cocker III DR, Fitz DR, Malkina IL, Bumiller K, Sauer CG, Pisano JT, Bufalino C, Song C (2005) A new environmental chamber for evaluation of gas-phase chemical mechanism and secondary aerosol formation, Atmos Environ, 39, 7768–7788, Copyright 2005, Reprinted with permission from Elsevier

Cocker D, Whitlock N, Collins D, Wang J, Flagan R, Seinfeld J (1999) Instrumentation for state-of-art aerosol measurements in smog chambers, Combined U.S./German Ozone/Fine Particle Science and Environmental Chamber Workshop, Riverside, CA, October 4–6

Cocker III DR, Flagan RC, Seinfeld JH (2001) State-of-art chamber facility for studying atmospheric aerosol chemistry, Environ Sci Technol, 35(12), 2594–2601

Crump JG, Seinfeld JH (1981) Turbulent deposition and gravitational sedimentation of an aerosol in a vessel of arbitrary shape, J Aerosol Sci, 12, 405–415

Dodge MC (2000) Chemical oxidant mechanisms for air quality modeling: Critical review, Atmos Environ, 34, 2103–2130

Edney EO, Kleindienst TE, Jaoui M, Lewandowski M, Offenberg JH, Wang W, Claeys M (2005) Formation of 2-methyl tetrols and 2-methylglyceric acid in secondary organic aerosol from laboratory irradiated isoprene/NO_x/SO_2/air mixtures and their detection in ambient $PM_{2.5}$ samples collected in the eastern United States, Atmos Environ, 39, 5281–5289

Finlayson B, Pitts Jr JN (1976) Photochemistry of the polluted troposphere, Science, 192, 111–119

Forstner HJL, Flagan RC, Seinfeld JH (1997) Secondary organic aerosol from the photooxidation of aromatic hydrocarbons: Molecular composition, Environ Sci Technol, 31, 1345–1358

Geiger H, Kleffmann J, Wiesen P (2002) Smog chamber studies on the influence of diesel exhaust on photosmog formation, Atmos Environ, 36, 1737–1747, Copyright 2002, Reprinted with permission from Elsevier

Gery MW, Fox DL, Jeffries HE, Stockburger L, Weathers WS (1985) A continuous stirred tank reactor investigation of the gas-phase reaction of hydroxyl radicals and toluene, Int J Chem Kinet, 17(9), 931–955

Gery MW, Corouse RR (2002) User's Guide for Executing OZIPR, US EPA home page (http://www.epa.gov/scram001/models/other/oziprdme.txt), Accessed in December 2002

Ghim YS, Moon KC, Lee SH, Kim YP (2005) Visibility trends in Korea during the past two decades, J Air Waste Manage Assoc, 55, 73–82, Reprinted with permission of JOURNAL of A&WMA

Grosjean D (1985) Wall loss of gaseous pollutants in outdoor Teflon chambers, Environ Sci Technol, 19, 1059–1065

Grosjean D, Seinfeld JH (1989) Parameterization of the formation potential of secondary organic aerosols, Atmos Environ, 23(8), 1723–1747

Heisler SL, Friedlander SK (1977) Gas-to-particle conversion in photochemical smog: Aerosol growth laws and mechanisms for organics, Atmos Environ, 11, 157–168

Hu D, Kamens RM (2007) Evaluation of the UNC toluene-SOA mechanism with respect to other chamber studies and key model parameters, Atmos Environ, 41, 6465–6477

Hu D, Tolocka M, Li Q, Kamens RM (2007) A kinetic mechanism for predicting secondary organic aerosol formation from toluene oxidation in the presence of NO_x and natural sunlight, Atmos Environ, 41, 6478–6496, Copyright 2007, Reprinted with permission from Elsevier

Hurley MD, Sokolov O, Wallington TJ, Takekawa H, Karasawa M, Klotz B, Barnes I, Becker KH (2001) Organic aerosol formation during the atmospheric degradation of toluene, Environ Sci Technol, 35(7), 1358–1366

Hynes RG, Angove DE, Saunders SM, Haverd V, Azzi M (2005) Evaluation of two MCM v3.1 alkene mechanisms using indoor environmental chamber data, Atmos Environ, 39, 7251–7262

Izumi K, Fukuyama T (1990) Photochemical aerosol formation from aromatic hydrocarbons in the presence of NO_x, Atmos Environ, 24A(6), 1433–1441

Jaimes JLL, Sandoval JF, González UM, González EO (2003) Liquefied petroleum gas effect on ozone formation in Mexico City, Atmos Environ, 37, 2327–2335

Jaimes JLL, Sandoval JF, González EO, Vázquez MG, González UM, Zambrano AG (2005) Effect of liquefied petroleum gas on ozone formation in Guadalajara and Mexico City, J Air Waste Manage Assoc, 55 (6), 841–846

Jang M, Kamens RM (2001) Atmospheric secondary aerosol formation by heterogeneous reactions of aldehydes in the presence of a sulfuric acid aerosol catalyst, Environ Sci Technol, 35, 4758–4766

Jang M, Carroll B, Chandramouli B, Kamens, RM (2003) Particle growth by acid-catalyzed heterogeneous reactions of organic carbonyls on preexisting aerosols, Environ Sci Technol, 37, 3828–3837

Jeffries H, Sexton K, Yu J (1998) Atmospheric photochemistry studies of pollutant emissions from transportation vehicles operating on alternative fuels, Report to National Renewable Energy Laboratory Under Contract No DE-AC36-83CH10093, NREL/TP-452-21426, University of North Carolina, Chapel Hill, NC

Ju OJ (2006) Effect of temperature on the photochemical reaction of toluene–NO_x–air mixture (in Korean), Master dissertation thesis, Seoul National University, Reprinted with permission from Ok Jung Ju

Kalberer M, Paulsen D, Sax M, Steinbacher M, Dommen J, Prevot ASH, Fisseha R, Weingartner E, Frankevich V, Zenobi R, Baltensperger U (2004) Identification of polymers as major components of atmospheric organic aerosols, Science, 303, 1659–1662

Kanakidou M, Seinfeld JH, Pandis SN, Barnes I, Dentener FJ, Facchini MC, Van Dingenen R, Ervens B, Nenes A, Nielsen CJ, Swietlicki E, Putaud JP, Balkanski Y, Fuzzi S, Horth J, Moortgat GK, Winterhalter R, Myhre CEL, Tsigaridis K, Vignati E, Stephanou EG, Wilson J (2005) Organic aerosol and global climate modelling: A review, Atmos Chem Phys, 5, 1053–1123

Kelly NA (1982) Characterization of fluorocarbon-film bags as smog chambers, Environ Sci Technol, 16(11), 763–770

Kelly NA, Olson KL, Wong CA (1985) Tests for fluorocarbon and other organic vapor release by fluorocarbon film bags, Environ Sci Technol, 19(4), 361–364

Kelly NA (1987) The photochemical formation and fate of nitric acid in the metropolitan Detroit area: Ambient, captive-air irradiation and modeling results, Atmos Environ, 21(10), 2163–2177

Kelly NA, Gunst RF (1990) Response of ozone to changes in hydrocarbon and nitrogen oxide concentrations in outdoor smog chambers filled with Los Angeles air, Atmos Environ, 24A(12), 2991–3005

Kleindienst TE, Smith DF, Hudgens EE, Snow RF, Perry E, Claxton LD, Bufalini JJ, Black FM, Cupitt LT (1992) The photo-oxidation of automobile emissions: Measurements of the transformation products and their mutagenic acitivity, Atmos Environ, 26A(16), 3039–3053

Lamorena RB, Jung SG, Bae GN, Lee W (2007) The formation of ultra-fine particles during ozone-initiated oxidations with terpenes emitted from natural paint, J Hazard Mater, 141, 245–251, Copyright 2007, Reprinted with permission from Elsevier

Lee S-B, Bae G-N, Moon K-C (2004a) Aerosol wall loss in Teflon film chambers filled with ambient air, J Korean Soc Atmos Environ, 20(E1), 35–41

Lee S-B, Bae G-N, Moon K-C, Choi M (2006) Effect of diesel exhaust on the photochemical reactions of ambient air (in Korean), Part Aerosol Res, 2(3–4), 127–140, Reprinted with permission of KAPAR

Lee SB, Bae GN, Moon KC (2007) Effect of diesel particles on the photooxidation of a diluted diesel exhaust-toluene mixture, SAE Technical 2007-01-0315

Lee S, Jang M, Kamens RM (2004b) SOA formation from the photooxidation of α-pinene in the presence of freshly emitted diesel soot exhaust, Atmos Environ, 38, 2597–2605

Lee Y-M, Bae G-N, Lee S-B, Kim M-C, Moon K-C (2005a) Effect of initial toluene concentration on the photooxidation of toluene–NO_x–air mixture – I. Change of gaseous species (in Korean), J Korean Soc Atmos Environ, 21(1), 15–26

Lee Y-M, Bae G-N, Lee S-B, Kim M-C, Moon K-C (2005b) Effect of initial toluene concentration on the photooxidation of toluene–NO_x–air mixture – II. Aerosol formation and growth (in Korean), J Korean Soc Atmos Environ, 21(1), 27–38

Lonneman WA, Buflini JJ, Kuntz RL, Meeks SA (1981) Contamination from fluorocarbon films, Environ Sci Technol, 15(1), 99–103

McDonald JD, Barr EB, White RK (2004) Design, characterization, and evaluation of a small-scale diesel exhaust exposure system, Aerosol Sci Technol, 38, 62–78

McMurry PH, Friedlander SK (1978) Aerosol formation in reacting gases: Relation of surface area to rate of gas-to-particle conversion, J Colloid Interface Sci, 64(2), 248–257

McMurry PH, Grosjean D (1985) Gas and aerosol wall losses in Teflon film smog chambers, Environ Sci Technol, 19(12), 1176–1182

McMurry PH, Rader DJ (1985) Aerosol wall losses in electrically charged chambers, Aerosol Sci Technol, 4, 249–268

McMurry P, Shepherd M, Vickery J (2004) Particulate Matter Science for Policy Makers – A NARSTO Assessment, Cambridge University Press, USA

Moon K-C et al. (2004a) A Study on the Smog Mechanism and Control Technology (in Korean), Report of Korea Institute of Science and Technology to Korean Ministry of Science and Technology, M1-0204-00-0049

Moon KC, Bae GN, Lee SB, Lee YM, Choi JE (2004b) Smog study using twin chambers filled with ambient air, 13th World Clean Air and Environmental Protection, IUAPPA, London, UK, August 22–27

Moon K-C, Bae G-N, Lee Y-M, Lee S-B (2006) Effect of the initial concentration ratio of toluene/NO_x on the photochemical reactions of ambient air, 15th IUAPPA, Lilli, France, September 5–8

Morrison GC, Nazaroff WW (2002) Ozone interactions with carpet: Secondary emissions of aldehydes, Environ Sci Technol, 36, 2185–2192

Nazaroff WW, Weschler CJ (2004) Cleaning products and air fresheners: Exposure to primary and secondary air pollutants, Atmos Environ, 38, 2841–2865

Odum JR, Hoffmann T, Bowman F, Collins D, Flagan RC, Seinfeld JH (1996) Gas/particle partitioning and secondary organic aerosol yields, Environ Sci Technol, 30, 2580–2585

Odum JR, Jungkamp TPW, Griffin RJ, Flagan RC, Seinfeld JH (1997) The atmospheric aerosol-forming potential of whole gasoline vapor, Science, 276, 96–99

Oh S, Andino JM (2000) Effects of ammonium sulfate aerosols on the reactions of the hydroxyl radical with organic compounds, Atmos Environ, 34, 2901–2908

Oh S, Andino JM (2001) Kinetics of the gas-phase reactions of hydroxyl radicals with C_1–C_6 aliphatic alcohols in the presence of ammonium sulfate aerosols, Int J Chem Kinet, 33, 422–430

Oh S, Andino JM (2002) Laboratory studies of the impact of aerosol composition on the heterogeneous oxidation of 1-propanol, Atmos Environ, 36, 149–156

Pandis SN, Paulson SE, Seinfeld JH, Flagan RC (1991) Aerosol formation in the photooxidation of isoprene and β-pinene, Atmos Environ, 25A, 997–1008

Pankow JF (1994a) An absorption model of gas/particle partitioning of organic compounds in the atmosphere, Atmos Environ, 28(2), 185–188

Pankow JF (1994b) An absorption model of gas/aerosol partitioning involved in the formation of secondary organic aerosol, Atmos Environ, 28(2), 189–193

Pitts Jr JN, Smith JP, Fitz DR, Grosjean D (1977) Enhancement of photochemical smog by N, N′ diethylhydroxylamine in polluted ambient air, Science, 197, 255–257

Prager MJ, Stephens ER, Scott WE (1960) Aerosol formation from gaseous air pollutants, Indust Eng Chem, 52(6), 521–524

Roberts PT, Friedlander SK (1976) Photochemical aerosol formation SO_2, 1-Heptene, and NO_x in ambient air, Environ Sci Technol, 10(6), 573–580

Saathoff H, Naumann K-H, Schnaiter M, Schöck W, Möhler O, Schurath U, Weingartner E, Gysel M, Baltensperger U (2003) Coating of soot and $(NH_4)_2SO_4$ particles by ozonolysis products of α-pinene, J Aerosol Sci, 34, 1297–1321, Copyright 2003, Reprinted with permission from Elsevier

Seinfeld JH, Pandis SN (1998) Atmospheric Chemistry and Physics, Wiley, New York, USA

Shibuya K, Nagashima T, Imai S, Akimoto H (1981) Photochemical ozone formation in the irradiation of ambient air samples by using a mobile smog chamber, Environ Sci Technol, 15(6), 661–665

Song C, Na K, Cocker III DR (2005) Impact of the hydrocarbon to NO_x ratio on secondary organic aerosol formation, Environ Sci Technol, 39, 3413–3419

Song C, Na K, Warren B, Malloy Q, Cocker III DR (2007) Impact of propene on secondary organic aerosol formation from m-xylene, Environ Sci Technol, 41, 6990–6995

Stephens ER, Hanst PL, Doerr RC, Scott WE (1956) Reactions of nitrogen dioxide and organic compounds in air, Indust Eng Chem, 48(9), 1498–1504

Stephens ER, Schuck EA (1958) Air pollution effects of irradiated auto exhaust as related to fuel consumption, Chem Eng Prog, 54(11), 71–77

Stroud CA, Makar PA, Michelangeli DV, Mozurkewich M, Hastie DR, Barbu A, Humble J (2004) Simulating organic aerosol formation during the photooxidation of toluene/NO_x mixtures: Comparing the equilibrium and kinetic assumption, Environ Sci Technol, 38, 1471–1479

Wang S-C, Paulson SE, Grosjean D, Flagan RC, Seinfeld JH (1992) Aerosol formation and growth in atmospheric organic/NO_x systems – I. Outdoor smog chamber studies of C_7- and C_8-hydrocarbons, Atmos Environ, 26A(3), 403–420

Wentzel M, Gorzawski H, Naumann K-H, Saathoff H, Weinbruch S (2003) Transmission electron microscopical and aerosol dynamical characterization of soot aerosols, J Aerosol Sci, 34, 1347–1370

Wiesen P (1999) Investigation of real car exhaust in the EUPHORE chamber, in Combined US/German Ozone/Fine Particle Science and Environmental Chamber Workshop, Riverside, CA, October 4–6

Wiesen P (2000) Diesel Fuel and Soot: Fuel Formulation and Its Atmospheric Implications, Final report of EU project contract ENV4-CT97-0390

Winer AM, Peters JW, Smith JP, Pitts Jr JN (1974) Response of commercial chemiluminescent NO–NO_2 analyzers to other nitrogen-containing compounds, Environ Sci Technol, 8(13), 1118–1121

Zafonte L, Rieger PL, Holmes JR (1977) Nitrogen dioxide photolysis in the Los Angeles atmosphere, Environ Sci Technol, 11(5), 483–487

Zielinska B, Samy S, Seagrave JC, McDonald J, Wirtz K, Vazquez MM (2007) Investigation of atmospheric transformations of diesel emissions in the European Photoreactor (EUPHORE), 5th Asian Aerosol Conference, Kaohsiung, Taiwan, August 26–29

Aerosol Concentrations and Remote Sources of Airborne Elements Over Pico Mountain, Azores, Portugal

Maria do Carmo Freitas, Adriano M.G. Pacheco, Isabel Dionísio and Bruno J. Vieira

Abstract Aerosol samples (PM_{10}) were collected using an aethalometer from 15 July 2001 to 18 April 2004 at the PICO-NARE site in Pico island, Azores, Portugal. The aethalometer is at an altitude of 2225 m AMSL, and sampled for 24 h in most cases, and for a few periods continuously. Samples were assessed through instrumental neutron activation analysis (k_0-variant), and concentrations of up to 15 airborne elements were determined. Concentrations are in the order of magnitude of a moderately polluted urban-industrial site. Elements are predominantly entrained by air masses from North-Central America, and to a lesser extent from Europe and North Africa. PCA and PMF assigned sources related to pollution (traffic, fossil-fuel combustion, mining, industrial processing) and to natural occurrences (crustal, Saharan episodes, marine). Although data uncertainties are relatively high due to the small masses collected in the filters and impurities in them, PMF – which includes the uncertainty – did not prove better than PCA when missing data are replaced by arithmetic means of the determined values for each element.

Keywords Aerosol concentrations · Airborne elements · Air-mass trajectories · Azores archipelago · Cluster analysis · Enrichment factors · Free troposphere · HYSPLIT model · k_0-INAA · Pico mountain · PICO-NARE observatory · PM_{10} · Positive matrix factorization (PMF) · Principal-components analysis (PCA) · Remote sources · Seven-wavelength aethalometer · Source attribution

1 Introduction

The Azores archipelago comprises nine islands ($2335 km^2$), geographically split into three groups – Eastern (Santa Maria, São Miguel), Central (Terceira, Graciosa, São Jorge, Pico, Faial) and Western (Flores, Corvo) – which span over 600 km in an overall ESE-WNW direction ($36°55'–39°43'$ N; $25°00'–31°15'$W), as outlined in Fig. 1. The climate is oceanic temperate, strongly influenced by the Azores anticyclone – often referred to as Azores High – and the North Atlantic Drift of the Gulf Stream, with high humidity and precipitation, and much milder temperatures (average- and amplitude-wise) than Lisbon or New York City, NY, both at roughly the same latitude.

Fig. 1 The Azores archipelago in the context of the North Atlantic Ocean

M. do Carmo Freitas (✉)
Reactor-ITN, Technological and Nuclear Institute; E.N. 10, 2686-953 Sacavém, Portugal
e-mail: cfreitas@itn.pt

Y.J. Kim et al. (eds.), *Atmospheric and Biological Environmental Monitoring*, DOI 10.1007/978-1-4020-9674-7_9, © Springer Science+Business Media B.V. 2009

All islands have volcanic origins, with some sedimentary (reef) contribution for Santa Maria, the oldest, formed some 7 million years ago. Geological-age estimates widely differ though, and an abundance of marine fossils in Santa Maria's rocks imply that, at some point, the island may have sunk and then resurfaced. The youngest island is Pico, only about 300 thousand years old. The archipelago lies on the Mid-Atlantic Ridge, near the so-called Azores triple-junction (Searle 1980), in an area of confluence of three major tectonic plates – the North American Plate, the Eurasian (or Euroasian) Plate and the African (or Nubian) Plate – complete with the local Azores Microplate. The two westernmost islands (Flores, Corvo) actually lie on the North American Plate. Tectonic patterns/geodynamics and geochemical aspects relating to the formation of the Azores plateau have been extensively discussed (Flower et al. 1976, Searle 1980, Fernandes et al. 2006, Silveira et al. 2006, Yang et al. 2006).

The archipelago sits directly on the pathway of long-range transport of airborne species over the North Atlantic Ocean, namely mineral particles from Africa's Sahara and Sahel (Chazette et al. 2001) – arguably, the world's largest sources of Aeolian soil dusts – and non-natural contaminants from the United States' eastern seaboard down to Central America and the Caribbean. The transport mechanisms and remote impacts of African dust have been clearly established (Prospero

1999a, and references therein). African-dust events have been shown not only to affect coastal North America and the Caribbean basin (Prospero and Nees 1986, Li et al. 1996, Li-Jones and Prospero 1998, Prospero 1999b, Prospero et al. 2001, Prospero and Lamb 2003), but also to reach deep into the continent, to the point that African-dust incursions could be discerned over much of the eastern half of the United States, including areas that make up the so-called "dust bowl" (Perry et al. 1997). In what concerns the Azores proper, contamination of African dust with polluted aerosols from Europe has been observed as well (Reis et al. 2002), in line with similar mixing processes detected in the North-Eastern Atlantic Ocean (Desboeufs and Cautenet 2005).

The archipelago is thus an ideal platform for watching significant deposition episodes in the area and, especially, for tracking the regular oceanic transit of air masses from the surrounding continents – Africa, Europe and North-Central America. Furthermore, the Pico mountain, rising up to 2351 m above mean sea level (AMSL) in Pico island – Fig. 2 – is high enough to enable land-based access to the lower free troposphere, which means that contaminant loads from afar can be detected over the altitude range of significant removal by (sinking into) the ocean, and that their trajectories may be appraised without accounting for the influence of the marine boundary layer. This work addresses the former issues, by looking into

Fig. 2 Pico mountain (2351 m AMSL) in Pico island (Azores' Central group)

the characteristics of aerosol samples collected through a seven-wavelength aethalometer at the PICO-NARE observatory.

2 Experimental

2.1 Instrumental Aspects

The PICO-NARE observatory (38.470 °N, 28.404 °W; 2225 m AMSL), which includes the seven-wavelength aethalometer (model AE31), is an automated, self-contained, experimental station located near the summit of the Pico mountain – Fig. 3. Relevant features that make this location unique for assessing regional and hemispheric impacts from measurements within the lower free troposphere or the marine boundary layer have already been highlighted (Honrath and Fialho 2001). In particular, measurements of aerosol absorption coefficients have been made at the observatory since July 2001. Details on the operation and calibration of the aethalometer system, as well as a few examples of its use on the characterization of biomass-burning plumes and Saharan-dust aerosols, can be found elsewhere (Honrath et al. 2004, Fialho et al. 2005, Fialho et al. 2006, Freitas et al. 2007).

All elemental determinations for the present study were carried out at the Portuguese Research Reactor (RPI; pool-type reactor; maximum nominal power: 1 MW) of the Technological and Nuclear Institute (ITN-Sacavém, Portugal), through k_0-standardised, instrumental neutron activation analysis – k_0-INAA (De Corte 1987, Freitas and Martinho 1989a, Freitas and Martinho 1989b, Freitas 1993, De Corte 2001). An excellent account of fundamentals and techniques of activation analysis with thermal (low energy) neutrons was given in a review by Erdtmann and Petri (1986).

In brief, this methodology uses the production of artificial radionuclides from stable elements for their identification and quantitative determination. Samples are irradiated with a neutron beam produced along the nuclear fission of ^{235}U nuclei in a nuclear reactor, thermalised by a moderator in order to induce the nuclear reaction type (n, γ). The latter consists in the capture of a thermal neutron by the nucleus of an atom with simultaneous emission of gamma radiation, that hits an electron of the atom of the material which makes up the counting detectors (germanium). The full energy is absorbed by the electron that gets ejected,

and its energy is collected as an electronic impulse with the same energy of the gamma radiation emitted. The result is a spectrum of several full-energy peaks, which enables an identification of elements whose atoms were involved in the former neutron captures. The peak areas provide the numerical bases for assessing the corresponding amounts of elements present in the irradiated sample.

As mentioned above, all calculations for this work were based on the k_0-variant of INAA, which includes a comparator – usually gold, to obtain the radionuclide ^{198}Au – to be irradiated together with the sample. Elemental concentrations are then calculated by comparing the full-energy peaks present in the spectra of the sample and comparator. Irradiation conditions and reactor parameters for the current implementation of the k_0-INAA methodology at ITN have been previously reported (Pacheco et al. 2006, Freitas et al. 2006).

The results herein refer to 425 aerosol samples with particles under $10\,\mu m$ aerodynamic diameter (PM_{10}), collected from July 15, 2001, until April 18, 2004, with an irregular periodicity. As a consequence, the air-intake volume varied between $0.12\,m^3$ and $81\,m^3$. Most frequent values were around $8\,m^3$, which may be seen to correspond to an average flow rate close to $5.5\,L \cdot min^{-1}$, on a 24 h sampling basis. The AE31 instrument has been operated with quartz-fibre filter tapes, reinforced with a non-woven polyester support layer as a strength binder (manufacturer: Pallflex®; type: Q250F) – Fig. 4a. The increase in tensile strength regarding similar, binder-free media has been found quite significant (Weingartner et al. 2003). Due to the remote setting of PICO-NARE and local weather conditions, instrumental maintenance has been performed once-twice a year.

Aerosol-laden spots were identified after removing the tape from the aethalometer – see Fig. 4b for an example – and packed together with unexposed portions of the tape (as blanks) for elemental analysis. Upon removal, each tape was first cut into segments of ca. 20 cm, which were photographed for archival backups, and then sealed into polyethylene sleeves for shipping purposes. At ITN, both samples and blanks were carefully cut from the quartz-fibre strips in circular shapes with an area of $0.950 \pm 0.086\,cm^2$, the former centred on the selected-spot area ($0.50 \pm 0.05\,cm^2$). The exposed side of each cut-out was protected with an identical piece of blank filter, to avoid contamination of the sample with the polyethylene

Fig. 3 **a** PICO-NARE experimental site (2225 m AMSL) near the summit of the Pico mountain (2351 m AMSL); and **b** a closer view of the automated observatory

top cover. The same procedure was also applied to every actual blank, to obtain an identical geometry. All sandwich-like sets were packed into aluminium sheets for further irradiation.

Each sample or blank set was irradiated for 5–7 h at a thermal-neutron flux of about $10^{13}\,\mathrm{n \cdot cm^{-2} \cdot s^{-1}}$, together with one disc (thickness: 125 μm; diameter: 0.5 cm) of an Al–0.1%Au alloy as comparator. After

Fig. 4 **a** Instrumental set-up for the quartz-fibre filter tape on site (continuous strip); and **b** an exposed section of the tape after removal at the laboratory (four identifiable spots)

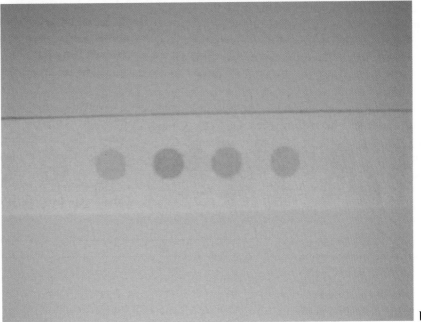

irradiation, sets were allowed to decay for 4–6 days and then measured; after this, they kept on decaying for an additional 2–3 weeks, and were measured again. All gamma spectra were acquired on a liquid nitrogen-cooled, ORTEC®-calibrated, high-purity germanium detector (1.85 keV resolution at 1.33 MeV; 30% relative efficiency), connected to a 4096 multi-channel analyser. The comparator disc was measured in the same detector 7 days after the end of an irradiation.

2.2 Data Analysis, Quality Control and Elemental Uncertainties

Elemental concentrations in loaded and blank media were assessed with the current version of the "k_0-IAEA" software, that has been developed by Blaauw (2007) with an endorsement by the International Atomic Energy Agency (IAEA). Standard packages for exploratory, correlation and multivariate data analyses were STATISTICA® 6.0 (StatSoft Inc.) and SPSS 11.0 for Windows® (SPSS Inc.). Backward trajectories of air masses reaching the Azores above PICO-NARE were produced in isentropic mode with HYSPLIT from NOAA ARL READY Website (Draxler and Rolph 2003, Rolph 2003).

Filters are made from quartz, and their dimensions are reduced. Their main constituent – silicon – does not activate that much with thermal neutrons; however, impurities are significant for a few elements. Unfortunately, aerosol reference materials with a quartz matrix are not commercially available, the only ones existing are polycarbonate membrane filters, such as the Nuclepore® brand. The latter are composed mainly of C and O, which do not activate, and they feature lower impurities than the quartz filters. For the reference material "Air Particulate on Filter Media" (NIST-SRM® 2783; $PM_{2.5}$ on a polycarbonate filter membrane) measured by INAA, the reproducibility was within 5–15% (Almeida et al. 2006a). Impurity amounts found in the present quartz filters were, in $\mu g\ cm^{-2}$: As: 0.02; Br: 0.002; Co: 0.007; Cr: 0.02; Fe: 0; Hf: 0.002; K: 5; La: 0.001; Mo: 0.2; Na: 6.8; Sb: 0.5; Sc: 0.0001; Sm: 0.0006; U: 0.003; Yb: 0; Zn: 0.07.

The ratio of an elemental (typical) mass determined in loaded filters to its occurrence (as an impurity) in blank filters is 2 for most elements, except for Br (ratio: 90), La, K, Sc and U (ratio: 4), and Na (ratio: 9). Iron and ytterbium are not present in the blanks. The uncertainties from the analysis proper, which arise mainly from the measurements

of full-peak areas, are, typically: 2% for Br and Sb; 10–15% for Mo, Na and U; 15–20% for Co, K and La; 25–35% for Cr, Fe, Hf, Sc, Sm, Yb and Zn; and 70% for As.

3 Results and Discussion

3.1 Concentrations of Elements at PICO-NARE

Table 1 shows the elemental concentrations found in the analysed samples. Only the elements detected in at least 25% of the cases are shown. Other elements with a ratio of determination under 25% were: As (5%; average: $2.0\ ng \cdot m^{-3}$), Cr (24%; average: $1.9\ ng \cdot m^{-3}$), K (18%; average: $376\ ng \cdot m^{-3}$), Se (8%; average: $1.3\ ng \cdot m^{-3}$), Tb (0.2%; average: $2.0\ ng \cdot m^{-3}$), Yb (5%; average: $ng \cdot m^{-3}$), and Th (3%; average: $0.4\ ng \cdot m^{-3}$). When compared by their absolute magnitude of concentrations, the average data can be grouped as (1) above $100\ ng \cdot m^{-3}$: Fe, Na; (2) between 100 and $10\ ng \cdot m^{-3}$: Br, Mo, Sb, Zn; (3) between 1 and $0.1\ ng \cdot m^{-3}$: Co, Hf, La, Sm, U; and (4) below $0.1\ ng \cdot m^{-3}$: Sc.

Comparing to total suspended particles at a remote site such as the Antarctic Peninsula (King Sejong station), the concentrations at PICO-NARE are three orders of magnitude higher for Zn and U, two orders of magnitude higher for Co, and one order of magnitude higher for Cr (Mishra et al. 2004). A similar comparison with PM_{10} from McMurdo station, Antarctica,

Table 1 Number of samples for an element (N), proportion of samples for an element to the total analysed samples (in %), mean, median, standard deviation (SD), minimum and maximum (all in ng m^{-3}), referring to the whole collection from the PICO-NARE site, July 15, 2001, through April 18, 2004 (425 aerosol samples)

	Br	Co	Fe	Hf	Mo	La
N	302	233	154	152	120	170
% Total	71	55	36	36	28	40
Mean	17.1	0.83	275	0.13	13.6	0.23
Median	1.34	0.49	85.8	0.090	8.94	0.094
SD	88.6	2.01	711	0.15	42.5	0.51
Minimum	0.015	0.00048	6.71	0.00093	0.023	0.0016
Maximum	1227	28.6	7978	1.35	469	4.40
	Na	Sb	Sc	Sm	U	Zn
N	186	154	136	127	256	145
% Total	44	36	32	30	60	34
Mean	477	22.8	0.054	0.14	0.78	10.4
Median	789	16.2	0.012	0.048	0.34	5.63
SD	1129	28.6	0.20	0.44	1.94	27.3
Minimum	1.40	0.0013	0.00037	0.0015	0.0022	0.059
Maximum	8341	257	2.23	3.81	20.0	276

results in three orders of magnitude for Fe, two orders of magnitude for As, and one order of magnitude for Co, Cr, K, Se and Zn (Mazzera et al. 2001). Therefore, the Pico summit has no characteristics of a remote clean area.

At Bobadela – an urban-industrial neighbourhood in the northern outskirts of Lisboa, Portugal, some 20 km straight from the open Atlantic Ocean – the following elemental concentrations, in $ng \cdot m^{-3}$, were found in PM_{10} (Almeida et al. 2006b): As: 0.43; Br: 4.3; Co: 0.2; Fe: 400; K: 270; La: 0.23; Na: 1500; Sb: 2.6; Sc: 0.061; Se: 4.4; Sm: 0.041; Zn: 36. The present results compare well for Co, Fe, La, Sc and Sm, which were crustal elements for the PM_{10} fraction (Almeida et al. 2006b), even if higher values could occur due to Saharan-dust episodes (Almeida et al. 2008).

As to (potential) marine elements, the present work found higher concentrations of Br (one order of magnitude) and lower concentrations of Na (one order of magnitude). This divergence may be due to an added anthropogenic component of Br (volatile element) in Pico (Vieira et al. 2006; see also further) and a stronger marine component of Na in Bobadela, since sea-salt advection inland over a flat terrain is likely easier than the vertical mixing or upslope flow of salt-laden air into the high reaches of Pico mountain. Actually, enrichment data for a major sea-salt tracer (Cl) in biomonitors from a high-altitude site in Pico island (Cabeço Redondo; 1000 m AMSL) strongly suggests that such an enrichment may be due to deposition of salt undergoing long-range transport through the upper troposphere (or lower stratosphere, for finer particles), rather than to the direct advection of locally-produced, marine aerosol (Pacheco and Freitas 2007).

The anthropogenic elements As and Sb (volatile elements) show higher concentrations at the Pico summit, while Zn has a higher concentration in Bobadela. Fugitive emissions from an urban-waste incinerator nearby may not be ruled out for Bobadela though, for Zn has been widely regarded as the prime indicator for refuse incineration (Rahn and Huang 1999). As a conclusion, the site at the Pico summit seems largely influenced by anthropogenic emissions as well as by natural sources, and thus may be viewed as close to an urban-industrial location in what concerns the more volatile elements and the African-dust occurrences.

In an earlier study (Freitas et al. 2007), 109 aerosol filters collected from July 2001 through July 2008 at PICO-NARE were dealt with, and a correlation analysis suggested the existence of 5 groups: G.I, split into groups G.Ia with Fe and Ce, and G.Ib with Sm, La and Sc; G.II, with Co, Hf, Sb and Th; G.III, with Br, W and Zn; G.IV, with Br and Na; and G.V, with Mo and U. The application of such grouping to the extended data-set of 425 aerosol filters is shown in Table 2.

Purely crustal before, G.I now contains a larger number of elements with good or even excellent correlation related to an anthropogenic origin, such as Br, Sb and Zn, as well as Co, Mo and U of probable crustal sources. Other than good correlations between Hf and Sb, and Co and Sb, G.II includes also good correlations for Mo, Na, U and Zn. The former high degree of association between Br and Zn in G.III still exists, and now Br appears highly correlated with Co, Mo and U too. There are no significant associations within G.IV and G.V though.

3.2 Enrichment Factors at PICO-NARE

To look into the status of relative pollution at the PICO-NARE site, crustal, marine and aerosol enrichment

Table 2 Significant results of the correlation-coefficient matrix ($r \geq 0.5$) for elements associated into the five identified groups at PICO-NARE (Freitas et al. 2007)

G.I	Sm	La	Fe	G.II	Hf	Co	G.III	Br	G.IV	Br	G.V	Mo
La	0.93	–	0.87	Na	–	0.51	–	–	Na	–	U	–
Sc	0.69	0.87	0.98	–	–	–	–	–	–	–	–	–
Fe	0.69	–	–	–	–	–	–	–	–	–	–	–
Co	0.96	0.86	–	–	–	–	Co	0.96	–	–	–	–
Zn	0.49	–	–	Zn	–	0.48	Zn	0.50	–	–	–	–
Br	0.97	0.85	–	–	–	–	–	–	–	–	–	–
Mo	0.99	0.86	–	Mo	–	0.97	Mo	1.00	–	–	–	–
Sb	0.52	–	–	Sb	0.68	0.56	–	–	–	–	–	–
U	0.80	–	–	U	–	0.70	U	0.81	–	–	–	–

Fig. 5 Crustal enrichment factors relative to scandium as the reference element. Marine enrichment factors of bromine and potassium relative to sodium as the reference element are also shown. Reference values were taken from Bowen (1979)

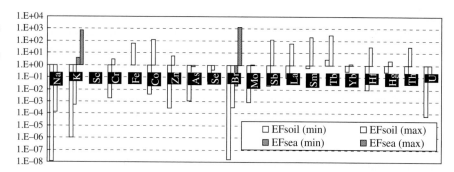

factors (EFs) were calculated. The EF value for an element X was computed according to $EF_X = [X/REF]_{sample} / [X/REF]_{ref.\ sample}$, where REF is Sc when "ref. sample" is a crustal reference value or an aerosol reference value, and REF is Na when "ref. sample" is a marine reference value. Sodium is a primary (cationic) sea-salt tracer. Even if Al is the historical and, arguably, most used element to account for soil-dust inputs or even correct for them (Zoller et al. 1974, Lantzy and Mackenzie 1979, Bargagli et al. 1995), there is no such thing as an established crustal reference (Manoli et al. 2002). Besides, Sc has already been deemed an optimal choice for that purpose (Cao et al. 2002). Values of "ref. sample" were taken from Bowen (1979). For crustal and marine cases, average data are available; for aerosols, data ranges (minima and maxima) are given. Therefore, the latter have two calculated values, using the extremes of the corresponding aerosol ranges. Considering that many EF values would be produced for each element, the criterion was to calculate minimum and maximum concentrations of each element in the whole set of concentrations for that element, and then compute the minimum and the maximum enrichment factors for the same element.

Figure 5 illustrates the results for crustal and marine enrichment factors, respectively referred to Sc and Na as reference elements. Concerning the crustal EFs, there is some depletion of Na, K, As, Se, Br, Mo and U relatively to the reference soil values, which seems inconclusive at best. On the contrary, EFs for K and Br of potential marine origin (relative to sodium, in grey colour) may span several decades. Even if EFs are crude indices, not to be taken strictly at face value, some concentrations of K and Br in the samples can hardly be ascribed to oceanic inputs, and likely are of anthropogenic origin.

The elements Cr and Zn show no significant enrichment, or are depleted relative to their soil-reference values; no enrichment is observed for Yb and Hg either. All these four elements are likely coming from crustal sources. Clearly enriched samples do exist for Fe, Co, Sb, La, Sm, Tb, Hf and Th, though, some of which have been associated with Saharan-dust events – Fe, La, Sm (Almeida et al. 2008, Fialho et al. 2006). When the present EFs relative to soil, with Sc as the crustal reference, are matched against former ones from Bobadela (Almeida et al. 2005), it results that Sb is enriched at both places, although to a lesser extent in Pico summit (one order of magnitude lower), while

Fig. 6 Aerosol enrichment factors relative to scandium as the reference element. Reference ranges were taken from Bowen (1979)

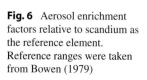
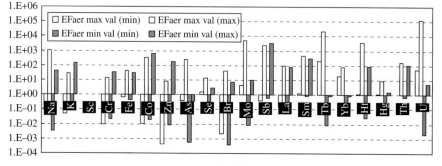

As, Br, and Zn present no enrichment relative to soil in Pico summit (EFs > 100 at Bobadela for all these elements).

Enrichment results relative to aerosol reference compositions, again using Sc as the normalising element, are summarised in Fig. 6. An identical trend is observed whether one takes the minimum or the maximum of the reference range, that is the aerosols from PICO-NARE are enriched in Na, K, Co, Zn, As, Mo, Sb, La, Sm, Tb, Hf, Th and U, when comparing to aerosol-reference data. The magnitude of enrichment is actually strong for all elements but Cr, Fe, Se, Br, Yb and Hg.

3.3 Multivariate Statistical Analysis of Aerosol Data from PICO-NARE

Source categories for PM_{10} constituents were tentatively identified through principal-component, factor analysis (PCA). This was carried out by using the orthogonal transformation method with Varimax rotation (Kaiser 1958, 1959), and retention of principal components whose eigenvalues were greater than unity. Factor loadings indicate the correlation of each pollutant species with each component, and are related to the emission-source composition. PCA has been commonly used in environmental studies (Almeida et al. 2005, Henry 1997, Buhr et al. 1996). Only the species assessed in more than 28% of the samples were considered for PCA. The whole data-set was included in the analysis; missing values were replaced by detection limits for the corresponding elements. Results are listed in Table 3, where four factors (principal components) can be seen to account for about 79% of the total variability in the data.

The first factor appears strongly loaded by traffic- (Br, Zn) and soil-derived elements (Co, La, Mo, Sm), suggesting that the former may have resulted from re-suspension processes (Artíñano et al. 2001). The representation of different source processes in a single factor may be explained by the relatively well-mixed nature of the oceanic troposphere, and by regional phenomena responsible for the aerosol mixing (Heidam 1985). Apart from refuse burning, Zn has been related to vehicle wear (Huang et al. 1994, Sternbeck et al. 2002), while Br, even if prevalent in the marine environment, has long been viewed as an elemental marker for mobile sources after the phasing-out of leaded gasoline (Huang et al. 1994).

The second factor stands for a crustal contribution, given that it shows high loadings for Fe and Sc, typical soil elements. African dust is likely the main contributor to this component, the highest concentrations of which may have been reached at PICO-NARE from October 31 to November 5, 2001 (Fialho et al. 2006). It should be recalled that the Sahara desert alone is responsible for the emission of 50% of the total mass of mineral aerosols to the atmosphere (Pacyna 1998).

The third component, with high loadings for Sb and Hf, suggests oil combustion. The lack of determination of Ni and V in the filters precludes an unequivocally confirmation of that source though. Still, the source was confirmed in the Azores atmosphere by the composition of organic matter in carbonaceous aerosols (Alves et al. 2007). Moreover, the same filters – sampled in Terceira island, at sea level – were recently re-analysed for V and U, and some degree of association is apparent between these elements, of the order of magnitude of the uranium loading in the factor (Freitas et al. 2008).

The fourth factor relates to marine sources. Other than sodium, there is a significant loading for uranium. This might be typical of high altitudes in oceanic environments, since no association of Na and U could be confirmed at sea level (Almeida et al. 2009).

As an alternative approach to the former analysis, and instead of replacing missing values by detection limits, replacements were made in terms of averages of determined data, which may seem more reasonable (see below). The alternative results are listed in Table 4.

The communality increased with an almost total explanation for all elements (only U is below 0.90). There are still four factors, yet a few changes did occur in between. Now, Sc, Fe, La and Sm – elements strongly associated with Saharan events (Almeida et al. 2008) – are grouped together. Antimony and hafnium still make up a likely oil-burning component, but now joined by cobalt, which is not unheard of from previous studies (Almeida et al. 2005, Freitas and Pacheco 2004). The marine factor included Na and U (Table 3). Sodium is now well correlated with Br and Zn (traffic tracers), and the fourth factor implies an association of U and Mo that may stem from uranium mining/milling and nuclear-waste processing (more further).

Table 3 Factor structure for elemental concentrations in aerosol samples from PICO-NARE (whole data-set), after extracting principal components from the correlation matrix and rotating axes to maximum variance for each factor. Missing values were replaced by detection limits. Bold figures indicate factor loadings higher than **0.60**

Variable (Element)	Factor 1	Factor 2	Factor 3	Factor 4	Communality
Br	**0.96**	−0.04	0.13	0.12	0.97
Co	**0.93**	−0.03	0.22	0.17	0.94
Fe	0.08	**0.92**	0.09	0.19	0.78
Hf	0.22	−0.03	**0.80**	−0.08	0.36
La	**0.70**	0.26	−0.11	0.22	0.54
Mo	**0.96**	−0.10	0.07	−0.03	0.92
Na	0.12	−0.03	−0.22	**0.81**	0.39
Sb	−0.01	0.22	**0.85**	0.06	0.50
Sc	0.08	**0.94**	0.10	0.02	0.79
Sm	**0.94**	0.11	0.18	−0.02	0.94
U	0.00	0.28	0.24	**0.69**	0.51
Zn	**0.60**	0.24	−0.08	−0.16	0.34
Eigenvalue	4.5	2.0	1.6	1.3	–
% Variance	39.70	17.81	12.16	8.90	78.57
Probable Source	Crustal Traffic	Crustal (Saharan)	Combustion (Oil)	Marine	–

Table 4 Factor structure for elemental concentrations in aerosol samples from PICO-NARE (whole data-set), after extracting principal components from the correlation matrix and rotating axes to maximum variance for each factor. Missing values were replaced by averages of determined values. Bold figures indicate factor loadings higher than **0.60**

Variable (Element)	Factor 1	Factor 2	Factor 3	Factor 4	Communality
Br	0.01	−0.01	**0.98**	−0.15	0.99
Co	0.30	**0.78**	0.22	0.24	0.87
Fe	**0.98**	0.05	0.16	−0.07	1.00
Hf	−0.10	**0.93**	−0.10	0.00	0.90
La	**0.98**	−0.06	0.16	0.04	1.00
Mo	0.01	0.31	−0.22	**0.83**	0.95
Na	0.09	−0.05	**0.79**	0.31	0.99
Sb	−0.10	**0.96**	−0.09	0.20	0.96
Sc	**0.99**	−0.01	0.02	0.08	1.00
Sm	**0.93**	0.03	−0.04	0.33	0.99
U	0.25	0.11	0.08	**0.87**	0.77
Zn	0.17	0.03	**0.68**	−0.39	0.97
Eigenvalue	4.0	2.5	2.2	1.9	–
% Variance	33.16	20.91	18.55	16.19	88.81
Probable source	Crustal (Saharan)	Combustion (Oil)	Marine Traffic	Uranium (Processes)	–

Generally speaking, this second approach to filling in the missing values with average data not only objectively decreases the unexplained variance, but may conceptually be closer to the actual reason why such missing data did occur in the first place. In fact, close values in blank and loaded samples may yield a null result for an element when accounting for its concentration in blanks, thus making justifiable a replacement by the limit of detection. There are, however, other missing data due to the radioactivity decay along the first week of measurements, and, for these ones, an average of the determined values could be more reasonable.

The results of cluster analysis are shown in Fig. 7, for the situation of filling in missing values with detection limits. Two main clusters are apparent: one with Na only, isolated from the other that contains the remaining elements, which is in itself an indication of the large oceanic influence at Pico summit. The mixed group is split into two parts. One subgroup

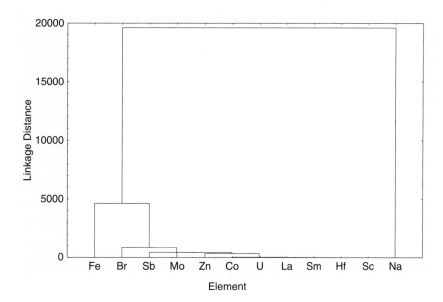

Fig. 7 Results of cluster analysis for elemental concentrations in aerosol samples from PICO-NARE (single linkage; Euclidean distances). Missing values were replaced by detection limits

has Fe isolated, indicating the high concentrations of this element during the Saharan events, whereas the anthropogenic elements Br, Sb, Mo, Zn, Co and U form another subgroup. There is also a crustal subgroup made up of La, Sm, Hf and Sc. Changing the replacement mode (detection limits by averaged data), just like before, would mainly affect the U positioning, putting it in aside the crustal elements La, Sm, Hf and Sc.

Since the error estimates of the present aerosol data broadly vary, and also to avoid the occurrence of negative loadings within the former factor structures (by PCA followed by rotations), a more robust approach may be convenient. Positive matrix factorization (PMF) is a multivariate receptor model that uses an inversely-weighted, least-squares methodology, and an iterative re-weighting algorithm, to provide a factor structure that is essentially non-negative. The factor-analysis solution by PMF is thus constrained so that both matrices of source contributions (scores) and source profiles (loadings) are required to be non-negative. In brief, that means neither negative source contributions to the samples, nor negative species concentrations at the sources – which, by the way, seems perfectly reasonable an assumption on phenomenological grounds.

The PMF technique was developed by Paatero and Tapper (1994) and Paatero (1997) from their own stance on factor analyses as least-squares-fit problems (Paatero and Tapper 1993). Since its

inception – and first application to environmental (precipitation) data by Juntto and Paatero (1994) – PMF has been extensively used in receptor-modelling and source-attribution studies (Anttila et al. 1995, Lee et al. 1999, Paterson et al. 1999, Xie et al. 1999, Chueinta et al. 2000, Polissar et al. 2001, Kim et al. 2003, 2004, Liu et al. 2005, 2006, Buzcu-Guven et al. 2007, Chan et al. 2008, Baumann et al. 2008, Yatkin and Bayram 2008). A comprehensive review of methods for apportioning sources of ambient particulate matter through the PMF algorithm – with an emphasis on procedural decisions and parameter selection – was recently given by Reff et al. (2007).

In practical terms, PMF is a trial-and-error procedure where data uncertainties play a key role for obtaining an optimal – and physically significant – solution with a given number of factors. Error estimates of the data points not only provide the residual sum of squares with (inverse) weights for the optimization (minimization) process of an objective function Q – a goodness-of-fit index – but also enable an adequate handling of missing and below-detection-limit values (Paatero 1997, Polissar et al. 1998). The criteria adopted here to fill in blanks and attribute uncertainties was based in Hopke et al. (1998). For determined values, the concentration stands as it is, with the uncertainty as it is plus one third of the limit of detection (LOD). For values below LOD, the concentration is half of LOD, and the uncertainty is an arithmetic mean of elemental LODs in the sampling

Table 5 Six-factor structure for elemental concentrations in aerosol samples from PICO-NARE (partial data-set: July 2001–July 2002; number of sampling days: 135), by positive matrix factorization. Bold figures indicate factor loadings ≥ **0.30**

Variable (Element)	Factor 1	Factor 2	Factor 3	Factor 4	Factor 5	Factor 6
As	0.00	0.16	0.00	0.00	0.01	**0.36**
Br	0.00	0.17	0.08	0.17	0.07	**0.30**
Co	0.06	0.06	0.00	**0.55**	0.01	0.00
Cr	0.00	0.00	0.00	0.26	0.10	0.19
Fe	**0.66**	0.00	0.00	0.10	0.00	0.16
Hf	0.00	0.00	0.05	**0.33**	0.02	**0.32**
K	0.13	0.00	0.00	0.17	**0.64**	0.00
La	**0.30**	**0.31**	0.19	0.05	0.04	0.01
Mo	0.00	**0.41**	**0.55**	0.01	0.00	0.28
Na	0.12	0.12	0.00	0.00	**0.32**	0.19
Sb	0.00	0.00	0.03	**0.53**	0.17	0.00
Sc	**0.82**	0.00	0.00	0.00	0.00	0.02
Sm	0.22	0.00	**0.67**	0.00	0.05	0.00
U	0.01	**0.82**	0.00	0.00	0.00	0.00
Zn	0.01	0.00	0.05	**0.33**	0.02	**0.32**
Probable source	Crustal (Saharan)	Uranium (Processes)	Crustal	Combustion (Oil)	Marine	Traffic

site, divided by 2, plus one third of LOD. For missing values, the concentration is the geometric mean of the determined values, and the uncertainty is 4 times the geometric mean of the determined values.

Table 5 shows the results of a six-factor PMF analysis upon data from aerosol samples collected between July 2001 and July 2002, accounting for the above criteria. This data-set had 15 chemical elements determined for 135 sampling days. The calculated Q value was 2172, which is fairly close to the {[number of chemical elements]·[number of sampling days]} = 2072 (theoretical, optimized Q). Still, comparing the PMF results to the ones given by PCA (Tables 3 and 4) – which does not include uncertainties, and either fills the missing data with arithmetic means and/or LODs – it may be concluded that they do not actually differ.

3.4 Directional Predominance of Air Masses and Continental Provenance of Airborne Elements at PICO-NARE

The air-mass trajectories for the studied period, as given by back analysis (HYSPLIT model), were associated in order to establish their main directions. The trajectories were requested 5 days backwards, for 100, 500 and 1000 m above the sampling site (AGL), in

an isentropic mode due to the complete oceanic situation of the sampling site. As no differences were found for the three altitudes, the 100 m AGL was selected thereafter. This trajectory height appears to be more than enough to ensure a transport pathway well above the marine boundary layer (MBL), that is within the (lower) free troposphere (FT), an essential attribute to assess long-range impacts of upwind source regions. It should be emphasized that the PICO-NARE station itself (2225 m AMSL) already lies above the regional MBL (Lapina et al. 2006), which is typically less than 1000 m high in summer (Owen et al. 2006, Val Martín et al. 2006). Subtropical MBLs are usually found between 800 and 1600 m AMSL, supporting the lower-stratocumulus cloud layer (Albrecht et al. 1995, Honrath et al. 2004) – and in Pico, more often than not, such cloud-topped, visually-apparent boundary is situated well below the mountain summit. Therefore, despite some possibility of marine upslope flows (Kleissl et al. 2006, 2007), trajectories whose end-point altitudes run at least above the mountain-top height most likely stand for transport phenomena within free tropospheric air.

The trajectories were split for seasonal study into summer – April to September – and winter – October to March. Figure 8 shows the directional predominance for summer and winter, as well as for the whole period.

Fig. 8 Directional prevalence of air masses reaching the Azores at 100 m above the PICO-NARE. **a**: annual; **b**: summer; **c**: winter

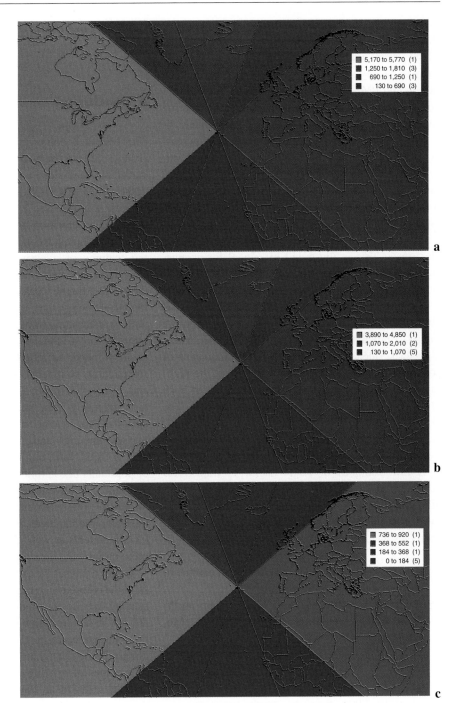

There are no significant differences between the three outputs by MapInfo Professional v7.5. Most air masses originate in North-Central America, followed by Europe, North Africa and Greenland. Less frequent are the ones coming from South America, Central Africa and South Africa. Such pattern is not surprising and likely explains why the area is under the influence of anthropogenic emissions from North America and Europe (Alves et al. 2007).

This is the first study on the concentrations of chemical elements at an altitude of more than 2000 m AMSL, in the middle of the North Atlantic Ocean. A previous study using the same filters dealt with black carbon and PM_{10} concentrations (Fialho et al. 2005). So the next question would be which trajectories bring As, Br, Sb, U, etc, detected in the analysed filters. To answer that, each elemental data-set was put in increasing order of concentrations, and the 20% highest values were selected. The latter corresponded to particular sampling times. Synoptic, 5-day backward trajectories were then computed in isentropic mode, for the highest concentrations of elements entrained by air masses arriving at 100 m above PICO-NARE at those particular times. This is illustrated by Fig. 9 for arsenic, antimony and uranium (drawings by MapInfo Professional v7.5).

Even if the assignment of remote sources is always debatable, arsenic clearly originates from the Canada/United States of America (USA) border, where high-stack smelters are in operation. Antimony is of continental origin: all trajectories that carry the highest concentrations of this element can be traced back to North-Central America or Western Europe. The highest concentrations of uranium come in from well-defined directions: Canada/USA, Caribbean basin and Central-Eastern Europe.

A few examples of 5-day backward trajectories (100, 500 and 1000 m AGL), ending at times when high concentrations of an element (or elements) was (were) observed, are given in Figs. 10 and 11. Such examples include:

October 6, 2001: from the North American continent, across Canada; 25 ng·m^{-3} of Mo.

April 25, 2002: from the European continent, north-south Iberian peninsula; 21.7 ng·m^{-3} of Mo, 50.8 ng·m^{-3} of Sb, 0.8 ng·m^{-3} of La, 0.3 ng·m^{-3} of Sm.

September 2, 2003: from the European continent, Ireland and United Kingdom; 0.3 ng·m^{-3} of Hf.

February 19, 2004: from the North American continent, across USA and Canadian/USA border; 20 ng·m^{-3} of U.

Some of the former values are considerably higher than those obtained almost a decade ago from PM_{10} at the vicinity of an oil-fired power plant in Setúbal, mainland Portugal (Freitas et al. 2008):

La: 0.3 ng·m^{-3}; Sb: 1.8 ng·m^{-3}; Sm: 0.07 ng·m^{-3}; U: 0.4 ng·m^{-3}.

4 Conclusion

The chemical composition of 425 aerosol samples (PM_{10}) collected on quartz-fibre filter tapes by a seven-wavelength aethalometer from 15 July 2001 to 18 April 2004 at the PICO-NARE observatory, Pico island, Azores, was assessed by k_0-standardised, instrumental neutron activation analysis (k_0-INAA). Generally speaking, elemental concentrations do not comply with the features of a clean, remote area, as would be expected for a high-altitude location in the middle of the North Atlantic Ocean. Instead, and for a few determined elements, they may even be seen to compare with similar data from moderately-polluted, urban-industrial areas in mainland Portugal. This is likely due to an influence of air masses originated in the three surrounding continents, which entrain anthropogenic and crustal elements from sources afar, the latter more intensely from North Africa when Saharan-dust episodes happen to occur. Still, the predominance of air masses reaching – and crossing over – the Pico-summit area is from North and Central America. Enrichment of pollutants like antimony, arsenic, bromine, uranium and zinc – relative to crustal/marine (gross) averages or aerosol (reference) data – may thus be assigned to those continental areas. Europe contributes with similar components as well, albeit not that often due to the preferential wind direction.

Factor analyses by extracting principal components from the correlation matrix for the whole data-set – and rotating axes to maximum variance for each factor – did not result in major differences between the factor solutions corresponding to different ways of handling missing data, that is filling in blanks with LODs or averages of existing values. The second approach, though, leads to an almost general enhancement of the proportion of the variables' (elements') variances accounted for by the four-factor structure, except for Co, with a substantial decrease of the total unexplained variance. Uranium appears strongly associated with molybdenum in a compositional profile that could fit a number of U-processing operations (mining, milling, recycling), instead of showing up in a marine profile – with Na, that has been clearly segregated by

Fig. 9 Directional prevalence of air masses with high elemental loads, reaching the Azores at 100 m above the PICO-NARE observatory. **a**: arsenic; **b**: antimony; **c**: uranium

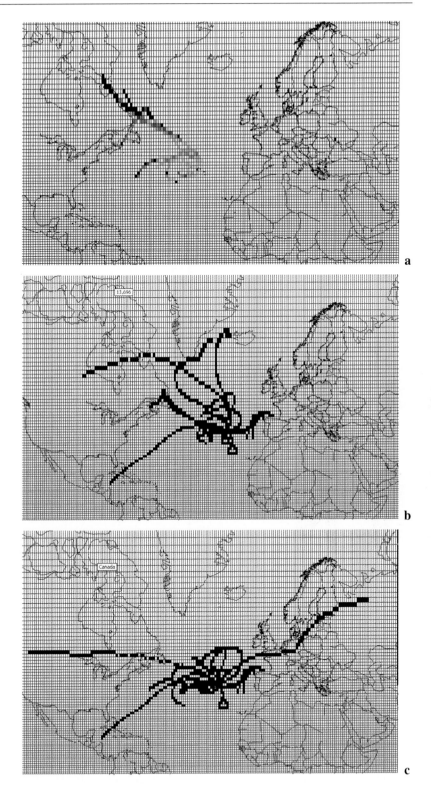

Fig. 10 Five-day synoptic back trajectories arriving at PICO-NARE on October 6, 2001, and April 25, 2002, at three different heights – 100, 500 and 1000 m AGL. Calculations were done in isentropic mode with the HYSPLIT model, via the NOAA ARL READY Website

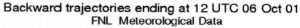

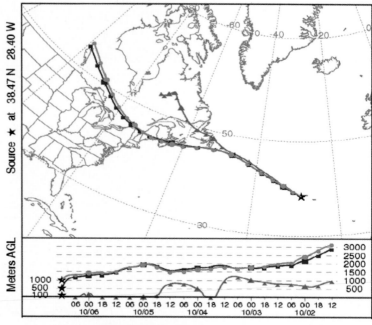

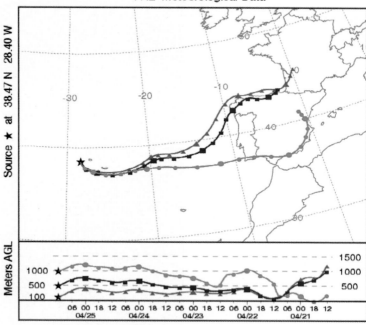

cluster analysis. Other than crustal/marine sources, and notwithstanding minor variations, both PCA analyses point to significant inputs from far-flung, traffic- and combustion-related sources to the Pico atmosphere.

Since uncertainties can be relatively high, yet readily available – 5–15% for the method alone, based on quality control of k_0-INAA through a surrogate reference material (NIST-SRM® 2783), plus individual

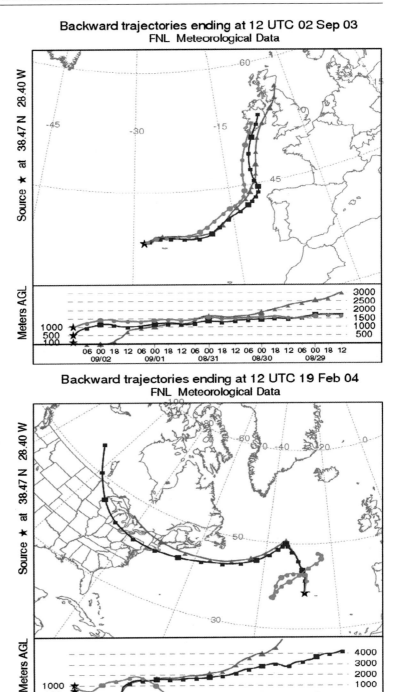

Fig. 11 Five-day synoptic back trajectories arriving at PICO-NARE on September 2, 2003, and February 19, 2004, at three different heights – 100, 500 and 1000 m AGL. Calculations were done in isentropic mode with the HYSPLIT model, via the NOAA ARL READY Website

(elemental) errors in the counting (spectral) statistics – a factor-analysis solution by PMF was sought. An optimal six-factor solution by PMF all but confirms the former PCA results. Again, an U-Mo association is clearly apparent in a factor that fits a compositional profile akin to U-processing operations; the remaining factors stand for probable sources already identified by PCA – soil dusts (atypical and Saharan), marine salts, traffic and oil combustion, both (the latter two) of remote origin.

Acknowledgments The authors gratefully acknowledge the NOAA Air Resources Laboratory (ARL) for the provision of the HYSPLIT transport and dispersion model and/or READY website (http://www.arl.noaa.gov/ready.html) used in this publication. The authors are indebted to Prof. Paulo Fialho and Dr. Filipe Barata (Group of Chemistry and Physics of the Atmosphere, Department of Agricultural Sciences, University of Azores, Portugal), respectively for filter availability and spot identification, and to Prof. Gürdal Tuncel (Department of Environmental Engineering, Middle East Technical University, Ankara, Turkey) for making available the mapping application MapInfo Professional v7.5.

Abbreviations

AGL	above ground level
AMSL	above mean sea level
ARL	Air Resources Laboratory (a part of the USA National Oceanic and Atmospheric Administration – NOAA)
EF	enrichment factor
FT	free troposphere
HYSPLIT (model)	hybrid single-particle lagrangian integrated trajectory (model)
IAEA	International Atomic Energy Agency
ITN	*Instituto Tecnológico e Nuclear* (official Portuguese designation of the Technological and Nuclear Institute)
k_0-**INAA**	k_0-standardised, instrumental neutron activation analysis
LOD	limit of detection
MBL	marine boundary layer
N	number of samples
NIST	National Institute of Standards and Technology (a non-regulatory federal agency within the USA Department of Commerce)
NOAA	National Oceanic and Atmospheric Administration (a federal agency within the USA Department of Commerce)
PCA	principal components analysis
PM$_{2.5}$	airborne particulate matter with an equivalent, mass-median aerodynamic diameter less than 2.5 μm
PM$_{10}$	airborne particulate matter with an equivalent, mass-median aerodynamic diameter of 10 μm or less
PMF	positive matrix factorization
READY	real-time environmental applications and display system (a web-based platform for accessing and displaying meteorological data, and running trajectory and dispersion models on the NOAA ARL web server)
REF	reference element
RPI	*Reactor Português de Investigação* (official Portuguese designation of the Portuguese Research Reactor)
SD	standard deviation
SRM®	standard reference material (from NIST)
USA	United States of America

References

Albrecht BA, Bretherton CS, Johnson D, Scubert WH, Frisch AS (1995) The Atlantic Stratocumulus Transition Experiment – ASTEX. Bull Am Meteorol Soc 76: 889–904

Almeida SM, Pio CA, Freitas MC, Reis MA, Trancoso MA (2005) Source apportionment of fine and coarse particulate matter in a sub-urban area at the Western European Coast. Atmos Environ 39: 3127–3138

Almeida SM, Freitas MC, Reis MA, Pio CA, Trancoso MA (2006a) Combined application of multielement analysis – k_0-INAA and PIXE – and classical techniques for source apportionment in aerosol studies. Nucl Instrum Meth A 564: 752–760

Almeida SM, Pio CA, Freitas MC, Reis MA, Trancoso MA (2006b) Source apportionment of atmospheric urban aerosol based on weekdays/weekend variability: Evaluation of road re-suspended dust contribution. Atmos Environ 40: 2058–2067

Almeida SM, Freitas MC, Pio CA (2008) Neutron activation analysis for identification of African mineral dust transport. J Radioanal Nucl Chem 276: 161–165

Almeida SM, Freitas MC, Repolho C, Dionísio I, Dung HM, Pio CA, Alves C, Caseiro A, Pacheco AMG (2009) Evaluating children exposure to air pollutants for an epidemiological study. J Radioanal Nucl Chem (accepted)

Alves C, Oliveira T, Pio C, Silvestre AJD, Fialho P, Barata F, Legrand M (2007) Characterisation of carbonaceous aerosols from the Azorean Island of Terceira. Atmos Environ 41: 1359–1373

Anttila P, Paatero P, Tapper U, Järvinen O (1995) Source identification of bulk wet deposition in Finland by positive matrix factorization. Atmos Environ 29: 1705–1718

Artíñano B, Querol X, Salvador P, Rodríguez S, Alonso DG, Alastuey A (2001) Assessment of airborne particulate levels in Spain in relation to the new EU-directive. Atmos Environ 35, Suppl 1: 43–53

Bargagli R, Brown DH, Nelli L (1995) Metal biomonitoring with mosses: Procedures for correcting for soil contamination. Environ Pollut 89: 169–175

Baumann K, Jayanty RKM, Flanagan JB (2008) Fine particulate matter source apportionment for the chemical speciation trends network site at Birmingham, Alabama, using positive matrix factorization. J Air Waste Manage 58: 27–44

Blaauw M (2007) Software for single-comparator instrumental neutron activation analysis – The k_0-IAEA program manual for version 3.21. International Atomic Energy Agency, Vienna, Austria, and Delft University of Technology, Delft, The Netherlands. http://www.tudelft.nl/live/binaries/8bba6542-6c38-468d-8f15-b98f0fc23a70/doc/k0IAEAmanual.pdf. Accessed 20 January 2008

Bowen HJM (1979) Environmental chemistry of the elements. Academic Press, London

Buhr MP, Hsu K-J, Liu CM, Liu R, Wei L, Liu Y-C, Kuo Y-S (1996) Trace gas measurements and air mass classification from a ground station in Taiwan during the PEM-West A experiment (1991). J Geophys Res 101: 2025–2035

Buzcu-Guven B, Brown SG, Frankel A, Hafner HR, Roberts PT (2007) Analysis and apportionment of organic carbon and fine particulate matter sources at multiple sites in the Midwestern United States. J Air Waste Manage 57: 606–619

Cao L, Tian W, Ni B, Zhang Y, Wang P (2002) Preliminary study of airborne particulate matter in a Beijing sampling station by instrumental neutron activation analysis. Atmos Environ 36: 1951–1956

Chan Y-C, Cohen DD, Hawas O, Stelcer E, Simpson R, Denison L, Wong N, Hodge M, Comino E, Carswell S (2008) Apportionment of sources of fine and coarse particles in four major Australian cities by positive matrix factorisation. Atmos Environ 42: 374–389

Chazette P, Pelon J, Moulin C, Dulac F, Carrasco I, Guelle W, Bousquet P, Flamant P-H (2001) Lidar and satellite retrieval of dust aerosols over the Azores during SOFIA/ASTEX. Atmos Environ 35: 4297–4304

Chueinta W, Hopke PK, Paatero P (2000) Investigation of sources of atmospheric aerosol at urban and suburban residential areas in Thailand by positive matrix factorization. Atmos Environ 34: 3319–3329

De Corte F (1987) The k_0-standardization method – A move to the optimization of neutron activation analysis (Aggrégé Thesis). Institute for Nuclear Sciences, University of Gent, Gent

De Corte F (2001) The standardization of standardless NAA. J Radioanal Nucl Chem 248: 13–20

Desboeufs KV, Cautenet G (2005) Transport and mixing zone of desert dust and sulphate over Tropical Africa and the Atlantic Ocean region. Atmos Chem Phys Discuss 5: 5615–5644

Draxler RR, Rolph GD (2003) HYSPLIT (HYbrid Single-Particle Lagrangian Integrated Trajectory) Model. NOAA Air Resources Laboratory, Silver Spring, MD; access via NOAA ARL READY Website. http://www.arl.noaa.gov/ready/hysplit4.html. Accessed 24 January 2008

Erdtmann G, Petri H (1986) Nuclear activation analysis: Fundamentals and techniques. In: Elving PJ, Krivan V, Kolthoff IM (eds) Treatise on analytical chemistry, Part I – Theory and practice (Volume 14, Section K), 2nd edn. Wiley Interscience, New York

Fernandes RMS, Bastos L, Miranda JM, Lourenço N, Ambrosius BAC, Noomen R, Simons W (2006) Defining the plate boundaries in the Azores region. J Volcanol Geoth Res 156: 1–9

Fialho P, Hansen ADA, Honrath RE (2005) Absorption coefficients by aerosols in remote areas: A new approach to decouple dust and black carbon absorption coefficients using seven-wavelength Aethalometer data. J Aerosol Sci 36: 267–282

Fialho P, Freitas MC, Barata F, Vieira B, Hansen ADA, Honrath RE (2006) The Aethalometer calibration and determination of iron concentration in dust aerosols. J Aerosol Sci 37: 1497–1506

Flower MFJ, Schmincke H-U, Bowman H (1976) Rare earth and other trace elements in historic azorean lavas. J Volcanol Geoth Res 1: 127–147

Freitas MC, Martinho E (1989a) Neutron activation analysis of reference materials by the k_0-standardization and relative methods. Anal Chim Acta 219: 317–322

Freitas MC, Martinho E (1989b) Accuracy and precision in instrumental neutron activation analysis of reference materials and lake sediments. Anal Chim Acta 223: 287–292

Freitas MC (1993) The development of k_0-standardized neutron activation analysis with counting using a low energy photon detector (PhD Thesis). Institute for Nuclear Sciences, University of Gent, Gent

Freitas MC, Pacheco AMG (2004) Bioaccumulation of cobalt in *Parmelia sulcata*. J Atmos Chem 49: 67–82

Freitas MC, Pacheco AMG, Dionísio I, Sarmento S, Baptista MS, Vasconcelos MTSD, Cabral JP (2006) Multianalytical determination of trace elements in atmospheric biomonitors by k_0-INAA, ICP-MS and AAS. Nucl Instrum Meth A 564: 733–742

Freitas MC, Dionísio I, Fialho P, Barata F (2007) Aerosol chemical elemental mass concentration at lower free troposphere. Nucl Instrum Meth A 579: 507–509

Freitas MC, Marques AP, Reis MA, Farinha MM (2008) Atmospheric dispersion of pollutants in Sado estuary (Portugal) using biomonitors. Int J Environ Pollut 32: 434–455

Heidam NZ (1985) Crustal enrichments in the Arctic aerosol. Atmos Environ 19: 2083–2097

Henry RC (1997) History and fundamentals of multivariate air quality receptor models. Chemometr Intell Lab 37: 37–42

Honrath RE, Fialho P (2001) The Azores Islands: A unique location for ground-based measurements in the MBL and FT of the central North Atlantic. IGACtivities Newsletter 24: 20–21

Honrath RE, Owen RC, Martín MV, Reid JS, Lapina K, Fialho P, Dziobak MP, Kleissl J, Westphal DL (2004) Regional and hemispheric impacts of anthropogenic and biomass burning emissions on summertime CO and O3 in the North Atlantic lower free troposphere. J Geophys Res 109: D24310 (17 pp)

Hopke PK, Paatero P, Jia H, Ross RT, Harshman RA (1998) Three-way (PARAFAC) factor analysis: Examination and comparison of alternative computational methods as applied to ill-conditioned data. Chemometr Intell Lab 43: 25–42

Huang X, Olmez I, Aras NK, Gordon GE (1994) Emissions of trace elements from motor vehicles: Potential marker elements and source composition profile. Atmos Environ 28: 1385–1391

Juntto S, Paatero P (1994) Analysis of daily precipitation data by positive matrix factorization. Environmetrics 5: 127–144

Kaiser HF (1958) The varimax criterion for analytic rotation in factor analysis. Psychometrika 23: 187–200

Kaiser HF (1959) Computer program for varimax rotation in factor analysis. Educ Psychol Meas 19: 413–420

Kim E, Hopke PK, Edgerton ES (2003) Source identification of Atlanta aerosol by positive matrix factorization. J Air Waste Manage 53: 731–739

Kim E, Hopke PK, Edgerton ES (2004) Improving source identification of Atlanta aerosol using temperature resolved carbon fractions in positive matrix factorization. Atmos Environ 38: 3349–3362

Kleissl J, Honrath RE, Henriques DV (2006) Analysis and application of Sheppard's airflow model to predict mechanical orographic lifting and the occurrence of mountain clouds. J Appl Meteorol Clim 45: 1376–1387

Kleissl J, Honrath RE, Dziobak MP, Tanner D, Val Martín M, Owen RC, Helmig D (2007) Occurrence of upslope flows at the Pico mountaintop observatory: A case study of orographic flows on a small, volcanic island. J Geophys Res 112: D10S35 (16 pp)

Lantzy RJ, Mackenzie FT (1979) Atmospheric trace metals: Global cycles and assessment of man's impact. Geochim Cosmochim Acta 43: 511–525

Lapina K, Honrath RE, Owen RC, Val Martín M, Pfister G (2006) Evidence of significant large-scale impacts of boreal fires on ozone levels in the midlatitude Northern Hemisphere free troposphere. Geophys Res Lett 33: L10815 (4 pp)

Lee E, Chan CK, Paatero P (1999) Application of positive matrix factorization in source apportionment of particulate pollutants in Hong Kong. Atmos Environ 33: 3201–3212

Li X, Maring H, Savoie D, Voss K, Prospero JM (1996) Dominance of mineral dust in aerosol light-scattering in the North Atlantic trade winds. Nature 380: 416–419

Li-Jones X, Prospero JM (1998) Variations in the size distribution of non-sea-salt sulfate aerosol in the marine boundary layer at Barbados: Impact of African dust. J Geophys Res 103: 16073–16084

Liu W, Wang Y, Russell A, Edgerton ES (2005) Atmospheric aerosol over two urban-rural pairs in the southeastern United States: Chemical composition and possible sources. Atmos Environ 39: 4453–4470

Liu W, Wang Y, Russell A, Edgerton ES (2006) Enhanced source identification of southeast aerosols using temperature-resolved carbon fractions and gas phase components. Atmos Environ 40 (Suppl 2): 445–466

Manoli E, Voutsa D, Samara C (2002) Chemical characterization and source identification/apportionment of fine and coarse air particles in Thessaloniki, Greece. Atmos Environ 36: 949–961

Mazzera DM, Lowenthal DH, Chow JC, Watson JG (2001) Sources of PM_{10} and sulfate aerosol at McMurdo station, Antarctica. Chemosphere 45: 347–356

Mishra VK, Kim K-H, Hong S, Lee K (2004) Aerosol composition and its sources at the King Sejong Station, Antarctic peninsula. Atmos Environ 38: 4069–4084

Owen RC, Cooper OR, Stohl A, Honrath RE (2006) An analysis of the mechanisms of North American pollutant transport to the central North Atlantic lower free troposphere. J Geophys Res 111: D23S58 (14 pp)

Paatero P, Tapper U (1993) Analysis of different modes of factor analysis as least squares fit problems. Chemometr Intell Lab 18: 183–194

Paatero P, Tapper U (1994) Positive matrix factorization: A non-negative factor model with optimal utilization of error estimates of data values. Environmetrics 5: 111–126

Paatero P (1997) Least squares formulation of robust non-negative factor analysis. Chemometr Intell Lab 37: 23–35

Pacheco AMG, Freitas MC, Ventura MG, Dionísio I, Ermakova E (2006) Chemical elements in common vegetable components of Portuguese diets, determined by k_0-INAA. Nucl Instrum Meth A 564: 721–728

Pacheco AMG, Freitas MC (2007) Trace-element enrichment in epiphytic lichens and tree bark at Pico island, Azores, Portugal. In: Proceedings of the A&WMA's 100th Annual Conference and Exhibition (ACE 2007; Pittsburgh PA, June 26–29, 2007; ISBN: 978-092-32049-5-2). Air & Waste Management Association, Pittsburgh

Pacyna JM (1998) Source inventories for atmospheric trace metals. In: Harrison RM, Van Grieken R (eds) Atmospheric. particles – IUPAC series on analytical and physical chemistry of environmental systems (Volume 5). Wiley, Chichester

Paterson KG, Sagady JL, Hooper DL, Bertman SB, Carroll MA, Shepson PB (1999) Analysis of air quality data using positive matrix factorization. Environ Sci Technol 33: 635–641

Perry KD, Cahill TA, Eldred RA, Dutcher DD, Gill TE (1997) Long-range transport of North African dust to the eastern United States. J Geophys Res 102: 11225–11238

Polissar AV, Paatero P, Hopke PK, Malm WC, Sisler JF (1998) Atmospheric aerosol over Alaska 2. Elemental composition and sources. J Geophys Res 103: 19045–19057

Polissar AV, Hopke PK, Poirot RL (2001) Atmospheric aerosol over Vermont: Chemical composition and sources. Environ Sci Technol 35: 4604–4621

Prospero JM (1999a) Long-range transport of mineral dust in the global atmosphere: Impact of African dust on the environment of the southeastern United States. Proc Natl Acad Sci USA 96: 3396–3403

Prospero JM (1999b) Long-term measurements of the transport of African mineral dust to the southeastern United States: Implications for regional air quality. J Geophys Res 104: 15917–15927

Prospero JM, Nees RT (1986) Impact of the North African drought and El Niño on mineral dust in the Barbados trade winds. Nature 320: 735–738

Prospero JM, Olmez I, Ames M (2001) Al and Fe in PM 2.5 and PM 10 suspended particles in south-central Florida: The impact of the long range transport of African mineral dust. Water Air Soil Poll 125: 291–317

Prospero JM, Lamb PJ (2003) African droughts and dust transport to the Caribbean: Climate change implications. Science 302: 1024–1027

Rahn KA, Huang SA (1999) A graphical technique for distinguishing soil and atmospheric deposition in biomonitors from the plant material. Sci Total Environ 232: 79–104

Reff A, Eberly SI, Bhave PV (2007) Receptor modeling of ambient particulate matter data using positive matrix factorization: Review of existing methods. J Air Waste Manage 57: 146–154

Reis MA, Oliveira OR, Alves LC, Rita EMC, Rodrigues F, Fialho P, Pio CA, Freitas MC, Soares JC (2002) Comparison of continental Portugal and Azores Islands aerosol during a Sahara dust storm. Nucl Instrum Meth B 189: 272–278

Rolph GD (2003) Real-time Environmental Applications and Display sYstem (READY). NOAA Air Resources Laboratory, Silver Spring, MD. http://www.arl.noaa.gov/ready/hysplit4.html. Accessed 24 January 2008

Searle R (1980) Tectonic pattern of the Azores spreading centre and triple junction. Earth Planet Sci Lett 51: 415–434

Silveira G, Stutzmann E, Davaille A, Montagner J-P, Mendes-Victor L, Sebai A (2006) Azores hotspot signature in the upper mantle. J Volcanol Geoth Res 156: 23–34

Sternbeck J, Sjödin AÅ, Andréasson K (2002) Metal emissions from road traffic and the influence of resuspension – results from two tunnel studies. Atmos Environ 36: 4735–4744

Val Martín M, Honrath RE, Owen RC, Pfister G, Fialho P, Barata F (2006) Significant enhancements of nitrogen oxides, black carbon, and ozone in the North Atlantic lower free troposphere resulting from North American boreal wildfires. J Geophys Res 111: D23S60 (17 pp)

Vieira BJ, Biegalski SR, Freitas MC, Landsberger S (2006) Atmospheric trace metal characterization in industrial area of Lisbon, Portugal. J Radioanal Nucl Chem 270: 55–62

Weingartner E, Saathoff H, Schnaiter M, Streit N, Bitnar B, Baltensperger U (2003) Absorption of light by soot particles: Determination of the absorption coefficient by means of aethalometers. J Aerosol Sci 34: 1445–1463

Xie Y-L, Hopke PK, Paatero P, Barrie LA, Li S-M (1999) Identification of source nature and seasonal variations of Arctic aerosol by positive matrix factorization. J Atmos Sci 56: 249–260

Yang T, Shen Y, van der Lee S, Solomon SC, Hung S-H (2006) Upper mantle structure beneath the Azores hotspot from finite-frequency seismic tomography. Earth Planet Sci Lett 250: 11–26

Yatkin S, Bayram A (2008) Source apportionment of PM_{10} and $PM_{2.5}$ using positive matrix factorization and chemical mass balance in Izmir, Turkey. Sci Total Environ 390: 109–123

Zoller WH, Gladney ES, Duce RA (1974) Atmospheric concentrations and sources of trace metals at the South Pole. Science 183: 199–201

Part II
Contaminants Control Process Monitoring

Removal of Selected Organic Micropollutants from WWTP Effluent with Powdered Activated Carbon and Retention by Nanofiltration

Kai Lehnberg, Lubomira Kovalova, Christian Kazner, Thomas Wintgens, Thomas Schettgen, Thomas Melin, Juliane Hollender and Wolfgang Dott

Abstract The increasing demand of potable water as well as process water is leading to scarcity of readily available water resources in many regions of the world. In order to preserve fresh water resources waste water reclamation can serve as a promising but technically challenging alternative to conventional water sources. Besides pathogens, organic trace pollutants cause major concerns in reclamation schemes. In this project, the removal of four micropollutants, more specifically bisphenol A, 17α-ethinylestradiol, 5-fluorouracil and cytarabine, from municipal wastewater treatment plant (WWTP) effluent was studied in laboratory scale experiments and in spiking experiments in a pilot plant combining sorption to powdered activated carbon (PAC) and retention by capillary nanofiltration (NF). Laboratory scale experiments evaluating adsorption isotherms and kinetics showed decreased adsorption capacities but increased affinities in the presence of natural organic matter. Spiking experiments showed that the combination of adsorption to PAC and NF retains micropollutants significantly better than direct NF. Retention of lipophilic micropollutants was increased up to 99.9%. Therefore, the PAC/NF process can be considered as an effective way to treat WWTP effluent for reuse.

Keywords Micropollutants · Cytostatics · Endocrine disrupting compounds · Water cycle · Nanofiltration · Adsorption · Powdered activated carbon

W. Dott (✉)
RWTH Aachen University, Institute of Hygiene and Environmental Medicine, Pauwelsstr. 30, D-52074 Aachen, Germany
e-mail: wolfgang.dott@post.rwth-aachen.de

1 Introduction

A major influence affecting the ecosystems worldwide is the introduction of artificial and anthropogenic substances which, if not degraded, pollute the environment. Pollution includes a broad variety of organic and inorganic contaminants but also pathogens such as viruses, bacteria or protozoa. When released to water, soil or air in amounts which cannot be degraded immediately, they potentially affect human health and the quality of ecosystems. Since the 1970s, widespread water pollution from soil contamination in industrial areas and disposal of untreated municipal and industrial wastewaters are considered as a threat for human health and the quality and fitness of ecosystems men rely on. Nitrate, pesticides, poly aromatic hydrocarbons, heavy metals, and emerging pollutants such as pharmaceuticals are found in surface and ground water even in remote places. Therefore, after ensuring the removal of pathogens, odorous and bad-tasting components, water treatment focuses nowadays increasingly on the removal of toxic agricultural, municipal or industrial contaminants. As the demand for potable water increases worldwide while fresh water resources are continuously decreasing, procedures and processes for preserving water resources by removing or preventing water pollution become ever more important. In a growing number of regions, natural water resources are depleting or cannot cope with the fresh water demand. Therefore, replenishment of natural sources and access to alternative sources for potable and process water are key features of modern water management strategies.

Y.J. Kim et al. (eds.), *Atmospheric and Biological Environmental Monitoring*, DOI 10.1007/978-1-4020-9674-7_10, © Springer Science+Business Media B.V. 2009

1.1 Micropollutants

With the availability of highly sensitive analytical methods for the detection of organic and inorganic pollutants in environmental samples the number of findings of anthropogenic compounds in waste water, surface water, groundwater and potable water increased and is still increasing (Ternes, 2007). With pre-concentration steps such as solid phase extraction, and high performance liquid chromatography (HPLC) or gas chromatography combined with mass spectrometer like triple quadrupole mass spectrometer or orbitrap mass spectrometer, detection limits in the range of picograms per liter for organic compounds in environmental samples are possible: plastic softeners are found in the atmosphere (Kamiura et al., 1997), flame retardants in the Canadian arctic (Alaee and Wenning, 2002), pharmaceuticals in potable water (Webb et al., 2003). Due to their low concentrations in the environment, these pollutants are referred to as micropollutants.

The findings of anthropogenic compounds nearly ubiquitous in the environment raise the question if they can affect human health or the quality of ecosystems (Dorne et al., 2007). For the risk assessment of these substances, it is a key issue to obtain information about their environmental behavior. Determination of the micropollutant concentrations in aquatic systems and the knowledge of their degradation kinetics – or their persistence – lead to predictions about their distribution and their life time in the environment. Sorption characteristics indicate path ways to the environment and the possibility to retain or immobilize compounds (Fig. 1). On the other hand, it has to be answered whether, and down to what concentration levels, micropollutants could affect the quality of aquatic ecosystems or human health need to be answered. For example, persistent organic pollutants (POPs) accumulate, alike heavy metals, in the food chain and are found in toxic amounts in the top consumers of the food chain. Some micropollutants are even without accumulation effective in very low dosages, due to of their continuous introduction into the environment (Petrovic et al., 2003). To predict an impact on human health or the quality of ecosystems, and to simplify risk assessment, the ratio between the measured or predicted environmental concentration (MEC, e.g. PEC) and the predicted non-effective

concentration (PNEC), PEC/PNEC, is calculated to indicate whether a certain substance should be considered harmful for ecosystems or not.

An important group within the micropollutants are endocrine disrupting compounds, which are "…defined as an exogenous agent that interferes with the synthesis, secretion, transport, binding, action or elimination of natural hormones in the body that are responsible for the maintenance of homeostasis, reproduction, development, and/or behavior." (EPA International Workshop on Endocrine Disruptors, 1997). Since 1960, effects of organic compounds on reproduction of vertebrates are known: dichlorodiphenyltrichloroethane (DDT) and dichlorodiphenyldichloroethylene (DDE) lead to thinning of egg shells, polychlorinated biphenyls (PCB) showed estrogenic behavior on marine mammals, tributyltin (TBT) lead to masculinization of snails, dicofol, an organo-chlorine pesticide, impaired the population of alligators in lake Apokpa, Florida USA. Since 1992 waste water treatment effluent is known to have estrogenic effects on fish. Endocrine disrupting compounds in waste water can be natural hormones like β-estradiol, synthetic hormones like 17α-ethinylestradiol, pesticides (see above), xenoestrogens like nonylphenol or bisphenol A or even phytoestrogens like coumestans and isoflavones. The increased occurrence of anthropogenic EDCs in the environment raised the question if these compounds have an effect on human health and this question is discussed quite controversially: Some scientists see a correlation between decreased fertility and sperm production, hyperactivity in children or testicular cancer and continuous intake of very low dosages of EDCs, others claim that the effects are diminished by the natural hormone levels in both male and female bodies but sound scientific prove of these hypotheses is quite difficult (Gore et al., 2006; Welshons et al., 2003, 2006; Sharpe and Skakkebaek, 1993). Unquestionable though is the fact that environmentally effective amounts of EDCs pass water treatment and are taken up by humans (Webb et al., 2003; Heberer et al., 2002; Drewes et al., 2002).

The mass balances for 17α-ethinylestradiol from Adler et al. (2001) and Andersen et al. (2003) in waste water treatment showed that due to its nonpolar characteristics, 17α-ethinylestradiol and its glucuronides are strongly adsorbed to particles and activated sludge. Most, but not all, of the 17α-ethinylestradiol is cycling

Fig. 1 Aquatic pathways of micropollutants through the environment to potable water and groceries. For the risk assessment of micropollutants it is a key feature to obtain information about their behaviour in environmental systems

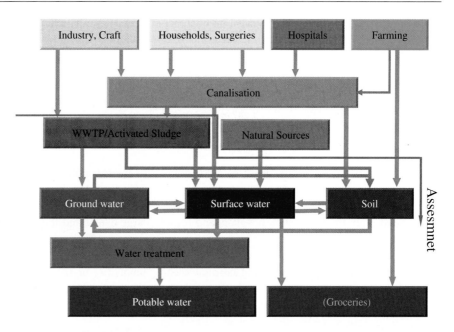

in the waste water treatment plant with the internal circulation and the return sludge, until it is chemically or biologically degraded (Fig. 2). Nevertheless, in 2001, every WWTP in Germany was dumping 70 mg of 17α-ethinylestradiol in average to the environment every day – enough to pollute 87500 m³ with an environmental effective concentration of 0.8 ng/L (Brown et al., 2007). Bisphenol A is found in concentrations of 3 μg/L in municipal WWTP influent and 0.275 μg/L in the WWTP effluent (Koerner et al., 2000). Tan et al. 2007 specified the concentrations for a WWTP influent in South East Queensland, Australia, with 0.14 μg/L

in the water, 0.271 μg/g in the solids and sludge, and 0.086 μg/L in the effluent water (Table 1). But also other groups of micropollutants like pharmaceutically active compounds (PhACs) including antibiotics or cytostatics are known or discussed to occur in effective concentrations in the environment. Antibiotic residues in the aquatic environment are suspected to lead to an increased resistance of pathogenic bacteria like Klebsiella-strains (Stelzer et al., 1985) or coliform bacteria (Cooke, 1976). Coliform bacteria with antibiotic resistance were also found in drinking water (Pandey and Musarrat, 1993). Due to their high pharmaceutical

Fig. 2 Mass balances for 17α-ethinylestradiol in a waste water treatment plant. Most of the nonpolar 17α-ethinylestradiol is adsorbed to the activated sludge and circles with the return sludge in the WWTP. Still, environmental effective concentrations leave the WWTP through the effluent. Graphic based on Andersen et al. 2003 and Adler et al. 2001

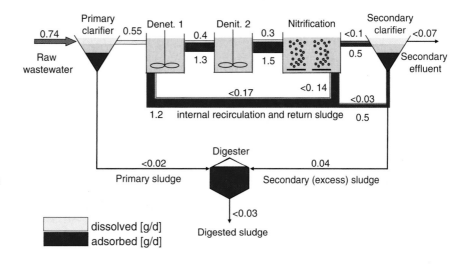

Table 1 Concentrations of 17α-ethinylestradiol and bisphenol A detected in WWTP effluent

Concentration of 17α-Ethinylestradiol	Concentration of Bisphenol A	
< 1 − 15 ng/L (Median 1 ng/L)	–	Ternes et al., 1999; Adler et al., 2001
0,3–0,5 ng/L	1 − 1000 ng/L (Median 83 ng/L)	Spengler et al., 1999
< 1 − 35 ng/L (Median < 1 ng/L)	< 0,1 − 700 ng/L (Median 70 ng/L)	Wenzel et al., 1998
–	30 − 2500 ng/L (Median 490 ng/L)	Hegemann et al., 2002, Bilitewski et al., 2002
–	247 ng/L	Koerner et al., 2000
–	86 ng/L	Tan et al., 2007

potential, cytostatics, used for cancer treatment, could have mutagenic, genotoxic, embryo toxic or teratogenic effects in very low dosages.

Cytostatics are mainly used in hospitals and found in hospital waste water in concentrations up to several micrograms per liter (Steger-Hartmann et al., 1997). Waste water treatment plants cleaning hospital waste water are known to contain cytostatics in their influent and effluent (Steger-Hartmann et al., 1996; Kümmerer et al., 1997; Ternes, 1998; Mahnik et al., 2007). In 2008, Johnson et al. (2008) predicted a concentration of 50 ng/L of 5-fluorouracil in the Aire and Calder catchment's area in North Yorkshire, United Kingdom, under low flow conditions.

To monitor organic micropollutants and predict their behavior in the environment, it is necessary, in respect to reasonable workloads and the size of samples, to determine certain substances, which indicate the distribution and behavior of trace compounds with similar physical, chemical or biological properties. These substances can then be monitored in the environment and, by modeling their behavior, predictions for other compounds can be made. Fig. 3 shows a selection

of micropollutants which could be used as indicators. These substances were selected because they are persistent against chemical or biological degradation to a large extent and they affect biological systems in low dosages. These selected micropollutants are mainly introduced to the environment through hospital and municipal or industrial wastewater but also through landfill leaching, livestock breeding and aqua farming.

1.2 Waste Water Treatment and Reuse

A major concern worldwide is the availability of clean potable water for a growing population in times of increasing scarcity of feasible water resources and the contamination of freshwater systems with anthropogenic organic and inorganic compounds (Kolpin et al., 2002; Schwarzenbach et al., 2006; Snyder et al., 2007; Ternes and Joss, 2006). Anthropogenic (micro-) pollutants found in aquatic environments are mostly introduced to the sewerage by industry and craft, households and surgeries, hospitals and farming. Pharmaceuticals and personal care products, are often

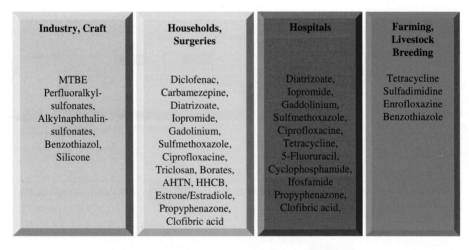

Fig. 3 Indicator substances grouped by sources (Body of experts, DECHEMA 2005)

Fig. 4 Scheme of the combination of adsorption to powdered activated carbon and retention by nanofiltration (PAC/NF)

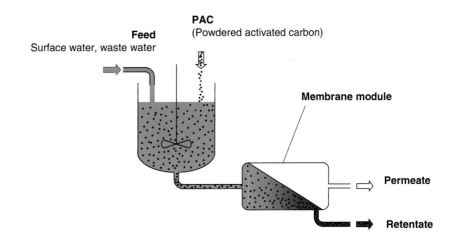

optimized and synthesized for enhanced stability and effectiveness, and restrictions for their use are difficult to establish. Once in the waste water system, they can pass the waste water treatment due to their persistence; and enter surface water, soil, ground water or potable water. To protect water resources for the use as process water and the production of potable water, waste water treatment is challenged to retain these substances. A possible way to preserve water resources is to use "polished" waste water treatment plant (WWTP) effluent as process water or for replenishment of ground water and surface water resources. A high quality of the water intended to be used for replenishment must be ensured, to meet high safety standards for employees who come into contact with process water and because natural and artificial barriers for pollutants may not be efficient enough to remove a growing number of POPs in the production of potable water (Vieno et al., 2007). Effluents from WWTP can be polished in different ways: degradation, filtration or adsorption of (micro-) pollutants. Degradation is performed either with ozone dosage or irradiation with ultra violet radiation (Snyder et al., 2006). The resulting products are mostly less toxic and better degradable (Goi et al., 2004); however in some cases degradation products are even more toxic than their educts (von Gunten, 2003). Filtration processes are common throughout the world but face membrane fouling which leads to the need of extensive pretreatments (Wintgens et al., 2005). Adsorption processes show good removal rates for nonpolar substances yet polar compounds are not retained efficiently. WWTP effluent containing natural organic matter show reduced adsorption capacities for small organic compounds due to blocked binding sites and pores.

In the project "Removal of Selected Organic Micropollutants from WWTP Effluent with Powdered Activated Carbon and Retention by Nanofiltration" funded by the Otto-von-Guericke Foundation, we evaluated the possibility to retain micropollutants with adsorption to powered carbon combined with retention by nanofiltration (PAC/NF, Fig. 4).

In laboratory scale experiments, adsorption isotherms and adsorption kinetics were evaluated. The gained knowledge helps to understand sorption processes of micropollutants in environmental systems and leads to an optimization of the PAC/NF process. Subsequent to the laboratory scale experiments, influent of a pilot plant situated at the waste water treatment plant in Aachen Soers, Germany was spiked with selected micropollutants and monitored for their removal. PAC/NF was already successfully tested for the treatment of severely contaminated waste water (Meier et al., 2002) and for waste water reclamation (Meier and Melin, 2005). The PAC/NF process showed reduced membrane fouling, a better permeate quality and in operation with very dense membranes, like reverse osmosis, a reduced osmotic pressure.

2 Material and Methods

2.1 Selected Micropollutants and Carbons

To evaluate the removal of micropollutants, four organic compounds with different physical, chemical and biological properties were selected. Two endocrine disrupting compounds, bisphenol A (BPA) and 17α-ethinylestradiol (EE2) as well as two cytostatics,

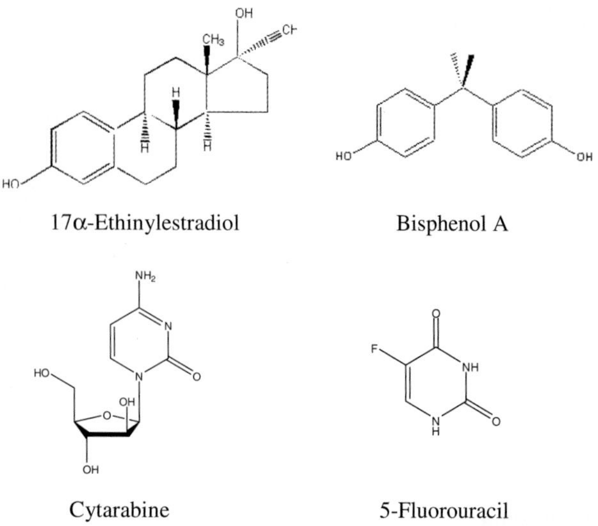

Fig. 5 Structures of
investigated substances

17α-Ethinylestradiol   Bisphenol A

Cytarabine   5-Fluorouracil

5-fluorouracil (5-Fu) and cytarabine (CytR) were
selected for laboratory scale and pilot plant experiments (Fig. 5). Bisphenol A is used as softener in
plastic products, as monomer for tin coatings, for the
production of polycarbonate e.g. for CDs and DVDs
and for thermal paper. 17α-ethinylestradiol is an active
compound of contraceptive pills and is also used for
the treatment of menopausal symptoms. 5-fluorouracil
and cytarabine are antimetabolites which mainly
inhibit enzymes necessary for the synthesis of the
DNA. They are mainly applied in hospitals for treatment of colorectal, pancreatic and hematological cancer. With a consumption of 1,629 kg (5-fluorouracil)
and 78.5 kg (cytarabine) in the year 2001, they belong
to the most prescribed cytostatics in Germany.

The molecular weights of bisphenol A, 17α-
ethinylestradiol and cytarabine are above the molecular weight cut-off (MWCO) of the nanofiltration membrane (Norit X-Flow 1.5 NF50 M10) of
200 Dalton. 5-fluorouracil has a molecular weight
below the MWCO (Table 2). Bisphenol A and

Table 3 Tested powdered carbons, information provided by
RWE Power and Norit

	Lignite coke dust	SAE Super
Manufacturer	RWE Power	Norit
Raw material	Char coal	Diverse
Inner surface, m^2/g	300	1300
D_{50}, μm	24	15
Price, €/t	350	1200

17α-ethinylestradiol are lipophilic with n-octanol-
water partition coefficients (log K_{OW}) of 3.2 and
4.1, respectively (Danielsson and Zhang, 1996).
The cytostatics are hydrophilic with log K_{OW} of
−0.9 for 5-fluorouracil and −2.2 for cytarabine.
The K_{OW} correlates with the association between
the compound and a solid surface (Byrns, 2001)
and is therefore used to predict sorption to solids
(Fuerhacker et al., 2001). The two different commercially available adsorbents were used for the laboratory
scale and pilot plant experiments: lignite coke dust provided by RWE Power and powdered activated carbon,
Norit SAE Super (Table 3).

Table 2 Physical-chemical properties of the four micropollutants

	17α-Ethinylestradiol	Bisphenol A	Cytarabine	5-Fluorouracil
MW	296.4 [1]	228.28 [1]	243.22 [3]	130.08 [3]
pK_a	10.2 [2]	9.7 [2]	4.4 [2]	8.0 [2]
Log K_{OW}	4.1 [1]	3.2 [1]	−2.2 [4]	−0.9 [4]
K_{oc}	6830 [2]	1750 [2]	2 [2]	9 [2]

MW: molecular weight, pK_a: acid dissociation constant (all values calculated), Log K_{OW}: n-octanol-water partition coefficient, K_{oc}:
water-organic-carbon partition coefficient. [1] European Commission DG ENV-report 6/2000, [2] calculated with Advanced Chemistry Development (ACD/Labs) Software V8.14 for Solaris, [3] EPI Suite, US EPA 2008, [4] Hansch, et al., 1995

2.2 Analytical Methods

Two analytical methods were employed for detection and quantification of target micropollutants in laboratory scale experiments: one method for the nonpolar endocrine disrupting compounds bisphenol A and 17α-ethinylestradiol, and another method for the polar cytostatics (Kovalova, 2008). Separation was performed with high performance liquid chromatography (HP1100, Agilent) on a reverse phase column (HyPurity C18, Thermo Fischer) for EDCs and a HILIC column (ZIC-HILIC, SeQuant) for cytostatics. Triple quadrupole mass spectrometer (API 3000, Applied Biosystems) was employed for detection of cytostatics, while EDCs were detected by a fluorescence detector (FLD G 1321 A, Agilent). For the pilot plant experiments, where the concentration range of selected micropollutants was two orders of magnitude lower comparing to lab scale tests, solid phase extractions (SPE) was used for sample preconcentration and clean-up prior to analysis. Oasis HLB cartridges were used for the EDCs, cytostatics were concentrated on ENV+. Samples were concentrated by a factor of 1000 for bisphenol A and 17α-ethinylestradiol, and by a factor of 100 for 5-fluorouracil and cytarabine. HPLC separation, as described for laboratory scale experiments, was followed by detection by tandem MS in case of all four analytes. Detection limits with pre-concentration in WWTP effluent were 2.0, 2.8, 4.6 and 0.9 ng/L for bisphenol A, 17α-ethinylestradiol, 5-fluorouracil and cytarabine respectively. For all samples analyzed by mass spectrometry, ^{13}C or ^{15}N labeled internal standards were used for quantification. In samples with low concentrations of cytarabine, it could not be distinguished from its isomer cytidine, naturally occurring in WWTP effluent as a DNA building block. Dissolved organic carbon (DOC) measurements were performed with a DIMA-TOC 100 total organic carbon analyzer (Dimatec Analysentechnik GmbH, Germany). DOC samples were pre-filtered with 0.45 μm Acrodisc filters from Pall Corporation.

2.3 Laboratory Scale Experiments

Adsorption kinetics and isotherms of the selected micropollutants in WWTP effluent were evaluated. All experiments were performed in two parallels and, were applicable, mean values were calculated. To provide a constant and comparable WWTP effluent matrix, 100 liters of effluent from the WWTP Aachen Soers (TOC 5 mg/l, pH 7.80) were frozen in 2 L Bottles at −20°C. Effluent used for experiments was melted at room temperature over night and filtered with glass fiber filters (0.7 μm). Experiments for the evaluation of the influence of natural organic matter were performed also with four times by nanofiltration concentrated effluent which was frozen and treated as WWTP effluent. For isotherms, 100 ml of filtered effluent in 250 ml Erlenmeyer beakers were spiked with 200 μg/L of each micropollutant individually and shaken at 100 rpm with different adsorbent doses for 16 h at 20°C. Kinetic experiments were performed in 1 liter bottles. These were also shaken at 100 rpm for 16 h at 20°C. At defined times, samples were taken and immediately filtered. In total, less than 5% of the total volume were taken for sampling.

2.4 Pilot Plant Experiments

After evaluating the adsorption behavior in laboratory scale experiments, a pilot plant combining adsorption to powdered carbon and retention by nanofiltration was used to evaluate the removal of selected micropollutants from WWTP effluent. The pilot plant is located at the waste water treatment plant Aachen Soers, Germany. This WWTP produces a high quality effluent with around 52 ±0.9 mg/l DOC, 15.6 ±2.3 mg/l COD, average conductivity of 0.97 ±0.9 mS/cm and a mean pH of 7.7±0.4 (Kazner et al., 2007). The effluent at least doubles the size of the receiving river Wurm which flows downstream through the popular recreational area "Wurmtal" of Aachen. The pilot plant uses the effluent which is pumped from the outlet of a sand filtration which is the last treatment step of the WWTP. It is pre-filtered with a 100 μm filter, shortly stored in a glass tank, mixed with powdered carbon slurry, stirred for approximately one hour in a continuous stirring tank reactor and is then pumped through an 8″ capillary nanofiltration (NF) module (Figs. 6 and 7).

This capillary NF module is equipped with an X-Flow NF50 M10 membrane from Norit with a molecular weight cut-off (MWCO) of 200 gram per mol and a membrane area of 20 m² (Futselaar et al., 2002).

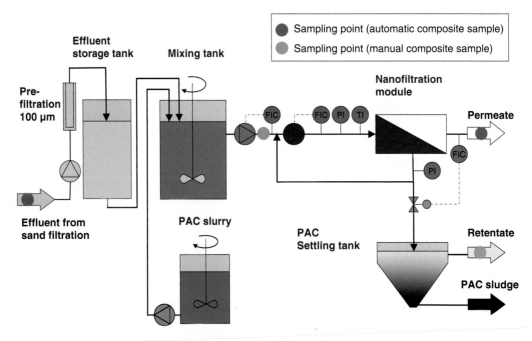

Fig. 6 Schematic view of the pilot plant combining retention by nanofiltration and adsorption to powdered (activated) carbon

The composite membrane has a negatively charged active layer of polyamide with a supporting layer of polyethersulfone.

For spiking experiments with the selected micropollutants, WWTP effluent was spiked with 1 μg/L of bisphenol A, 1 μg/L 17α-ethinylestradiol, 2 μg/L 5-fluorouracil and 2 μg/L Cytarabine. Samples of the pilot plant influent and permeate were taken with auto sampling units (MAXX GmbH). The sampling rate was 200 mL in 30 min over a period of 24 hours. For the feed (pilot plant influent mixed with carbon and stirred for 1 h) and retentate, manually produced composite samples were produced. Therefore, six times in 24 hours, samples were taken and immediately filtered with a glass fiber filter with a pore size of 0.7 μm. For the evaluation of the removal of micropollutants, over a period of five months, 105 samples were analyzed for 17α-ethinylestradiol and bisphenol A and 15 samples were analyzed for 5-fluorouracil and cytarabine/cytidine. For determination of the effluent organic matter (EfOM) matrix, dissolved organic carbon (DOC) was measured for every sample. For

Fig. 7 Pilot plant at the waste water treatment plant (WWTP) in Aachen-Soers, Germany. *Left picture*: 40ft-container housing the pilot plant and working space situated at the tertiary effluent channel of the WWTP (foreground). *Right picture*: Nanofiltration module of the pilot plant with operating terminal in the background

evaluating the membrane performance, the sulfate retention was measured two times a week. The pilot plant was also continuously monitored for trans-membrane pressure, temperature and pH.

3 Results and Discussion

3.1 Sorption Behavior of Selected Micropollutants

As mentioned above, 17α-ethinylestradiol and bisphenol A show significant sorption to particles and sludge. These findings lead to the assumption, that sorption processes are an effective way to remove nonpolar micropollutants from WWTP effluent. Polar substances show weak sorption to particles and activated sludge (Heberer and Stan, 1995; Scheytt et al., 2001).

The capacity of powdered (activated) carbon for the removal of trace compounds depends on the inner surface and the pore size distribution of the carbon as well as on the background matrix. In the case of waste water treatment plant effluent, this matrix is dominated by inorganic compounds, humic acids, proteins, carbohydrates and other biological building blocks. All these compounds compete with the micropollutants for binding sites on the carbon surface and decrease availability of the binding sites for trace compounds. The chemical properties of the carbon surface and trace compounds influence the affinity for sorption (Sontheimer et al., 1985). To evaluate the behavior of micropollutants in WWTP effluent, we performed laboratory scale experiments with four selected micropollutants (see above).

3.2 Adsorption Isotherms of Selected Micropollutants to Powdered Carbon

In the laboratory scale experiments the selected micropollutants showed a quite diverse adsorption behavior according to their physical and chemical properties (Fig. 8). For the evaluation of adsorption experiments, the mass of the substance adsorbed (adsorbate) per mass of adsorbing substance (adsorbent) is plotted against the adsorbate concentration

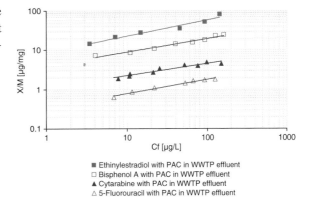

■ Ethinylestradiol with PAC in WWTP effluent
□ Bisphenol A with PAC in WWTP effluent
▲ Cytarabine with PAC in WWTP effluent
△ 5-Fluorouracil with PAC in WWTP effluent

Fig. 8 Freundlich adsorption isotherms of the chosen micropollutants in WWTP effluent with PAC. Micropollutants were used in concentrations of 200 μg/L, carbon was dosed in concentrations between 4.5 mg/L and 180 mg/L. C_f: concentration of micropollutant, M: amount of powdered activated carbon (PAC), X: amount of sorbed micropollutant

remaining in solution. The relationship between the amount adsorbed and the concentration remaining in the solution at equilibrium at a constant temperature is referred to as the adsorption isotherm. Adsorption isotherms are categorized in one of three sorption mechanisms, represented by characteristic shapes. These isotherms are referred to as the linear isotherm, the Freundlich isotherm and the Langmuir isotherm (Wiedemeier et al., 1999). The Freundlich equation (Equation 1), where C_0 is initial concentration of adsorbate, C_f is concentration of adsorbate, M is concentration of adsorbents, K stands for Freundlich coefficient and n^{-1} is Freundlich exponent, can be plotted in double logarithmic scale which allows performing a linear regression. The Freundlich coefficient (K) is then read as its Y-axis interception and the Freundlich exponent (n^{-1}) as the reciprocal of its slope. The Freundlich coefficient indicates the capacity of the adsorbent for the adsorbate, and the Freundlich exponent indicates the affinity of the adsorbate to the adsorbent.

$$\frac{C_0 - C_f}{M} = K \cdot C_f^{\frac{1}{n}} \qquad (1)$$

For the removal of dissolved organic carbon (DOC) with PAC as adsorbents, representing the equivalent background compound (EBC) (Crittenden et al., 1985; Graham et al., 2000), $K = 17.8\,\mathrm{L.g^{-1}}$ and $n^{-1} = 1.3$ were determined (data not shown). For 17α-ethinylestradiol and bisphenol A, a capacity ($K = 9.0\,\mathrm{L.mg^{-1}}$ and $K = 4.1\,\mathrm{L.mg^{-1}}$, respectively) and an affinity ($n^{-1} = 0.41$ and $n^{-1} = 0.35$, respectively) was found. In comparison, the cytostatics cytarabine and 5-fluorouracil show lower adsorption capacities ($K = 1.12\,L.mg^{-1}$ and $K = 0.35\,L.mg^{-1}$, respectively) than the EDCs but a similar affinity ($n^{-1} = 0.31$ and $n^{-1} = 0.36$, respectively) was found. These findings just confirm that nonpolar substances are better adsorbed to carbon than polar, hydrophilic substances.

The adsorption capacities of the lipophilic micropollutants correlate with the K_{OC} and K_{OW} which indicate good adsorption for 17α-ethinylestadiol followed by bisphenol A and much lower adsorption capacities for the cytostatics, as Fuerhacker et al. predicted in 2001. Within the group of cytostatics, the K_{OW} could not predict the observed adsorption behaviour, as here we are comparing partially charged 5-fluorouracil with uncharged cytarabine.

The comparison between lignite coke dust (LCD) and PAC showed that higher concentrations are necessary for LCD to reach the adsorption capacity of powdered activated carbon (Fig. 9). For 17α-ethinylestradiol, bisphenol A and cytarabine, Freundlich coefficients for LCD ($K = 1.8\,L.mg^{-1}$, $K = 1.2\,L.mg^{-1}$ and $K = 0.43\,L.mg^{-1}$, respectively) are approximately three to four times lower than for PAC. For 5-fluorouracil, the Freundlich coefficient of LCD ($K = 0.07$) is six times lower. The adsorption capacities correspond with the inner surfaces of LCD and PAC, except for the adsorption of 5-fluorouracil. Contrary to the adsorption capacities, adsorption affinities for 17α-ethinylestradiol, bisphenol A and cytarabine to LCD ($n-1 = 0.31$, $n-1 = 0.31$ and $n-1 = 0.36$, respectively) are twice as high for the selected substances to LCD compared with PAC. Again, 5-fluorouracil shows a differing behavior: The adsorption affinity to LCD ($n-1 = 0.53$) is lower than to PAC. The increased affinity of 17α-ethinylestradiol, bisphenol A and cytarabine could indicate an improved adsorption of micropollutants due to a different pore size distribution which is not influencing the selected micropollutant with the smallest molecular size.

To evaluate the influence of effluent organic matter (EfOM), experiments with different EfOM concentrations, represented by different DOC concentrations, were performed. To buffer pH-shifts induced by the powdered carbon, experiments with no EfOM-background were performed with 5 mM phosphate

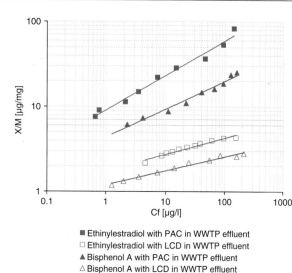

■ Ethinylestradiol with PAC in WWTP effluent
□ Ethinylestradiol with LCD in WWTP effluent
▲ Bisphenol A with PAC in WWTP effluent
△ Bisphenol A with LCD in WWTP effluent

Fig. 9 Comparison of Freundlich isotherms for 17α-ethinylestradiol and bisphenol A with two different carbons: powdered activated carbon (PAC) and lignite coke dust (LCD). For both micropollutants, the adsorption capacity of Norit SAE-Super is higher than the capacity of lignite coke dust. Micropollutants were used in concentrations of 200 μg/L, carbon was dosed in concentrations between 4.5 mg/L and 300 mg/L. C_f: concentration of micropollutant, M: amount of PAC, X: amount of adsorbed micropollutant

buffer. The pH was adjusted to 7.8, which was the pH of the applied WWTP effluent. Results were compared to experiments with WWTP effluent (5 mg/L DOC) and via nanofiltration four times concentrated WWTP effluent (20 mg/L) (Fig. 10). As expected and reported before (Ding et al., 2006), an increase of background-EfOM, and therefore a stronger concurrency for binding sites, decreases the capacity of powered carbon for the selected micropollutants. For 17α-ethinylestradiol, the Freundlich coefficient is reduced from $K = 5.3\,L.mg^{-1}$ in 5 mM phosphate buffer to $K = 1.8\,L.mg^{-1}$ in WWTP effluent and $K = 0.58\,L.mg^{-1}$ in four times concentrated WWTP effluent, while the Freundlich exponent is reduced from $n^{-1} = 0.28$ to $n^{-1} = 0.19$ and $n^{-1} = 0.15$, respectively. The reduction of adsorption capacities can be explained by blocked binding sites due to adsorption of EfOM components and through blocked pores which reduces the available inner surface significantly (Ebie et al., 2001). The selected micropollutants show also an increase of affinity to powdered carbon with EfOM background present, shown by the reduced Freundlich exponent.

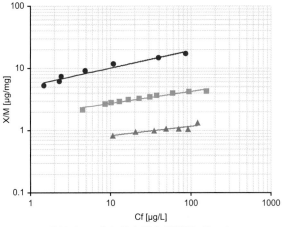

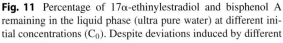

Fig. 10 Comparison of Freundlich isotherms for 17α-ethinylestradiol (200 μg/L) with LCD and different amounts of natural organic matter (EfOM). With increasing EfOM-background, the capacity of the powdered carbon decreases. 5 mM phosphate buffer: 0 mg/L DOC, WWTP effluent: 5 mg/L DOC, concentrated WWTP effluent: 20 mg/L DOC. Carbon was dosed in concentrations between 1.5 mg/L and 150 mg/L C_f: concentration of micropollutant, M: amount of PAC, X: amount of sorbed micropollutant

The laboratory scale experiments were performed with much higher initial concentrations of the selected micropollutants than those found in WWTP effluent. To validate the significance of the obtained results for the treatment of WWTP effluent, experiments were performed with a constant amount of carbon and different initial concentrations of the selected micropollutants. As Fig. 11 shows, despite a deviation due to

variations of the carbon dosage, the relative removal of 17α-ethinylestradiol and bisphenol A is constant in the tested range of initial concentrations (200 μg/L to 200 ng/L). For 5-fluorouracil and cytarabine, similar results were obtained (data not shown). Studies showed, consistent with the ideal adsorbed solution theory, a direct link is given between relative removal and carbon dosage even at different initial concentrations of trace components in presence of natural organic matter (Knappe et al., 1998; Qi et al., 2007; Crittenden et al., 1985; Graham et al., 2000; Matsui et al., 2003). This direct link is given, as long as the concentration of the organic trace components is significantly smaller than the concentration of the equivalent background component. In the case of the laboratory scale experiments in this project, it could be shown that the used concentration of 200 μg/L is small enough to fulfil this requirement. Therefore, laboratory scale experiments can provide evidence for the adsorption behaviour of micropollutants in waste water treatment plant effluent or similar matrices.

3.3 Adsorption Kinetics of Selected Micropollutants

The knowledge of the adsorption kinetics of micropollutants is essential for dimensioning of processes for the treatment of WWTP effluent. The limited contact time of micropollutants to adsorbent in treatment plants limits their removal efficiency. To obtain information about the adsorption kinetics, experiments were performed with carbon concentrations adsorbing 80%

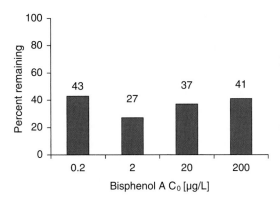

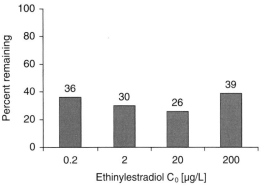

Fig. 11 Percentage of 17α-ethinylestradiol and bisphenol A remaining in the liquid phase (ultra pure water) at different initial concentrations (C_0). Despite deviations induced by different amounts of carbon, the removed percentage is not changed by changes in the concentration of micropollutant. LCD concentration was on average 50 mg/L

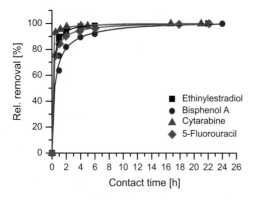

Fig. 12 Relative removal of bisphenol A, 17α-ethinylestradiol, cytarabine and 5-fluorouracil with lignite coke dust in dependence of the contact time. Carbon concentrations were chosen for 10% to 20% of initial concentration of micropollutants remaining at equilibrium: 100 mg/L for bisphenol A, 17α-ethinylestradiol and 500 mg/L for cytarabine and 5-fluorouracil

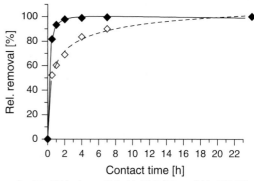

◆ 17α-Ethinylestradiol with PAC (20 mg/L) in WWTP effluent
◇ 17α-Ethinylestradiol with PAC (5 mg/L) in WWTP effluent

Fig. 13 Relative removal of 17α-ethinylestradiol in dependence of different adsorbent concentrations. Powdered activated carbon concentrations were chosen for 20% and 80% of initial concentration of micropollutants remaining at equilibrium

to 90% of the initial concentration of the micropollutant. Beside the different adsorption kinetics of the selected micropollutants, influences of initial concentration of adsorbate, presence of other micropollutants, concentration of EFOM-background, type of powdered carbon and the influence of the carbon concentration were investigated. Fig. 12 shows the relative removal for all four selected micropollutants plotted against the contact time with the adsorbent. All selected micropollutants show fast kinetics: for 17α-ethinylestradiol and bisphenol 88%, and 71% respectively, of the removal at equilibrium could be achieved within 1 hour contact time, which is approximately the contact time within the CSTR of the pilot plant used in this project. Despite their lower adsorption capacities, the cytostatics 5-fluorouracil and cytarabine show similar kinetics. Within one hour, 85% and 95%, respectively of the equilibrium concentration are reached.

For the adsorption kinetics of all four selected trace components, the adsorbent concentration proved to be the most important parameter. Figure 13 shows adsorption kinetics for 17α-ethinylestradiol with 5 mg/L and 20 mg/L PAC. Within one hour of contact time, for 17α-ethinylestradiol and bisphenol A, 63% and 60%, respectively, of equilibrium concentration was reached with 5 mg/L powdered activated carbon, within the same contact time, 20 mg/L of powdered activated carbon reached 89% and 80%, respectively, of equilibrium concentration.

The cytostatics showed a similar behavior at different carbon concentrations: within one hour, 40 mg/L and 220 mg/L PAC adsorbed 55% and 72% of the equi-

librium concentration of 5-fluorouracil; and 20 mg/L and 100 mg/L powdered activated carbon absorbed 54% and 79% of cytarabine. The comparison of LCD with PAC showed that at a concentration of 20 mg/L, the higher capacity of the powdered activated carbon leads to a faster adsorption for the EDCs. If high carbon concentrations with comparable adsorption capacities (500 mg/L LCD for both compounds compared with 220 mg/L for 5-fluorouracil and 100 mg/L for cytarabine) were used for the adsorption of the cytostatics, the LCD showed a faster adsorption (data not shown).

EFOM-background showed an influence on the lipophilic, non polar EDCs but no significant influence on the hydrophilic, polar cytostatics. To reach 80% of the equilibrium concentration in presence of EFOM, 20 mg/L LCD were necessary for 17α-ethinylestradiol and 500 mg/L for 5-fluorouracil. Bisphenol A and cytarabine show similar kinetics in presence of EFOM.

The difference in carbon dosages, caused by the weak adsorption capacities for 5-fluorouracil and cytarabine, requiring high carbon dosages, could lead to a reduced difference in sorption behavior in different background matrices. In the low concentration range, the presence of EFOM shows greater influence most presumably due to competition between EFOM and micropollutant for binding sites.

The other parameters showed no significant influence on the relative removal of micropollutants from WWTP effluent. The evaluation of the adsorption kinetics showed that within one hour with dosages of 100 mg/L LCD 89% of 17α-ethinylestradiol 80%

of bisphenol A and with dosages of 20 mg/L of PAC 93% of 17α-ethinylestradiol and 87% of bisphenol A can be removed from WWTP effluent. For the polar hydrophilic compounds 5-fluorouracil and cytarabine, higher carbon dosages are needed to reach similar removal rates: with 500 mg/L LCD 95% of cytarabine and 84% of 5-fluorouracil can be removed. With 220 mg/L of PAC, at least 79% of cytarabine and 72% 5-fluorouracil are adsorbed.

3.4 Pilot Plant Experiments

The evaluation of the wastewater treatment pilot plant spiking experiments showed that operation of the pilot plant without carbon dosage (direct nanofiltration) leads to retention for the non polar, lipophilic compounds BPA of 43% to 58% and for EE2 of 52% to 68%. This is in compliance with literature (Bellona et al., 2004). No significant correlation was found between membrane flux and permeate recovery rate and the retention of the nonpolar, lipophilic micropollutants, but significant losses in the mass balance are observed (Fig. 14). Most probably, this is due to adsorption to pilot plant parts, membrane material or fouling layers. The mass balances showed losses of the spiked micropollutants within the pilot plant after 21 days of continuous spiking: 18% (17α-ethinylestradiol) and 26% (bisphenol A) of the spiked amount was lost within 24 hours.

For the polar, hydrophilic compound cytarabine/cytidine a retention of 94% was achieved. 5-fluorouracil showed, despite its molecular weight below the cut-off of the membrane, highest retention by the pilot plant up to 100%. But also for these polar, hydrophilic substances, significant losses in the mass balance were observed. 79% of cytarabine/cytidine and 95% of 5-fluorouracil of the amount spiked within 24 hours are lost after 21 days of continuous spiking. Most probably, this is due to adsorption to polar polyamide membrane material or to fouling layers combined with degradation. Further research is necessary to evaluate these findings. Because of these mass balances, no spiking experiments with carbon dosage and 5-fluorouracil and cytarabine/cytidine were performed as permeate and retentate concentrations of direct nanofiltration were already close to the limit of detection.

Spiking Experiments with the EDCs and with dosage of carbon were performed with LCD in dosages of 100 mg/L, 200 mg/L and 350 mg/L and PAC in concentrations of 10 mg/L, 25 mg/L, 50 mg/L and 100 mg/L. Figure 15 shows the total removal of the pilot plant together with the adsorptive removal by adsorption to dosed carbon in the pilot plant. For the calculation of the removal rates, mass flows for each substance were used.

DOC was retained by the membrane between 91 and 97%. In operation with the lowest tested PAC concentration of 10 mg/l retention of 88% bisphenol

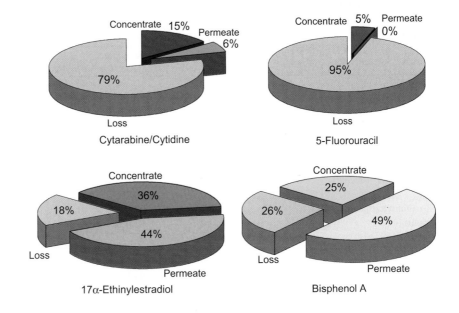

Fig. 14 Mass balances of cytarabine/cytidine, 5-fluorouracil, 17α-ethinylestradiol and bisphenol A in the pilot plant after 21 days of continuous spiking. All substances show significant losses within the pilot plant. These losses are greatest for 5-fluorouracil, followed by cytarabine/cytidine, bisphenol A and 17α-ethinylestradiol

Fig. 15 Adsorptive and total removal of 17α-ethinylestradiol (EE2) and bisphenol A (BPA) by the pilot plant. *Upper left graph*: Removal of 17α-ethinylestradiol with dosage of Norit SAE Super powdered activated carbon (PAC), *upper right graph*: 17α-ethinylestradiol with dosage of lignite coke dust (LCD). *Lower left graph*: bisphenol A with dosage of PAC, *lower right graph*: bisphenol A with dosage of (LCD). With 25 mg/l of PAC or 100 mg/l of LCD 99% of 17α-ethinylestradiol and 95% of bisphenol A are retained by the pilot plant

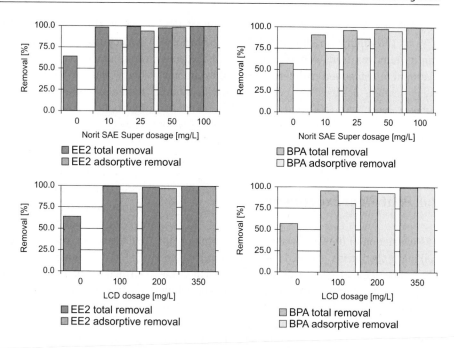

A and 98% 17α-ethinylestradiol could be achieved. With a dosage of 25 mg/L of powdered activated carbon, 95% of bisphenol A and 99% of 17α-ethinylestradiol could be removed from WWTP effluent. With 100 mg/L lignite coke dust, 95.4% and 99.9% were removed (Table 4). While the total removal of the selected hydrophilic micropollutants increased significantly with the dosage of carbon to the WWTP effluent, the removal of DOC, representing removal of natural organic matter, was not increased. With a dosage of 25 mg/L powdered activated carbon and 100 mg/L LCD, 90% and 91% of DOC, respectively, were retained by the pilot plant. This indicates that the dosage of carbon in concentrations less than 100 mg/L enhances the removal of micropollutants but does not increase EFOM removal.

The results indicate that almost complete retention of micropollutants can be achieved by the combination of NF membrane filtration and adsorption to powdered activated carbon. However; adsorption effects should be investigated more detailed for micropollutants.

Table 4 Measured concentrations and removal rates of the pilot plant for the selected micropollutants and DOC

	WWTP Effluent [μg/L]	Feed with carbon [μg/L]	Retention by carbon	NF Permeate [μg/L]	Retention by pilot plant
Direct NF					
DOC	5735	–	–	495	91.4%
17α-Ethinylestradiol	0.98	–	–	0.46	69.0%
Bisphenol A	1.18	–	–	0.66	49.6%
5-Fluorouracil	1.60	–	–	0.11	100%
Cytarabine/Cytidine	2.45	–	–	0.89	94%
NF with LCD 100 mg/L					
DOC	5115	3545	30.7%	448	91.3%
17α-Ethinylestradiol	0.74	0.06	92.3%	0.001	99.9%
Bisphenol A	0.78	0.15	81.0%	0.036	95.4%
NF with Norit 25 mg/L					
DOC	4870	3455	29.1%	475	90.2%
17α-Ethinylestradiol	0.99	0.06	94.1%	0.010	99.0%
Bisphenol A	0.98	0.13	86.4%	0.050	95.0%

4 Summary

Today, mankind worldwide faces scarcity of feasible water resources for potable and process water. Therefore, it is a key issue to replenish existing water resources. Reclamation of effluent from waste water treatment for groundwater recharge is a promising way to achieve this (Meier and Melin, 2005). A key feature of this reclamation process is the removal of persistent organic trace compounds like pharmaceuticals and compounds of personal care products (Schwarzenbach et al., 2006; Snyder et al., 2007; Ternes and Joss, 2006). The removal of four micropollutants with different compound properties, more specifically bisphenol A (BPA, log K_{OW} 3.2, pKa 9.7, MW 228.3), 17α-ethinylestradiol (EE2, log K_{OW} 4.1, pKa 10.2, MW 296.4), 5-fluorouracil (5-Fu, log K_{OW} -0.9, pKa 8.0, MW 130.1) and cytarabine (CytR, log K_{OW} −2.2, pKa 4.4, MW 243.2) from municipal wastewater treatment plant Aachen Soers, Germany (WWTP) effluent (DOC 5 mg/L) was studied in a pilot plant combining sorption to powdered activated carbon and retention by capillary nanofiltration (NF). These substances were selected because of their known or predicted effects on aquatic ecosystems and possibly on human health in low dosages. According to their chemical, physical and biological properties the four substances were divided in two groups: bisphenol A and 17α-ethinylestradiol as lipophilic endocrine disrupting compounds and 5-fluorouracil and cytarabine as hydrophilic cytostatics. For the evaluation of their removal by the process combination of adsorption to powdered activated carbon and retention by nanofiltration, samples were pre-concentrated by solid phase extraction and analyzed by HPLC-MS/MS. Preliminary laboratory scale experiments were performed to investigate the adsorption isotherms and kinetics of the selected micropollutants. These experiments showed that the substances have different adsorption behavior according to their physical-chemical properties. The comparison between two different commercial available powdered carbons, LCD and PAC, showed that four times higher concentrations are needed for LCD to reach the adsorption capacity of PAC. This corresponds to the difference in the area of the inner surface. The presence of natural organic matter as it is found in WWTP effluents decreases the adsorption capacity for the micropollutants of both carbons. In compliance with findings of Knappe et al., 1998, the relative removal of the selected compounds is not influenced by their initial concentration. The evaluation of the adsorption kinetic showed a fast adsorption for all four substances. At high carbon dosages of more than 200 mg/L, the presence of natural organic matter has little influence on the adsorption kinetic of the selected hydrophilic compounds. The adsorption kinetic of lipophilic compounds with low dosages of carbon (20 mg/L) showed a significant slowing down in the presence of natural organic matter. The dosage of carbon was found to have the strongest influence on the adsorption kinetic.

A waste water treatment pilot plant located at the waste water treatment plant Aachen Soers, Germany, equipped with a capillary NF membrane module with a molecular weight cut-off (MWCO) of 200 Dalton, was used to evaluate the removal of spiked micropollutants from WWTP effluent. WWTP effluent was used as raw water for the pilot plant and was spiked with 1–2 µg/L of the chosen micropollutants. Samples were taken as 24 h composite samples from pilot plant influent and permeate and as manually produced composite samples from the membrane feed and the retentate. The evaluation of the spiking experiments shows that operation without carbon dosage leads to retention by the pilot plant for bisphenol A of up to 58%, for 17α-ethinylestradiol of up to 68%, and for cytarabine/cytidine 94%. 5-Fu showed, despite its molecular weight below the cut-off of the membrane, retention by the pilot plant up to 100%. Dissolved organic carbon was retained up to 91%. The mass balances for the selected and spiked micropollutants showed that at least one third of the amount of the spiked lipophilic substances was lost within the pilot plant. For the hydrophilic substances, up to 96% of the spiked amount was lost in the pilot plant. The high losses within the pilot plant lead to the assumption that the substances were adsorbed by parts of the pilot plant, by the membrane or by fouling layers combined with degradation processes. In operation with dosage of powdered activated carbon or LCD, lipophilic organic trace compounds were highly retained. With 100 mg/L LCD, 95% of bisphenol A and 99.9% of 17α-ethinylestradiol were removed from WWTP effluent. To reach similar levels of removal, 25 mg/L PAC had to be applied. The selected lipophilic micropollutants showed much higher increase in removal at dosage of low concentrations of powdered carbon with regard to direct nanofiltration then was the increase in removal of natural organic matter. This project proved the applicability of the combination of adsorption to powdered activated carbon and retention by nanofiltration for the removal of organic trace compounds from WWTP effluent.

Acknowledgments This study was supported by budget funds from the German Federal Ministry of Economics and Technology through the German Federation of Industrial Research Associations "Otto von Guericke" (AiF project number 14773 N1). Norit X-Flow B.V. is gratefully acknowledged for the donation of the NF50 M10 membranes used for this study. The powdered carbons were donated by RWE Power AG – Rheinbraun Brennstoff GmbH and Norit Deutschland GmbH. We thank Daniela Schmitz, Anne Schneider, Jochen Herr, Markus Linden and Daas Jabbour for their valuable work for the project.

Abbreviations

5-Fu	5-fluorouracil
AHTN	1-(5,6,7,8-tetrahydro-3,5,5,6,8,8-hexamethyl-2-naphthalenyl)ethanone
BPA	Bisphenol A
C_0	Initial concentration of adsorbate
C_f	Concentration of adsorbate
CDs	Compact discs
CytR	Cytarabine
DDT	Dichlorodiphenyltrichloroethane
DDE	Dichlorodiphenyldichloroethylene
DNA	Deoxyribonucleic acid
DOC	Dissolved organic carbon
DVDs	Digital versatile discs
EBC	Equivalent background compound
EDCs	Endocrine disrupting compounds
EE2	17α-ethinylestradiol
EfOM	Effluent organic matter
EPA	Environmental protection agency
HILIC	Hydrophilic interaction liquid chromatography
HHCB	1,3,4,6,7,8-hexahydro-4,6,6,7,8,8-hexamethylcyclopenta-gamma-2-benzopyran
HPLC	High performance liquid chromatography
K	Freundlich coefficient
K_{OC}	Water-organic-carbon partition coefficient
K_{OW}	n-octanol-water partition coefficient
LCD	Lignite coke dust
M	Concentration of adsorbents
MEC	Measured environmental concentration
mM	Millimolar
MTBE	Methyl tertiary butyl ether
MW	Molecular weight
MWCO	Molecular weight cut-off
n^{-1}	Freundlich exponent
NF	Nanofiltration
PAC	Powdered activated carbon
PCB	Polychlorinated biphenyls
PEC	Predicted environmental concentration
PhACs	pharmaceutically active compounds
PNEC	Predicted non-effective environmental concentration
pK_a	Acid dissociation constant
POPs	Persistent organic pollutants
SPE	Solid phase extraction
TBT	Tributyltin
TOC	Total organic carbon
WWTP	Municipal wastewater treatment plant

References

Adler A, Steger-Hartmann T et al. (2001) Vorkommen natürlicher und synthetischer östrogener Steroide in Wässern des süd- und mitteldeutschen Raumes (Distribution of Natural and Synthetic Estrogenic Steroid Hormones in Water Samples from Southern and Middle Germany). Acta hydrochim hydrobiol 29(4):227–241.

Alaee M, Wenning RJ (2002) The significance of brominated flame retardants in the environment: current understanding, issues and challenges. Chemosphere 46:579–582.

Andersen H, Siegrist H et al. (2003) Fate of estrogens in a municipal sewage treatment plant. Environ Sci Technol 37(18):4021–4026.

Bellona C, Drewes JE et al. (2004) Factors affecting the rejection of organic solutes during NF/RO treatment – a literature review. Water Res 38(12):2795–2809.

Bilitewski D, Weltin P, Werner P (2002) Endokrin wirksame Substanzen in Abwasser und Klärschlamm – Neueste Ergebnisse aus Wissenschaft und Technik. Beiträge zur Abfallwirtschaft/Altlasten, Band 23 TU Dresden.

Brown KH, Schultz IR et al. (2007). Reduced embryonic survival in rainbow trout resulting from paternal exposure to the environmental estrogen 17α-ethynylestradiol during late sexual maturation. Reproduction 134(5):659–666.

Byrns G (2001) The fate of xenobiotic organic compounds in wastewater treatment plants. Water Res 35(10):2523–2533.

Cooke MD (1976) Antibiotic resistance in coliform and faecal coliform bacteria from natural waters and effluents. NZ J Mar Freshwater Res 10(3):391–397.

Crittenden JC, Luft P, Hand DW (1985) Prediction of multicomponent adsorption equilibria in background mixtures of unknown composition. Water Res 19(12):1537–1548.

Danielsson LG, Zhang Y-H (1996) Methods for determining n-octanol-water partition constants Trends Anal Chem 15(4):188–196.

Ding L, Marinas BJ et al. (2006) Competitive effects of natural organic matter: parametrization and verification of the three-component adsorption model COMPSORB. Environ Sci Technol 40(1):350–356.

Dorne J, Skinner L et al. (2007) Human and environmental risk assessment of pharmaceuticals: differences, similarities, lessons from toxicology Anal Bioanal Chem 387(4):1259–1268.

Drewes, JE, Heberer T, Reddersen K (2002) Fate of pharmaceuticals during indirect potable use. Water Sci Tech 46(3):73–80.

Ebie K, Li F et al. (2001) Pore distribution effect of activated carbon in adsorbing organic micropollutants from natural water. Water Res 35(1):167–179.

Fent K (2003) Ecotoxicological problems associated with contaminated sites. Toxicol Lett 1(13).

Fuerhacker M, Durauer A et al. (2001) Adsorption isotherms of 17[beta]-estradiol on granular activated-carbon (GAC). Chemosphere 44(7):1573–1579.

Futselaar H, Schonewille H, van der Meer W (2002) Direct capillary nanofiltration – a new high-grade purification concept. Desalination 145:75–80.

Goi D, de Leitenburg C et al. (2004) Catalytic wet oxidation of a mixed liquid waste: Cod an aox abatement. Environ Tech 25(12):1397–1403.

Gore AC, Heindel JJ, et al. (2006) Endocrine disruption for endocrinologists (and Others). Endocrinology 147(6):1–3.

Graham MR, Summers RS et al. (2000) Modeling equilibrium adsorption of 2-methylisoborneol and geosmin in natural waters. Water Res 34(8):2291–2300.

Hansch C, Leo A, Hoekman D (1995) Exploring OSAR – Hydrophobic, Electronic and Steric Constants. American Chemistry Society, Washington, D.C.

Heberer T, Reddersen K, Mechlinski A (2002) From municipal sewage to drinking water: fate and removal of pharmaceutical residues in the aquatic environment in urban areas. Water Sci Tech 46(3):81–88.

Heberer T, Stan, HJ (1995) Polar environmental contaminants in the aquatic system. Occurrence and identification by GC-MS. GIT Fachz Lab 39(8):718.

Hegemann W, Busch K et al. (2002) Einfluss der Verfahrenstechnik auf die Eliminierung ausgewählter Estrogene und Xenoestrogene in Kläranlagen- ein BMBF Verbundprojekt. Wasser – Abwasser 143:422–428.

Johnson AC, Jurgens MD et al. (2008) Do cytotoxic chemotherapy drugs discharged into rivers pose a risk to the environment and human health? An overview and UK case study J Hydrol 348(1–2):167–175.

Kamiura T, Tajima Y, Nakahara T (1997) Determination of bisphenol A in air. J Environ Chem 7:275–279.

Kazner C, Fink G et al. (2007) Removal of organic micropollutants by nanofiltration in combination with adsorption on powdered activated carbon for artificial groundwater recharge with reclaimed wastewater. Proc 5th IWA Micropol & Ecohazard 2007 conference, Frankfurt/Main, 259–265.

Knappe DRU, Matsui Y et al. (1998) Predicting the capacity of powdered activated carbon for trace organic compounds in natural waters. Environ Sci Technol 32(11):1694–1698.

Koerner W, Bolz U et al. (2000) Input/output of estrogenic active compounds in a major municipal sewage plant in Germany. Chemosphere 40(9–11):1131–1142.

Kolpin DW, Furlong ET et al. (2002) Pharmaceuticals, hormones, and other organic wastewater contaminants in U.S. Streams, 1999–2000: A national reconnaissance. Environ Sci Technol 36(6):1202–1211.

Kovalova L, McArdell CS et al. (2009) Challenge of high polarity and low concentrations in analysis of cytostatics and metabolites in wastewater by hydrophilic interaction chromatography/tandem mass spectrometry. Journal of Chromatography 1216(7):1100–1108.

Kümmerer K, Steger-Hartmann T et al. (1997) Biodegradability of the anti-tumour agent ifosfamide and its occurrence in hospital effluents and communal sewage. Water Res 31(11):2705–2710.

Mahnik SN, Lenz K et al. (2007) Fate of 5-fluorouracil, doxorubicin, epirubicin, and daunorubicin in hospital wastewater and their elimination by activated sludge and treatment in a membrane-bio-reactor system. Chemosphere 66(1):30–37.

Matsui Y, Fukuda Y et al. (2003) Effect of natural organic matter on powdered activated carbon adsorption of trace contaminants: characteristics and mechanism of competitive adsorption. Water Res 37(18):4413–4424.

Meier J, Melin T (2005) Wastewater reclamation by the PAC-NF process. Desalination 178:27–40.

Meier J, Melin T, Eilers LH (2002) Nanofiltration and adsorption on powdered adsorbent as process combination for the treatment of severely contaminated waste water. Desalination 146:361–366.

Pandey S, Musarrat J (1993) Antibiotic resistant coliform bacteria in drinking water. J Environ Biol 14(4):267–274.

Petrovic M, Gonzales S, Barcelo D 2003 Analysis and removal of emerging contaminants in wastewater and drinking water. Trends Anal Chem 22(10):685–696.

Qi S, Schideman L et al. (2007) Simplification of the IAST for activated carbon adsorption of trace organic compounds from natural water. Water Res 41(2):440–448.

Scheytt T, Grams S et al. (2001) Pharmaceuticals in groundwater: clofibric acid beneath sewage farms south of Berlin, Germany. ACS Sympos Ser 791:84–99.

Schwarzenbach RP, Escher BI et al. (2006) The challenge of micropollutants in aquatic systems. Science 313(5790):1072–1077.

Sharpe RM, Skakkebaek NE (1993) Are oestrogens involved in falling sperm counts and disorders of the male reproductive tract. The Lancet 341(8857):1392–1396.

Snyder SA, Adham S et al. (2007) Role of membranes and activated carbon in the removal of endocrine disruptors and pharmaceuticals. Desalination 202(1–3):156–181.

Snyder SA, Wert EC et al. (2006) Ozone oxidation of endocrine
disruptors and pharmaceuticals in surface water and wastewater. Ozone Sci Eng 28(6):445–460.

Sontheimer H, Frick R et al. (1985) Adsorptionsverfahren in der
Wasseraufbereitung, DVGW-Forschungsstelle am Engler-
Bunte-Institut, Karlsruhe.

Spengler P, Körner W, Metzger JW (1999) Schwer abbaubare
Substanzen mit östrogenartiger Wirkung im Abwasser von
kommunalen und industriellen Kläranlagenabläufen. (Hardly
degradable substances with estrogenic activity in effluents
of municipal and industrial sewage plants). Vom Wasser
93:141–157.

Steger-Hartmann T, Kümmerer K et al. (1996) Trace analysis
of the antineoplastics ifosfamide and cyclophosphamide in
sewage water by two-step solid phase extraction and gas
chromatography and mass spectrometry. Journal of chromatography A 726(1–2):179–184.

Steger-Hartmann T, Kümmerer K, et al. (1997) Biological degradation of cyclophosphamide and its occurrence in sewage
water. Ecotox Environ Safety 36(2):174–179.

Stelzer W, Ziegert E, Schneider E (1985) The occurrence
of antibiotic-resistant Klebsiellae in wastewater. Zentralbl
Mikrobiol 140(4):283–291.

Ternes TA (1998) Occurrence of drugs in German sewage treatment plants and rivers. Water Res. 32:3245–3260.

Ternes TA (2007) The occurrence of micopollutants in the
aquatic environment: a new challenge for water management.
Water Sci Tech 55(12):327–332.

Ternes TA, Joss A (eds.) (2006) Human Pharmaceuticals,
Hormones and Fragrances: The Challenge of Micropol-
lutants in Urban Water Management. IWA Publishing,
London.

Ternes TA, Stumpf M et al. (1999) Behavior and occurrence
of estrogens in municipal sewage treatment plants I. Investigations in Germany, Canada and Brazil. Sci Total Environ
225(1–2):81–90.

US EPA. (2008) Estimation Programs Interface Suite™
for Microsoft® Windows, v3.20. United States Environmental Protection Agency, Washington, DC, USA.
http://epa.gov/oppt/exposure/pubs/episuite.htm.   Accessed
10 April 2008.

Vieno N, Harkki MH et al. (2007) Occurrence of pharmaceuticals in river water and their elimination in a pilot-scale
drinking water treatment plant. Environ Sci Technol 41(14):
5077–5084.

von Gunten U (2003) Ozonation of drinking water: Part II. Dis-
infection and by-product formation in presence of bromide,
iodide or chlorine. Water Res 37(7):1469–1487.

Webb S, Ternes T et al. (2003) Indirect human exposure to
pharmaceuticals via drinking water. Toxicol Lett 142(3):
157–167.

Welshons WV, Nagel SC et al. (2006) Large effects from small
exposures. III. Endocrine mechanisms mediating effects of
bisphenol A at levels of human exposure. Endocrinology
147(6):56–69.

Welshons WV, Thayer KA, et al. (2003) Large effects from
small exposures. I. Mechanisms for endocrine-disrupting
chemicals with estrogenic activity. Environ Health Perspect
111(8):994–1006.

Wenzel A, Küchler T et al. (1998) Konzentrationen
östrogen wirkender Substanzen in Umweltmedien
(Concentration of estrogenic active substances in the
environment). Umweltforschungsplan des Bundesmin-
isters für Umwelt Naturschutz und Reaktorsicher-
heit. Umweltchemikalien/Schadstoffwirkungen.   IUCT,
Schmallenberg.

Wiedemeier TH, Wilson JT et al. (1999) Technical protocol for
implementing intrinsic remediation with long-term monitoring for natural attenuation of fuel contamination dissolved
in groundwater. U.S. Air Force Center for Environmental
Excellence, v.1&2, A324248, A324247a, A324247b.

Wintgens T, Melin T et al. (2005) The role of membrane
processes in municipal wastewater reclamation and reuse.
Desalination 178(1–3):1–11.

Development of Vertically Moving Automatic Water Monitoring System (VeMAS) for Lake Water Quality Management

Dongil Seo and Eun Hyoung Lee

Abstract A more complex water quality model requires more detailed data for calibration and validation. Without proper information of physical, chemical and biological transformations of water quality constituents, proper water management of lakes and reservoirs would not be possible. In general, in-lake water quality related data are less abundant compared to data for rivers. This is especially true for vertical profile data for lake stratification. An integrated lake water quality monitoring system, VeMAS, was developed to assist in such problems. The system is composed of sensors for water quality monitoring, motor and winch for vertical movement of the sensors, solar panel and rechargeable batteries for power supply, and wireless data communication system using a commercially available mobile phone network to send data and to control the system remotely. A software and interface are developed to monitor and control the entire system. It is expected that this system will enhance accuracy of water quality modeling by providing timely data and efficiency of other water quality management alternatives by providing real time data of inflow and in-lake conditions such as temperature, conductivity and turbidity.

Keywords Water quality monitoring · Modeling · Management · VeMAS · Stratification · Real time data

D. Seo (✉)
Department of Environmental Engineering,
Chungnam National University, Daejeon, 305–764, Korea
e-mail: seodi@cnu.ac.kr

1 Introduction

A development of water quality management strategies of a water body requires accurate predictions of future water status including its quality and quantity. In general, such predictions require large number of data that are not often available. Korean government operates 1,982 water quality monitoring points, 37 automatic water quality monitoring stations, 689 rainfall stations and 477 water level stations as of 2007. Weather in Korea is heavily affected by Monsoon effect that makes 2/3 of precipitation concentrated during the summer season. Therefore, it is essential to include rainfall effect or nonpoint source effect when analyzing water quality data in surface water bodies. To estimate accurate waste load of pollutant from a basin, it is necessary to collect water quantity data and water quality data simultaneously. Recently, major drinking water reservoirs in upstream areas in Korea have been experiencing persistent turbidity problems due to soil erosion from increased high land farming areas located mostly on steep mountainous areas. Due to low temperature in high altitude areas there are fewer problems from insect and this makes possible pesticide-less farming. However, since surfaces of high land area are mostly covered by rock or gravel, it is necessary to transport soil from low land area. To grow products, excessive amount of fertilizers are applied. Figure 1 shows transported soil and rows of pile of fertilizers to be applied over the top of soil. Figure 2 shows turbid water discharge from penstock of Soyang Dam due to density current resulted from soil erosion of its basin.

Figure 3 shows monthly TN and TP concentrations in Soyang Lake. The Soyang Lake is one of the most pristine water resources in Korea. However

Y.J. Kim et al. (eds.), *Atmospheric and Biological Environmental Monitoring*,
DOI 10.1007/978-1-4020-9674-7_11, © Springer Science+Business Media B.V. 2009

Fig. 1 High land farming practice in steep area in Gangwon Province, Korea

total phosphorus concentrations in the lake was over 0.16 mg/L in two consecutive months in the summer of 2007 due to soil erosion from catchments area.

In turbidity predictions in reservoirs, 2-Dimensional and 3-Dimensional boundary data and calibration data are essential for successful application of models. However, it is more often that such data are not available and turbidity predictions are made based on artificially formulated input conditions or limited field measurements. Kim et al. (2001) simulated movement

Fig. 2 Turbid water discharge from Soyang Dam, Korea (Jeon and Choi, 2004)

Fig. 3 Variations in total nitrogen and total phosphorus concentrations in Soyang Lake, Korea (Seo, 2007)

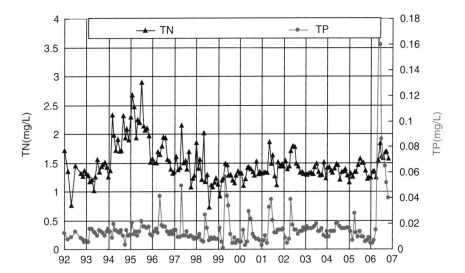

of turbidity water in Soyang Lake using CE-QUAL-W2 model. They suggested the timely withdrawal of density current of Soyang Lake is the effective method to control turbidity problem in the lake. Chung et al. (2005) applied the same model to Daechung Lake, Korea and reported that temperature drop due to storm is the major factor that affects density current in the lake.

Lee et al. (2002b) pointed out the need of continuous profile data of a reservoir when calibrating 2-Dimensional hydrodynamic water quality model. Seo and Yu (2007) applied 3-D hydrodynamic model to predict 2-Dimensional water quality variations in the Han River, Korea. They reported serious water quality differences in lateral direction of the river due to incomplete mixing especially when there was tributary input. This study reports that sampling for water quality monitoring in a fixed location may not provide representative data for the river.

2 Advances in Water Quality Modeling

Water quality models are effective tools to diagnosis the status of water, predict water quality and evaluate water quality management actions. Water quality model is a set of computer programs that solved systems of differential or algebraic equation that described interactions of factors affecting water quality in the water body of interest. After pioneering DO-BOD

modeling in a river by Streeter and Phelps (1958), water quality modeling has been evolved greatly as hardware and software of computer technique has been evolved. Brown and Barnwell (1987) developed a steady state river water quality modeling system, QUAL2E-UNCAS. QUAL2E model has been one of the most popular models for river water quality predictions in last couple of decades. Chapra et al. (2007) developed QUAL2K model. USEPA replaced their official stream water quality model from QUAL2E to QUAL2K. However these models only can apply hydraulically steady one dimensional situation.

Dynamic and multidimensional modeling was first suggested by Di Toro et al. (1971). Later Di Toro et al. (1983) developed the original version of WASP. WASP4 (Ambrose et al., 1988) and WASP5 (Ambrose et al., 1993) were two of the most widely used water quality models in recent decades that considers spatial and temporal variations of water quality in surface waters. However, the entire above model does not have module to simulate the movement of water. Therefore, a user have to assume hydraulic transport characteristics when use these models. Cole and Buchak (1995) developed CE-QUAL-W2 model that can consider hydrodynamics and water quality simultaneously. This model was applied to many reservoir and stream water quality simulations (Cole and Wells, 2007). However, this model assumes that lateral components of a river segment are homogeneous that is often not true especially when there is a tributary inflow or pollution input as discussed by Seo and Yu (2007).

Wool (personally communication, 2008) stated that USEPA has been developing EFDC-WASP that considers 3-dimensional hydrodynamic transport, complex water quality interactions and sediment diagenesis. EFDC was originally developed by Hamrick (1992). USEPA has been trying to link hydrodynamic part of EFDC and WASP model. This combination makes users possible to simulate hydrodynamic characteristics and water quality kinetics simultaneously. Water quality models are essential engine in TMDL development in US and reports for total waste load management act in Korea.

However, as model becomes more complex, its data requirements get increased greatly. For example, to model the movement of turbidity in a reservoir, it would be necessary to monitor continuous input turbidity and related data. Without proper input data, there will be no advantage of using a more complex model.

3 Automatic Water Quality Monitoring System in Korea

Korean government began to operate its nationwide automatic water quality monitoring stations since 1990 after accidental spill of phenolic compound in the Nakdong River. As of 2007, Korea operates 49 automatic monitoring stations and plans to construct 7 additional stations as shown in Fig. 4.

Current stations are equipped with Total Organic Carbon (TOC) analyzer, Volatile Organic Carbon (VOC) analyzer and Biomonitoring System that sense activities of Daphnia Magna inside the systems, DO, pH and Temperature sensors, automatic water samplers as shown in Fig. 5. Since TOC analyzer and VOC analyzer is unable to accept particulate material, sampled water needs to flow through membrane filtration system before any analysis. However, VOC compounds tend to have low density and flow only at surface of water body. Current submerged-type intake system may not be effective in collecting VOC samples from river since VOC material may have lighter density than water. Daphnia Magna is a biological organism that needs appropriate temperature, food and other requirement for its survival.

Dynamic nature of river water conditions and inside station may not be favorable to Daphnia Magna. As a result, Daphnia Magna often found dead or became inactive and issued false alarms and this biological monitoring system has been found difficult to maintain.

Seo (Lee et al., 2002a) reported that current sampling withdrawal location and system might not be effective due to incomplete mixing. Yotsukura (1968) suggested the following equation for a length required for complete mixing in a river when there is a side stream discharge.

$$L = 8.52U \frac{B^2}{H}$$

where,

L = mixing length (m)
U = stream velocity (m/sec)
B = stream width (m)
H = stream depth (m)

For example, if velocity is 0.1 m/s, width is 100 m and depth 1 m for a stream, then the length required for a complete mixing is calculated to be 8.52 km. This example indicates that if a toxic material was introduced from the other side of the river bank, it would be difficult to detect such incident from the other side of stream.

On the other hand, Lee et al. (2006) developed automatic water quality monitoring system for Yudeung-cheon, Daejeon Korea to investigate causes of fish kills in the stream. The system composed of sensors to detect DO, specific conductivity, ORP, turbidity, temperature and water depth, wireless data transmission system using embedded cell phone and computer server to integrate and analyze data. The system was successful to detect DO depletion due to combined sewer overflow. However, the system was not maintained and removed from the site due to unknown reason.

Korea Water Resources Corporation operates automatic water quality monitoring stations and turbidity monitoring stations at several places including Imha Dam, Yongdam Dam, Soyang Dam and Daechung Dam Sites (Chung et al., 2006a,b). Imha Dam site has the greatest number of automatic monitoring stations.

Fig. 4 Automatic water quality monitoring stations in Korea (http://www.emc.or.kr/measure/water_01.asp)

Fig. 5 Inside of automatic water quality monitoring stations in Korea

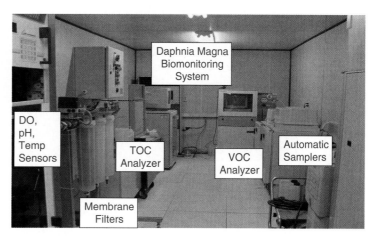

4 Development of VeMAS (Vertically Moving Automatic Water Monitoring System)

Persistent turbidity problem has become a serious issue for water resources, especially for drinking in Korea. Reservoirs or lakes show very different water qual-ity characteristics in vertical direction especially in the summer or where there is density current due to storm water as shown in Fig. 6.

Figure 7 shows continuous measurements of temperature and dissolved oxygen profile in Daeheong Lake. As shown in the figure, temperature difference in upper and lower layer of the lake is greater than 20° showing clear stratification. Due to the stratification,

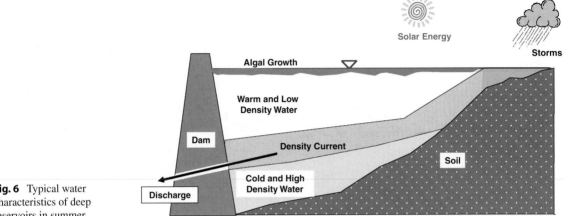

Fig. 6 Typical water characteristics of deep reservoirs in summer

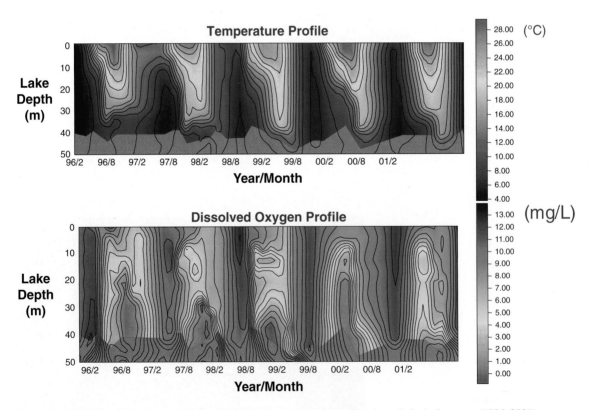

Fig. 7 Vertical profile of temperature and dissolved Oxygen concentration in Daechung Lake in five years (1996–2001)

Fig. 8 The first vertical water quality monitoring system in Daechung Lake

DO concentrations in the hyploimnion are nearly zero in every summer.

Seo et al. (1997a,b) developed automatic water quality monitoring system in Daechung Lake, Korea by moving sensors up and down along the depth of reservoir as shown in Fig. 8. The first system used AC and rechargeable battery. This system was installed in Sihwa Lake, Korea to monitor water quality changes due to water gate operation. Figure 9 show algorithm of movement of the system. Users can select depth interval and time interval between measurements dependent upon their needs. When a sensor reached lake bottom tension sensor sends command to rewind the system and the measurements can start from the initial state.

Authors believe this was the first moving automatic water quality monitoring system that users that can monitor vertical profile of a water body continuously. Later, several companies adopted the same idea and

commercialized the system that does the same job but with different vertical moving mechanism.

Vertical moving system in this research has kept being evolved since its first development as shown in Fig. 10 (Lee et al., 2004). In 2004, authors developed VeMAS (vertically moving automatic water monitoring system) that is advanced version of its first development in 1997 as shown in Fig. 11. The VeMAS consists of a barge, sensors, mechanical system including motor and wince, power system including solar panel and rechargeable battery and controller and wireless communication system using CDMA technique. Figure 12 shows installation of VeMAS in Daechung Lake, Korea.

The data generated from sensors are continuously sent to the server at the remote location as shown in Fig. 13. The computer server processes the received information and is designed to transfer data for real time web-service as shown in Fig. 14. A user or client may download or view the data dependent on their needs as shown in Fig. 15.

5 Future Improvements

Though the VeMAS has been improved for nearly 10 years, there still are aspects for improvement.

Firstly, it will be necessary improve physical data transfer line between sensors and data transmission system. When depth of reservoir becomes deeper this line has to become longer and this consequently

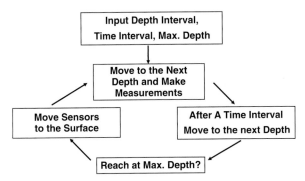

Fig. 9 Conceptual operation procedure of vertical automatic water quality monitoring system

Fig. 10 Evolution of VeMAS
(1997–2007)

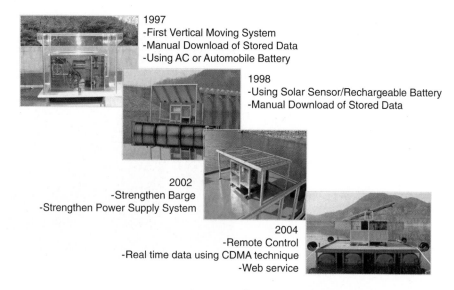

1997
-First Vertical Moving System
-Manual Download of Stored Data
-Using AC or Automobile Battery

1998
-Using Solar Sensor/Rechargeable Battery
-Manual Download of Stored Data

2002
-Strengthen Barge
-Strengthen Power Supply System

2004
-Remote Control
-Real time data using CDMA technique
-Web service

Fig. 11 Components of
VeMAS

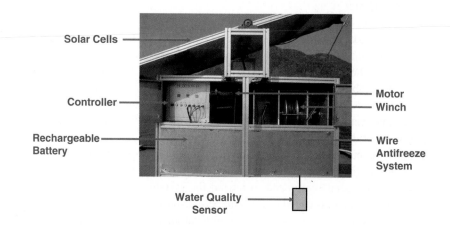

Solar Cells
Controller
Rechargeable Battery
Motor
Winch
Wire Antifreeze System
Water Quality Sensor

Fig. 12 Installation of
VeMAS at Daechung Lake

Fig. 13 Data processing structure for real time service of VeMAS

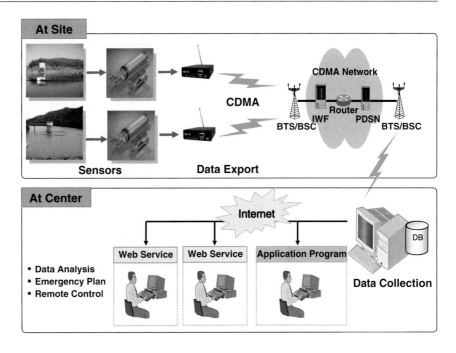

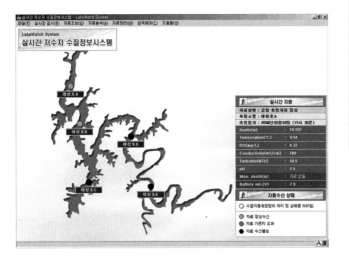

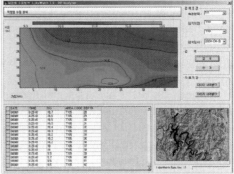

Fig. 14 Client server system using VeMAS

Fig. 15 Vertical Temperature profile in Daechung Lake, Korea using VeMAS

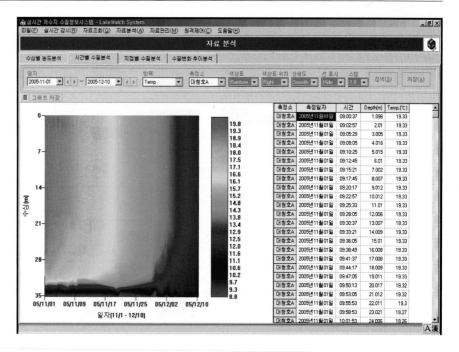

requires greater motor and also more power consumption. Therefore, it is desirable to reduce the size of communication line. If this system can be improved to be wireless such as using blue tooth technique, it would be a major break through to reduce the size of whole monitoring system.

Secondly, it is necessary to improve detection methods or sensors. Currently only conventional parameters such as DO, specific conductivity, pH and etc can be monitored due to detection limit. However, it is known that nutrient concentrations are important in reservoirs especially due to eutrophication. Alternative idea for this problem is that the monitoring system can be modified to withdraw samples from different depths and processed either at site using separately installed analyzing equipment or can be transferred periodically to laboratory.

Thirdly, VeMAS can be designed to control its mechanical system at site from the remote location for emergency such as hardware malfunctioning due to foreign object. The VeMAS can be designed to measure 3-Dimensional water velocity using Acoustic Doppler Profiler (ADP) system. This feature is very important when 3-D water movement needs to be calibrated.

Fourthly, this monitoring system needs to be integrated with water quality management actions of the water body such as artificial aeration system or early warning system. Song and Seo (2004) developed the method to effectively operate artificial aeration system in small drinking water reservoirs. They reported that it is necessary to integrate the artificial aeration system with automatic water quality monitoring system.

6 Conclusions

There have been many automatic water quality monitoring systems in treatment plants and fixed locations. Authors believe that we developed the first automatic water quality monitoring system in freshwater that moves vertically continuously. Due to development of information technology (IT), other techniques including environmental technology (ET) have been developed at a faster rate. VeMAS is the typical example of hybridation between two academic fields. The success of water quality monitoring system should be determined not by its technical complexity but by its usefulness in water quality management. This paper reports successful development of VeMAS that can be used for real time water quality information and water quality management actions. It should be noted that VeMAS was developed by field engineers like authors who have monitored water quality in a number of water body manually and thus felt necessity of its automation.

The technique used in development of VeMAS can be applied in many other water resources management actions including turbidity predictions using 2-D or 3-D hydrodynamic and water quality models, TMDL development. However, it should be noted that there will be no perfect automation in such system. Sensors require periodic maintenance and calibrations. Samples may have to be withdrawn and to be transported to the lab where more sophisticated analysis will be possible.

Acknowledgments Authors wish to thank, Jay Hoon Choi and Hyundong Hwang in M-Cubic, Inc., Daejeon, Korea for their support and contribution for the development of the VeMAS. Authors also wish to thank many graduate students of Department of environmental engineering in Chungnam National University for their sweat and effort to make VeMAS today.

Abbreviation (Dongil Seo)

ADP   Acoustic Doppler Profiler
BOD   Biochemical Oxygen Demand
CDMA   Code Division Multiple Access
DO   Dissolved Oxygen
EFDC   Environmental Fluid Dynamics Code
ORP   Oxygen Reduction Potential
TMDL   Total Maximum Daily Loads
TN   Total Nitrogen
TOC   Total Organic Carbon
TP   Total Phosphorus
USEPA   United States Environmental Protection Agency
VeMAS   Vertically moving automatic water monitoring system
VOC   Volatile Organic Carbon
WASP   Water Quality Analysis and Simulation Program

References

Ambrose R B, Wool T, Connolly J P and Schanz R W (1988) "WASP4, A Hydrodynamic and Water Quality Model: Model Theory, User's Manual, and Programmer's Guide," Environ. Res. Lab., EPA 600/3-87/039, Athens, GA.

Ambrose R B, Wool T A and Martin J L (1993) The Water Quality analysis Simulation Program, WASP5, Part A: Model Documentation, USEPA, Athens, GA.

Brown L C and Barnwell T O, Jr (1987) The Enhanced Stream Water Quality Models QUAL2E and QUAL2E-UNCAS: Documentation and User Manual, EPA/600/3-87/007.

Chapra S C, Pelletier G J and Tao H (2007) QUAL2K: A Modeling Framework for Simulating River and Stream Water Quality, Version 2.07: Documentation and Users Manual. Civil and Environmental Engineering Dept., Tufts University, Medford, MA, USA.

Chun M and Choi S (2004) "Present and Efficient Use of the Han River Waterahed Management Fund", Proceedings of Chuncheon Water Forum. (In Korean).

Chung S W, OH J K and Ko I H (2005) "Simulation of Temporal and Spatial Distribution of Rainfall-Induced Turbidity Flow in a reservoir using CE-QUAL-W2", Korea Water Resources Association, 38(8), 665–664. (In Korean with English Abstract).

Chung S W, Yoon S W and Ko I H (2006a) "Development of a Real-time Turbidity Monitoring and Modeling System for a Reservoir", The Proceeding of 7th International Conference on Hydroinformatics HIC 2006, 4–8 September, 2006, Nice, France, pp. 1495–1502.

Chung S W, Yoon S W and Ko I H (2006b) "A Real-Time Turbidity Monitoring and Modeling System for a Reservoir using CE-QUAL-W2 Model", The Proceeding of 5th IWA World Water Congress and Exhibition, 10–14 September, 2006, Beijing, China, p. 6.

Cole T and Buchak E (1995) "CE-QUAL-W2: A Two-Dimensional, Laterally Averaged, Hydrodynamic and Water Quality Model, Version 2.0," Tech. Rpt. EL-95-May 1995, Waterways Experiments Station, Vicksburg, MS.

Cole, T M, and Wells S A (2007) CE-QUAL-W2: A Two-Dimensional, Laterally Averaged, Hydrodynamic and Water Quality Model, Version 3.5 User Manual, Environmental Laboratory U.S. Army Corps of Engineers Waterways Experiment Station Vicksburg, MS 39180–6199 and Department of Civil and Environmental Engineering Portland State University Portland, OR 97207-0751.

Di Toro D M, O'Connor D J and Thomann R V (1971) A Dynamic Model of the Phytoplankton Population in the Sacramento San Joaquin Delta. Adv. Chem. Ser. 106, Am Chem Soc., pp. 131–180.

Di Toro D M, Fitzpatrick J J and Thomann R V (1983) Water Quality Analysis Simulation: Program (WASP) and Model Verification Program (MVP)-Documentation. Hydroscience, Inc., Westwood, NY doe USEPA, Duluth, MN Contract No. 68-01-3872, 1981, erv. 1983.

Hamrick J M (1992) A Three-Dimensional Environmental Fluid Dynamics Computer Code: Theoretical and Computational Aspects. The College of William and Mary, Virginia Institute of Marine Science. Special Report 317. 63pp.

Jeon M S and Choi S E (2004) Effective use of the Han River Basin Management Fund, Chuncheon Water Forum.

Kim Y, Choi S, Kim B and Seo D I (2001) "Modeling of thermal stratification and transport of density flow in Lake Soyang using the 2-D hydrodynamic water quality model, CE-QUAL-W2", J. Korean Soc. Water Wastewater, 15(1), 40 ~ 49 (In Korean with English Abstract).

Lee J, Choi J, Hwang H and Seo D (2004) "New Generation Automatic Lake Water Quality Monitoring System", The 24th International Symposium of North American Lake Management Society, Victoria, Canada.

Lee S, Seo D and Baek D (2002a) "Feasibility Stud of Automatic Water Quality Monitoring Stations in the Nakdong River, the Geum River and the Youngsan River", Report of Environmental Management Corporation. (In Korean).

Lee E H, Seo D, Hwang H, Yoon J and Choi J (2006) "Causes of Fish Kill in the Urban Stream and Prevention Methods II – Application of Automatic Water Quality Monitoring System and Water Quality Modeling", J. Korean Soc. Water Wastewater, 20(4), 585–594 (In Korean with English Abstract).

Lee S, Seo D, Kim J and Oh H J (2002b) "Automatic Monitoring and CE-QUAL-W2 Modeling of Stratification Characteristics in Daechung Lake, Korea", The 22th International Symposium of North American Lake Management Society, Anchorage, AK.

Seo D (2007) "Suggestions on Total Waste Load Management Act for Water Quality Management in Korea", Special Presentation in the Joint National Conference of Korea Water and Wastewater Society and Korea Society of Water Environment, Ilsan, Korea. (In Korean).

Seo D and Lee J (2005) "Prediction of sediment transport characteristics using 3-D Hydrodynamic Model, EFDC", IWA International Conference on Particle Separation 2005, Seoul, Korea.

Seo D and Yu H (2007) "Error Analysis of QUAL2E Water Quality Model using EFDC-Hydro and WASP7.2 – A Case Study of the Lower Han River", Korea Water and Wastewater Association and Korea Water Environment Association Joint Symposium, Seoul. (In Korean).

Seo D, Choi J, Jung S, Cha M and Song H (1997b) "Analysis of Stratification Characteristics in the Daechung lake by Using Continuous Water Quality Monitoring System", The 17th International Symposium of North American Lake Management Society, Houston, TX.

Seo D, Lee E H and Lim S (1997a) "Development of Automatic Vertical Moving System for Water Quality Sensors with Internal Memory to Monitor Stratification Characteristics of Lakes and Reservoirs", The 17th International Symposium of North American Lake Management Society, Houston, TX.

Song M and Seo D (2004) "Investigation of the Flow Generated by an Intermittent Hydraulic Jet Circulator", J. Ocean Sci. Technol., 1(2), 119–123.

Streeter H W and Phelps E B (1958), A Study of the Pollution and Natural Purification of The Ohio River, III, Factors Concerned in the Phenomena of Oxidation and Reseration, U.S. Public Health Serv., Pub. Health Bulletin Number. 146, February 1925, 75pp. Reprinted by US, HHEW, PHA.

Yotsukura N (1968) As referenced in "Surface Water Quality Modeling", Chapra S C, Mc-Graw Hill, New York, p. 247, 1997.

Part III
Environmental Toxicity Monitoring and Assessment

Toxicity of Metallic Nanoparticles in Microorganisms- a Review

Javed H. Niazi and Man Bock Gu

Abstract Recent advances in the synthesis and development of nanoparticles (NPs) for wide applications has lead to a serious threat to both human and environmental health. NPs are highly reactive and catalytic in nature compared to their ions or bulk counterparts and thus applicable in various fields including drug delivery, electronics, optics, and therapeutics. Due to these applications, many varieties of NPs in massive amounts are being industrially produced. These NPs are discharged in to the environment and thus providing a path to enter into food chain via microorganisms and eventually disturbs the ecological balance. The NPs exhibit toxicity to living organisms mainly because of their small size (>100 nm), large surface-to-volume ratio and highly reactive facets. The microorganisms including bacteria present in the natural ecosystem are the primary targets that get exposed to NPs. Before these NPs enter into the food chain, it is imperative to evaluate the toxicity associated with NPs in microorganisms. The most convenient and rapid way is to perform toxicity analysis using microorganisms such as bacteria. Toxicity of nanomaterials using microorganisms such as *E.coli, Pseudomonas, Bacillus* as models for prokaryotes gives an insight into the toxic impacts of NPs. Toxicities associated with NPs in microorganisms is mainly related to their nano-size that cause membrane disorganization, generation of reactive oxygen species (ROS) and in some cases, oxidative DNA damage. In this review article we describe the toxicity of various nanoparticles in bacteria and provide a rationale for assessing nanotoxicity and discuss the current status on toxicity impacts on microorganisms.

Keywords Nanoparticles · Nanotoxicity · Membrane damage · Reactive oxygen species · Oxidative toxicity

1 Introduction

Metallic nanomaterials are among the most important catalysts, the smaller the metal particles (<100 nm), the larger the fraction of the metal atoms that are exposed at surfaces, where they are accessible to reactant molecules and available for catalysis. Due to this the nanoparticles have unusual physical and chemical properties that differ substantially from those conventional bulk materials of the same composition. The unique characteristics of metallic nanoparticles (NPs) have drawn a lot of attention for their promising applications in optical, electrical, mechanical, chemical and medical uses. However, it is not currently clear whether these nanostructures present harmful effects on the human health and environment. Therefore, exploitation of the full potential of the nanotechnologies requires close attention to the toxicities of nanoparticles on the living cells.

Currently, nanomaterials that have been found to be toxic can be classified into four types: (i) carbon based nanomaterials that are mostly made of carbon in the form of hollow spheres, ellipsoids (fullerenes), or tubes (nanotubes). These are found to accumulate in living cells and cause cytotoxicity and pulmonary toxicity (Lam et al. 2004; Magrez et al. 2006; Porter et al. 2006; Wei et al. 2007); (ii) the metal based nanomaterials

Prof. M.B. Gu (✉)
College of Life Sciences and Biotechnology, Korea University, Anam-dong, Seongbuk-Gu, Seoul 136-701, South Korea
e-mail: mbgu@korea.ac.kr

including quantum dots, various metallic NPs, such as Au, Ag, Pt, and FePt NPs (Lengke et al. 2006; Maenosono et al. 2007; Morones et al. 2005), and metal oxides, such as TiO_2, ZnO_2, Fe_3O_4, Al_2O_3, CrO_3, and SiO_2. These have adverse effects mainly causing oxidative stress, apoptosis, inflammation of endothelial cells, and ecotoxicity (Borm et al. 2006; Gojova et al. 2007; Heinlaan et al. 2008; Jeng and Swanson 2006); (iii) the nanomaterials based on met-allodendrimers are composed of nanosized, metal conjugated with organic polymers, used in molecular electronics and catalysis. For example, water-soluble star-shaped organo-iron redox catalysts have been used for nitrate and nitrite cathodic reduction in water, and electron-reservoir serving as molecular batteries and the example of C_{60} (Astruc et al. 2003; Caminade and Majoral 2004; Partha et al. 2007). Lastly, (iv) metal NPs composites that combine two different NPs, which form bulk-type materials such as Fe-Pt NPs. Nanoma-terials composites, such as nanosized clays were used to enhance mechanical, thermal, barrier, and flame-retardant properties (Subramanian et al. 2003). Such composite metal NPs (Fe-Pt) causes mutagenicity in bacteria (Maenosono et al. 2007). Both metalloden-drimers and metallic NP composites present toxicities that are associated with metal NPs as seen in above two cases (i and ii).

Nanomaterials, such as fullerenes (C_{60}) and carbon nanotubes have many interesting and unique proper-ties potentially useful in a variety of biological and biomedical systems and devices and finally find their way into the environment (Seetharam and Sridhar 2007). The insoluble carbon nanotubes in aqueous phase have buoyancy and therefore float on top of the aqueous layer and are mistaken for food and ingested by aquatic organisms. There have been attempts to modify carbon nanotubes for improved bioapplica-tions, especially for the aqueous solubility and bio-compatibility of carbon nanotubes. As a result, var-ious methodologies for the aqueous dispersion and solubilization of carbon nanotubes, such as modifi-cation by biofunctionalization, functionalization with hydrophilic polymers, and non-covalent stabilization have been reported (Reviewed by Lin et al. 2004). Car-bon based nanoparticles including fullerene (C60) and single-walled carbon nanotubes (SWCNT) are taken up by aquatic organisms and that these are known to induce changes in biochemical or gene expression lev-els (Oberdorster 2004; Zhu et al. 2006).

Despite of the vast applications of these nanomate-rials for the benefit of human, there is increasing con-cern that NPs affect human and environmental health. Studies have shown that NPs lead to an increase of bioavailability and toxicity (Nel et al. 2006). Currently, a complete understanding of the size, shape, composi-tion and aggregation-dependent interactions of nanos-tructures with biological systems is lacking (Fischer and Chan 2007). Only, few studies have investigated the toxicological and environmental effects of direct and indirect exposure to nanomaterials and no clear guidelines exist to quantify these effects (Colvin 2003). Hence, an area of nanotechnology called nanotoxicol-ogy has emerged (Oberdorster et al. 2005b, 2007). There is a keen interest in nanotoxicology research because the processing of nanosturctures in biologi-cal systems could lead to unpredictable effects because of their distinct properties compared with their ions or bulk counterparts (Fischer and Chan 2007).

The potential hazard of manufactured NPs, their release into the aquatic environment and their harm-ful effects remain largely unknown (Moore 2006). Existing reports on nanoparticles show that they may conjugate with biological molecules in natural aque-ous environments making them gain soluble proper-ties that may have adverse effect on bacteria and other aquatic organisms. Interaction of carbon nanotubes or fullurenes with biological systems is well docu-mented, especially with biological macromolecules, such as DNA, RNA, proteins as well as lysophospho-lipids (reviewed in Ke and Qiao 2007). A first evi-dence of direct contact with purified SWNT aggregates induces damage to bacterial cell membrane and thus cell death indicating the strong antimicrobial activity of SWNTs (Kang et al. 2007). Similarly, *E. coli* under-goes severe membrane damage and subsequent loss of viability due to SWNTs. However, very little informa-tion is currently available with regard to the cytotoxic mechanisms of SWNTs (Kang et al. 2007). Studies on toxicity of carbon nanotubes using *Staphylococcus aureus* and *Staphylococcus warneri* showed antimicro-bial activity and inhibition of microorganism attach-ment and biofilm formation (Narayan et al. 2005). It is assumed that SWNTs induce significant morpho-logical changes, that included elongation and similar changes that have been shown in bacterial cells under extreme conditions, such as that under high tempera-ture (Rasanen et al. 2001), pressure (Ritz et al. 2001), and changes in surface-to-volume ratio and exposure to chemical agents (Veeranagouda et al. 2006). Recently, Ghafari et al. reported that SWNTs are internalized by a protozoan *T. thermophilia* and acquire inability to ingest and digest their prey bacteria species allowing nanotubes to move up the food chain (Ghafari et al. 2008). This suggests that presence of carbon nanotubes as contaminant in the aquatic environment may have deleterious effects, which eventually lead to ecological imbalance.

Thus toxicity testing of NPs should be performed in an environmentally relevant mode to avoid misleading information on toxicity of NPs (Oberdorster et al. 2006). The effect of nanoparticles on microorganisms is much more extensive and diverse than for plants, invertebrates and vertebrates (Oberdorster et al. 2007). Nanoparticles including TiO_2 and silver have been used as antibacterial agents regardless of the particle size, but this activity is enhanced when delivered in a nanoparticulate form. One material which is not inherently antimicrobial is carbon, however C_{60} fullerens have recently been found to inhibit the growth of *Escherichia coli* and *Bacillus subtilis* (Fortner et al. 2005). However, there is insufficient evidence to suggest that all nanoparticles have antimicrobial effects, or in fact that all nanoparticles are toxic to any organism exposed in an environment. The impact of nanoparticles with respect to toxicity on microorganisms is still in its infancy stage. Unicellular microorganisms, such as bacteria and yeast can serve as the model organisms to study the toxicology of NPs. Before the toxicity of any nanomaterial to be tested on living organisms, it is imperative to understand the physicochemical properties of a given nanomaterial.

2 Physical and Chemical Properties of Nanoparticles

Nanoparticles exhibit unique physical and chemical properties compared to same material without nanoscale features. It is mainly because of following reasons; (a) their size in nanoscale measuring 100 nm or less with one or more dimensions, surface characteristics and morphology of their sub-structure, (b) their properties differ significantly from those of larger size as a result of manipulation at atomic, molecular and macromolecular scales, (c) several nanoparticles have nanostructures at the nanoscale levels. Nanomaterials have complex interrelation between the structure and the composition of the materials. Thus they acquire novel properties derived from atomic and molecular origin in a complex way along with features of its native bulk counterpart.

Nanomaterials exist in various shapes and structures such as spheres, needle, tubes, plates, sheets etc. The size and shape of nanomaterials contributes to onset of cytotoxicty, for example, single-wall nanotubes are more toxic than multi-wall nanotubes (Jia et al. 2005; Kang et al. 2007). Understanding of important physicochemical properties of nanoparticles in order for characterizing nanoparticle's toxicity to biological systems comprise (a) size distribution, (b) nature of agglomeration/aggregation, (c) shape, (d) structure of nanomaterial, (e) surface area, (f) surface chemistry, (h) surface charge, and (i) porosity (Oberdorster et al. 2005a). Various physicochemical methods have been employed to characterize the nanomaterials, including Transmission and/or Scanning Electron Microscopy (TEM or SEM), X-Ray Diffraction (XRD), Dynamic Light Scattering (DLS), Zeta potential, Isothermal adsorption, and Spectroscopic techniques (UV vis, IR, Raman, NMR) (Oberdorster et al. 2005a).

Under ambient conditions, some nanoparticles form aggregates or agglomerates. Nanoparticles also tend to aggregate by fusing and deposition that form bulk components. Nanoparticles suspended in gas tend to stick to each other more rapidly than in liquids. The primary free nanoparticle may form agglomerated primary particles (agglomerates) by interparticle interaction, which forms a collection of particles that are attached together by both weak and strong forces, including van der Waals, electrostatic forces and sintered bonds (Oberdorster et al. 2005a). The particle-particle interaction at the nanoscale level is governed by weak van der Waals forces, stronger polar and electrostatic or covalent interactions. The interparticle interaction is also influenced by viscosity and polarisability of the aqueous environment in order to form nanoparticle's aggregation. The forces involved in nanoparticle – nanoparticle interaction and nanoparticle – aqueous solution interactions are the basis for physical and chemical processes. The attractive or repulsive forces of nanoparticles crucially determine the fate of individual and collective nanoparticles. This interaction between nanoparticles results in aggregates and/or agglomerates, which greatly influences on their physical and chemical nature.

Most of nanoparticles belonging to this category are modified chemically or engineered by surface modification to avoid agglomeration. The nanoparticles in the presence of chemical agents (surface active agents), the surface and interfacial properties may be modified and such agents can inderctly stabilize against coagulation (agglomeration) by preserving charge on the nanoparticles. The properties of nanoparticles can be significantly altered by surface modification and the distribution of nanoparticle that mainly depends upon the surface characteristics. Engineering nanoparticles by surface modification, addition or modification of surface functional groups and chemical composition to maintain the characteristics of nanoparticles that are often stable and prevent agglomeration or aggregation (reviewed in Oberdorster et al. 2005a, 2007). The behavior of nanoparticles will be dependent on their solubility and susceptibility to degradation and that neither the chemical composition nor particle size is to remain constant over time. This makes increasingly difficult to study and understand the cytotoxicity of any nanoparticle on biological systems. Therefore, the current review will provide some highlights and conclusions based on the existing information on the toxicological studies of nanosized particles, specific mechanisms underlying NPs' effects particularly focusing on the microorganisms with special attention to metallic nanoparticles as models.

3 Nanoparticles Pose Potential Threat to Bacteria

A majority of toxicity concerns that has been so far addressed is related to human cells or human health. Nevertheless, it is important to test the impact of NPs on other living organisms that exist in the natural environment including prokaryotes, such as bacteria, and other unicellular microorganisms. These unicellular microorganisms can also serve as model organisms for NP-toxicity analysis. It is most interesting that the bacteria are more sensitive than human fibroblast (Brunner et al. 2006; Limbach et al. 2005). The microorganisms are the primary targets for being

exposed to the man made NPs after they are discharged into the environment. As a result, the microbial interactions and uptake of NPs would lead to entering of persistent NPs into the food chain, which eventually disturbs the ecological balance. NPs gain entry into the living cells through various means including physical rupturing of cell membrane or wall or endocytosis and cause cellular toxicities at various levels. Studies have confirmed that the metallic NPs can pass through or remain attached to the cell membrane (Borm and Kreyling 2004; Kashiwada 2006). A number of studies have examined the uptake and effects of NPs at a cellular level to evaluate their impact on humans. It may not be extrapolated to other species, such as unicellular microorganisms (bacteria or yeast) based on the conclusions of these studies, but more research is needed to confirm this assumption. Therefore, there is a need to assess the toxic impacts of various types of NPs not only on human or higher organisms but also on the microorganisms. It is important to investigating effect of NPs on bacteria because of the potential impact on microorganisms that serve as the basis of the food chain and as primary agents for biogeochemical cycles.

NPs exhibit different toxicities, which is dependent on the two major factors: (i) nature of NPs, such as size, morphology, and chemical nature; (ii) interaction with different microbial species and underlying potential mechanisms that should be investigated, which include cell wall damage and the role of NPs in disruption of membrane integrity, oxidative stress via reactive oxygen species (ROS) formation, organic radicals generated in the absence of light, and possible genotoxicties exhibited. A summary of a range of nanoparticles, their size, effective concentrations, and potential toxicity mechanisms currently available for Gram-negative and Gram-positive bacteria is summarized in Tables 1 and 2.

4 Nanoparticles Disrupt the Integrity of Cell Membrane

Nanoparticles interact with the bacterial cell membrane by adsorption or electrostatic interactions (Thill et al. 2006). Large thickness in outer membranes of some bacteria such as *E. coli* certainly plays a crucial

Table 1 Summary of different metallic and metal oxide nanoparticles and their toxic effects in Gram-positive bacteria

Bacteria	NPs	Size (nm)	Effective conc.	Toxicity action	Reference
Escherichia coli	SiO$_2$, TiO$_2$, ZnO,	205–480	10–1000 mg/L	Light induced ROS generation, cellular internalization; oxidative toxicity, antibacterial activity, membrane disorganization	(Adams et al. 2006a, b; Brayner et al. 2006; Fu et al. 2006; Reddy et al. 2007; Rengifo-Herrera et al. 2007; Tsuang et al. 2008)
	Ag	1–40	25–100 mg/L	Increased membrane permeability, cellular internalization, perforation of membrane; membrane damage and cell death	(Baker et al. 2005; Gogoi et al. 2006; Morones et al. 2005; Pal et al. 2007; Ruparelia et al. 2007; Sondi and Salopek-Sondi 2004)
	C60	25–500	0.4–4 mg/L	Decreased CO2 production; cytotoxicity, mechanical stress on the cell wall or membrane	(Fortner et al. 2005; Lyon et al. 2005; Tang et al. 2007)
	FePt	9	2.5 mg/ plate	Mutagenicity; DNA damage	(Maenosono et al. 2007)
	MgO	4	ND	Damage cell membrane; cell wall leakage	(Stoimenov et al. 2002)
	CeO$_2$	7	1.2–37 mg/L	Interacts outer membrane and cell-membrane damage	(Thill et al. 2006)
Pseudomonas putida	Fullerene/C60	50–200	0.09–0.5 mg/L	Oxidative stress (ROS generation), decrease levels of unsaturated fatty acids, increase cyclopropane fatty acids, altering membrane lipid composition, Phase transition temperature (by ROS), membrane fluidity	(Fang et al. 2007)
Pseudomonas aeruginosa	Ag	1–10	25–100 mg/L	Interact with cell membrane and sulfur- and phosphorous containing compunds such as DNA; Damage cell membrane and DNA	(Morones et al. 2005)
	TiO$_2$	20	10 mg/mL	Photoactivation of TiO2 induced loss of viability; Bactericidal	(Tsuang et al. 2008)
Salmonella typhimurium	C60	50–200	300 mg/L	Generate single oxygen, ROS generation and mutagenicity Oxidative toxicity, oxidative DNA damage, mutagenicity	(Sera et al. 1996)
	FePt	9	2.5 mg/ plate	Mutagenicity; DNA damage	(Maenosono et al. 2007)
	Ag	1–10	25–75 mg/L	Interact with cell membrane and sulfur- and phosphorous containing compunds such as DNA; Damage cell membrane and DNA	(Morones et al. 2005)
Shewanella oneidensis	Pd(0)	1:1 to 1:10 size of cell to NP	∼50 mg/L	Bioreduction; Cytotoxicity	(De Windt et al. 2006)
	C60	ND	0–80 mg/L	Mechanical stress on the cell wall or membrane	(Tang et al. 2007)
Vibrio fischeri	TiO$_2$, CuO, ZnO	50–70	1.1–79 mg/L	Oxidative stress (extracellular ROS generation); acute toxicity, membrane damage, impaired growth,	(Heinlaan et al. 2008)
Vibrio cholera	Ag	1–10	25–75 mg/L	Interact with cell membrane and sulfur- and phosphorous containing compounds such as DNA; Damage cell membrane and DNA	(Morones et al. 2005)
Bacteroides fragilis	TiO$_2$	20	10 mg/mL	Photoactivation of TiO2 induced loss of viability; bactericidal	(Tsuang et al. 2008)

Table 2 Summary of different metallic and metal oxide nanoparticles and their toxic effects in Gram-negative bacteria

Bacteria	NPs	Size (nm)	Effective conc.	Toxicity action	Reference
Bacillus subtilis	C60	25–500	0.01–0.75 and 0.4–4 mg/L	Oxidative stress (ROS generation); increased iso- and anteiso-branched fatty acids, altering membrane lipid composition, Phase transition temperature (by ROS), membrane fluidity	(Fang et al. 2007; Fortner et al. 2005; Kai et al. 2003; Lyon et al. 2006, 2005)
	SiO_2, TiO_2, ZnO	205–480	10–5000 mg/L	Light induced ROS generation; oxidative toxicity, antibacterial activity	(Adams et al. 2006a)
	MgO	4	ND	Bactericidal effects, cell wall damage; cell wall disruption, desiccation	(Stoimenov et al. 2002)
	Ag	3	ND	Bactericidal effects, cell wall damage	(Ruparelia et al. 2007)
	CuO	9	ND	Bactericidal effects, cell wall damage	(Ruparelia et al. 2007)
Bacillus megaterium	MgO	4	ND	Bactericidal effects, cell wall damage; cell wall disruption, desiccation	(Stoimenov et al. 2002)
Streptococcus agalactiae	ZnO	60–150	<837 mg/L	Cellular internalization, membrane disorganization, increase membrane permeability; cytotoxic, bactericidal	(Huang et al. 2008)
Streptococcus pyogenes	Fe_3O_4-TiO_2	>100	2.57 mg/mL	Photokilling of bacteria	(Chen et al. 2008)
Staphylococcus aureus	ZnO	60–150	<837 mg/L	Cellular internalization, membrane disorganization, increase membrane permeability; cytotoxic, bactericidal	(Huang et al. 2008; Reddy et al. 2007)
	Ag	3	ND	Bactericidal effects, cell wall damage	(Ruparelia et al. 2007)
	CuO	9	ND	Bactericidal effects, cell wall damage	(Ruparelia et al. 2007)
	TiO_2	20	10 mg/mL	Photoactivation of TiO_2 induced loss of viability; bactericidal	(Tsuang et al. 2008)
	Fe_3O_4-TiO_2	>100	2.57 mg/mL	Photokilling of bacteria	(Chen et al. 2008)
Staphylococcus saprophyticus	Fe_3O_4-TiO_2	>100	2.57 mg/mL	Photokilling of bacteria	(Chen et al. 2008)
Enterococcus hirae	TiO_2	20	10 mg/mL	Photoactivation of TiO_2 induced loss of viability; bactericidal	(Tsuang et al. 2008)

role in the very high level of adsorption which is observed as already suspected by researchers (Chatellier et al. 2001). No clear evidence of the NPs passage inside the cells can be obtained by techniques such as transmission electron microscopy (TEM). This is possibly because of strong electrostatic interaction between the NPs and the membrane that might block them at the surface for very long time. However, it was found that NPs are found to be mainly located on the surface of the bacteria using adsorption isotherms and TEM images (Morones et al. 2005; Thill et al. 2006). Further, this adsorption onto the surface is linked to an oxidative stress for the bacteria.

Interaction of nanoparticles with cell membrane was found to be different in Gram positive and Gram negative bacteria because of their distinct membrane compositions. Exposure of nC_{60} with *Pseudomonas putida* (Gram-negative) and *Bacillus subtilis* (Gram-positive) result in altering membrane lipid composition, phase transition temperature, and membrane fluidity (Fang et al. 2007). It is suspected that lipid peroxidation is an important toxicity mechanism in bacteria, since bacterial lipids are mainly monounsaturated and thus unreactive to the lipid peroxidation reaction (Bielski et al. 1983; Imlay 2003). However, bacteria also tend to adapt physiologically by altering the membrane fatty acid compositions to cope up with the damage caused by the nanoparticles. It was found that Gram-positive bacteria exposed to nC_{60} nanoparticles increased the levels of iso- and anteiso- membrane fatty acids by 5–32%. Whereas, Gram-negative bacteria decreased the levels of unsaturated fatty acids and increased the cyclopropane fatty acids proportions (Fang et al. 2007). The distinct response by the different bacteria explains the differential responses associated with cell membrane integrity with respect to the toxicity of nC_{60}. However, it is to be noted that the nanoparticles also exist in a variety of different size, morphology, chemical nature that also contribute to the different ways of inducing cell-membrane damage. So far no detailed mechanisms of adaptation to the damage caused by NPs have been reported except for C_{60} (Fang et al. 2007). Only physical disruption of cell membrane with a range of metallic and metal oxide nanoparticles is evident from the literature. Recently size dependent silver nanoparticles were found to be located in the cell membrane as a result of direct interaction leading to bactericidal effects (Morones et al. 2005; Pal et al. 2007).

Smaller particles with a larger surface to volume ratio provide a more efficient means for antibacterial activity (Baker et al. 2005). *E. coli* cells exposed to ZnO NPs showed increase of membrane permeability leading to accumulation of ZnO NPs in the bacterial membrane and also cellular internalization of these NPs (Brayner et al. 2006). A substantial loss of cell viability/membrane integrity (~30%) was also observed in the *E. coli* following treatment with ZnO NPs (Reddy et al. 2007). A range of metal oxide NPs including ZnO, SiO2, TiO2, and MgO have shown to cause membrane disorganization, increased membrane permeability as a result of perforation, and finally leading to cell death (Adams et al. 2006a; Brayner et al. 2006; Reddy et al. 2007; Stoimenov et al. 2002; Tsuang et al. 2008). Large amount of CeO2 NPs measuring 7 nm sizes has been shown to be adsorbed on the *E. coli* outer membrane and undergo reduction bringing significant bacterial cytotoxicity. The toxicity effect of CeO2 NPs is brought on by interaction with *E. coli* via adsorption followed by oxidoreduction (Thill et al. 2006). Metallic nanoparticles such as nanosilver (Ag NPs), NPs of FePt, Pd, and C60 also found to cause membrane disruption in both Gram-positive and -negative bacteria (De Windt et al. 2006; Gogoi et al. 2006; Maenosono et al. 2007; Morones et al. 2005; Ruparelia et al. 2007; Sondi and Salopek-Sondi 2004).

Recently, studies have shown that the silver nanoparticles caused toxicity via protein/membrane and oxidative damage, but do not result in DNA damage. However, gold nanoparticles do not cause any damage to *E. coli* (Hwang et al. 2008). In addition, these findings and that of other groups, the silver nanoparticles appear to disrupt the cell membrane, which results in a synergistic toxicity effect to the cells (Lok et al. 2006; Sondi and Salopek-Sondi 2004).

5 Nanoparticles Induce Oxidative Toxicity by Generating Reactive Oxygen Species (ROS)

The mechanism by which NPs induce toxicity is thought to be via oxidative stress that damages lipids, carbohydrates, proteins and DNA (Fang et al. 2007; Kohen and Nyska 2002). Lipid peroxidation is considered most deleterious that leads to alterations in cell membrane properties which in turn disrupt vital cellular functions (Rikans and Hornbrook 1997; Sera et al. 1996). ROS production has been found to be with NPs as diverse as C_{60}, fullerenes, single walled nanotubes (SWNTs), quantum dots, and ultrafine particles (UFPs). These nanomaterials have shown to generate ROS especially under concomitant exposure to light, UV, or transition metals (Brown et al. 2001; Derfus et al. 2004; Fang et al. 2007; Joo et al. 2004; Li et al. 2003; Lu et al. 2004; Oberdorster et al. 2005b; Yamakoshi et al. 2003). An antibacterial activity of metal oxide nanoparticles is associated with light induced oxidative stress. For example, TiO2 and SiO2 was toxic to both *E. coli* and *B. subtilis* under both light and dark conditions, and cell growth inhibition appeared higher in the presence of light (Adams et al. 2006a, b). Oxidative stress mechanisms leading to membrane damage and antibacterial properties has been demonstrated for ZnO in *E. coli* (Zhang et al. 2007). Potential mechanisms of oxidative stress via ROS formation, organic radicals generated in the absence of light, and the role of other nanomaterials in disruption of membrane integrity have been investigated using fullerene in *Pseudomonas putida* (Fang et al. 2007), TiO_2 in *Psuedomonas aeruginosa* (Tsuang et al. 2008). Metal oxide NPs, such as SiO2, TiO2, ZnO dependent on light for inducing ROS generation in *Bacillus subtilis* and *E. coli* that eventually lead to oxidative toxicity, membrane disorganization, and antibacterial activities (Adams et al. 2006a, b; Brayner et al. 2006; Fu et al. 2006; Reddy et al. 2007; Tsuang et al. 2008). The antibacterial properties of silver is long been known for over the decades. However, the mechanism of bactericidal actions of silver is still not well understood. The action of silver nanoparticles is thought to be broadly similar to that of silver ion (Pal et al. 2007). It is speculated that a bacterial cell exposed to silver nanoparticles takes in silver ions, which inhibit a respiratory enzyme(s), facilitating the generation of reactive oxygen species and consequently damage the cell (Hwang et al. 2008).

Metal oxides are highly reactive to light because the particulate metal oxides such as MnO, WO_3, $SrTiO_3$, Fe_2O_3, ZnS, ZnO and TiO_2 absorb sufficient light/UV energy and result in the formation of electron-hole pairs through a process of electronic excitation between the valence and conduction band (Beydoun et al. 1999; Hoffmann et al. 1995). Photo-

generated electrons and holes undergo reaction with dissolved molecular oxygen, surface hydroxyl groups, and adsorbed water molecules to form hydroxyl (•OH) and superoxide ($O_2\bullet^-$) radicals, as shown in equations (1), (2), (3), and (4):

$$\xrightarrow{ZnO} \quad hv \quad h_{vb}^+ + e_{cb}^- \qquad (1)$$

$$h_{vb}^+ + e_{cb}^- \longrightarrow ZnO + hv \text{ (or heat)} \qquad (2)$$

$$O_2 + e^- \longrightarrow O_2\bullet^- \qquad (3)$$

$$H_2O + h^+ \longrightarrow \bullet OH + H^+ \qquad (4)$$

where, h_{vb}^+ is the valence-band hole, and e_{cb}^- is a conducting band electron. It is proposed that this type of reactions occur when metal oxide NPs come in contact with bacteria and exposed to light source. Studies in recent years on light induced oxidative toxicity by NPs, such as C_{60} or fullerene, TiO_2, SiO2, ZnO, and MgO in bacteria have surfaced, and the toxicity is thought to be associated with ROS generation (Adams et al. 2006a; Brayner et al. 2006; Fu et al. 2006; Tsuang et al. 2008). Recent studies showed that composite irradiated $Fe_3O_4@TiO_2$ nanoparticles induce photokilling in bacterial species such as *Streptococcus pyogenes, Staphylococcus saprophyticus*, and *Staphylococcus aureus*, (Chen et al. 2008). For example, induction of oxidative stress by ROS generation by fullerene exposure to *Pseudomonas putida* and *Bacillus subtilis* following alterations in membrane composition as a defense mechanism (Fang et al. 2007). *E. coli* exposed to C_{60} has been shown to decrease CO_2 production, membrane damage, and cytotoxicity is associated with oxidative toxicity (Fortner et al. 2005; Lyon et al. 2005; Tang et al. 2007). C_{60} oxidative toxicity in other bacteria, such as *Salmonella, Shewanella* sp. has also shown by inducing singlet oxygen, and oxidative DNA damage related mutagenicity (Sera et al. 1996; Tang et al. 2007). NPs interaction is likely to be unique to Gram-positive and –negative bacteria that may have different potential to induce toxicities because of the differing compositions in their cell membranes (Fang et al. 2007).

ROS and other radicals are involved in a variety of biological phenomena, such as mutation, and carcinogenesis (Kohen and Nyska 2002). It is not entirely surprising that the ROS generation by NPs can also lead to oxidative DNA damage or mutagenicity. FePt and fullerene C_{60} nanoparticles were mutagenic to bacteria belonging to *Salmonella* sp. (Maenosono et al. 2007; Sera et al. 1996). The mutagenicity of C_{60} NPs is thought to be due to the indirect action of singlet oxygen and lipid peroxidation of linoleate that causes oxidative DNA damage (Sera et al. 1996). Silver nanoparticles interact with cell membrane and sulfur- and phosphorous containing compounds such as DNA and induce DNA damage in *V. cholera* and *S. typhus* (Morones et al. 2005). Although no clear evidence has been reported regarding the toxicity mechanisms of silver nanoparticles by generating ROS in bacteria. The ROS generation through singlet molecular oxygen production was seen by interaction between bacteria (*E. coli*) and the photo-catalytic TiO_2 nanoparticles (Adams et al. 2006b; Rengifo-Herrera et al. 2007). Light induced ROS generation appear to be common in metal oxide nanoparticles, such as TiO_2, SiO_2, and ZnO mediated cytoxicities, and thus these NPs are known to possess effective bactericidal effects (Adams et al. 2006a; Brayner et al. 2006; Fu et al. 2006; Reddy et al. 2007; Tsuang et al. 2008). A previous study demonstrated that metallic silver nanoparticles led to the production of silver ions and, subsequently, superoxide radicals. This damage is linked with the size of the particles because larger silver particles, i.e., micro-sized particles showed no toxicity to *E. coli* (Hwang et al. 2008).

6 Current Understanding on Toxicity of Nano-Sized Particles to Bacteria

Few microorganisms grow in the presence of high metal concentrations that might result from specific mechanisms of resistance. Such mechanisms include efflux systems, alteration of solubility and toxicity by changes in the redox state of the metal ions, extracellular complexation or precipitation of metals, the lack of specific metal transport systems, and the changes in membrane composition (Beveridge et al. 1997; Fang et al. 2007; Silver 1996). Recently, Fang et al. (2007) have demonstrated that Gram-positive and Gram-negative bacteria have separate ways of adaptation to fullerene nanoparticles toxicity, but both by changing membrane composition in order to cope up with the toxicity, a first ever evidence for adaptation to metal NPs by aerobic bacteria, although bacteria were not

resistant to NPs. However, more research is needed to explain the similar mechanisms underlying adaptation to other nanoparticles in aerobic microorganisms. A number of researchers reported the cytotoxicity of a range of NPs in both Gram-positive and –negative bacteria. A majority of these findings conclude that NPs induce oxidative toxicity by generation of ROS, and in some instances this ROS was triggered by the light exposure on NPs (Tables 1 and 2). A most common toxicity effect of NPs is associated with physical membrane damage leading to fatal effects as a result of perforation and membrane fluidity and/or disorganization. Metal or metal oxide NPs also release soluble ions that also contribute to the chemical toxicity to bacteria (Heinlaan et al. 2008). Studies have shown that some bacteria belonging to *Pseudomonas* sp. can solubilize bulk NPs, such as ZnO NPs, into Zn ions that exhibited bactericidal effects (Fasim et al. 2002).

The fact that presence of metal/metal oxide nanoparticles is toxic to aerobic bacteria, which is most certainly due to the reactivity of metal/metal oxide NPs with molecular oxygen and/or light, followed by ROS generation. However, there are no reports on the toxicity of NPs in absence of oxygen, or under anaerobic conditions. It is assumed that light-induced metal/metal oxide NPs toxicity may have detrimental effects on anaerobic bacteria. Nevertheless, thorough experimental evidence is required on these lines to confirm the hypothesis. But, it is well documented that anaerobic bacteria (i.e., metal reducing bacteria) unlike aerobic bacteria adapt to excess metal ions by reduction of metal ions and produce metal/metal oxide nanoparticles (Mandal et al. 2006). Anaerobic bacteria tend to change the environment of their outer membrane in presence of metal ions, creating electrochemical conditions favorable for metal ion precipitation, which is most likely be associated with an organic matrix and produce a broad size-distributed nanoparticles (Frankel 1987). For example, magnetite particles with a narrow size distribution around 40–50 nm are produced by iron-reducing bacteria and these particles are enveloped by bacterial membranes (Balkwill et al. 1980; Gorby et al. 1988). Synthesis of metal/metal oxide nanoparticles from external high metal ion concentrations is an adaptation process of anaerobic bacteria to cope up with the metal ion toxicity (reviewed in Mandal et al. 2006; Nies 2003). It is unclear that the nano-sized particles produced by anaerobes exhibit toxicity to themselves or to the co-existing microorganisms. The ability of metal-reducing bacteria to produce copious amounts of extra-cellular nanoparticles is a process of biogeochemical cycling of metal, carbon, nitrogen, phosphate, and sulfur in natural and contaminated subsurface environments which is well documented in the literature (Fredrickson et al. 2001; Liu et al. 1997; Lovely et al. 1987; Lovley 1995). It is also now important to assess the toxicity of nanomaterials on anaerobic bacteria that may have distinct cytotoxic mechanism and gives an insight into the impact of nanomaterials on both aerobic and anaerobic mesocosms.

7 Toxicity Assays of NPs Using Bacteria as Models

Toxicity assays using specific microorganisms can be used to assess the detrimental effects of various NPs on living organisms and understand its impact, mode of action or mechanism. As evidenced from the literature that NPs exhibit toxicity to bacteria (Tables 1 and 2). The detailed toxicity action of NPs in the cells or interaction with cellular proteins/enzymes and other components seems to be overlooked in most cases. Researchers have found only two major effects of NPs in bacteria; (i) NPs induce oxidative toxicity and (ii) cell-membrane/wall damage. A detailed study on a particular NP, impact of its size, chemical nature would allow us to understand the mode of NP action to cause cytotoxicity to bacteria. The outcome of this study also gives an insight into the toxicity action of NPs to other living organisms including effects on humans and environment. Only little information is currently available from the literature explaining the mechanism of toxicity, interaction with biological systems and environment (Nel et al. 2006).

Assessing toxicity of NPs using bacteria as model organism have many advantages, including the fact that bacterial assays are faster, sensitive, less expensive and easy to handle when compared to the cells of mammalian origin. Recently, toxicity of silver nanoparticles in bacteria has been studied using recombinant bacterial biosensors and elucidated the potential mode of toxic action by silver nanoparticles (Hwang

et al. 2008). Similarly, a quite a few number of studies have shown the toxicity modes of few nanomaterials, although not in detail, for example, toxicity fullerene, metal oxide NPs including ZnO, CuO, SiO$_2$, and TiO$_2$ (Fang et al. 2007; Heinlaan et al. 2008; reviewed in Oberdorster et al. 2005b, 2007). However, the toxicity mode of action deduced using these nanoparticles may not be the same with the other nanomaterials. It may be because of their variable size, surface chemistry, or chemical nature of nanomaterials. Likewise, the NPs may also have different effect on different types of cells, which depends on cell-wall composition (Fang et al. 2007). Therefore the scientific committee on emerging and newly identified health risk (SCENIHR) of Europe has concluded that there is insufficient data available at the present time to allow the identification of any systematic rules that govern the toxicological characteristics of all products of nanotechnology (SCENIHR 2006). Further, a guideline has been proposed that the risk assessment needs to be made on a case by case basis.

8 Summary and Future Outlook

It is most probable that production of nanomaterals and use for the benefit of human will lead to its entering in the environment as a result of disposal. So far there is no clear consent among the regulatory bodies and the manufacturers to examine ecotoxicological impacts of NPs. Until recently, toxicities of most nanomaterials have only focused on human cells and it is still continued to do so in future. However, very little is known about their potential adverse effects on aerobic or anaerobic microorganisms. Developing resistance properties to NPs by these microorganisms in the environment can be an evolutionary process which might take decades or centuries. It is important to assess these NPs for their toxicity on different microorganisms which provides a means for possible measures needs to be taken for safety.

Despite of the preliminary knowledge regarding NPs toxicity on humans and microorganisms, their detailed toxic effects still remained unknown at large. Limited information available on toxicity of NPs either in human cells or bacteria consistently points out that the greater surface to volume ratio or small size of

NPs is a main cause for their biological activity than larger-sized particles of the same composition. Secondly, NPs tend to induce membrane disorganization as a result of adsorption by electrostatic interaction and oxidative stress by generation of ROS in aerobic bacteria. Most importantly, the metal oxide NPs in particular are highly susceptible to light and oxygen in order for production of ROS and thus oxidative toxicity. ROS generation by the NPs is also likely to induce mutagenicity by oxidative DNA damage. This has been implicated to be occurring with few NPs, such as C_{60} and FePt in bacteria. However, no detailed mechanism for mutagenicity has yet been elucidated. It is well known that ROS generation in cells is also linked to indirect oxidative DNA damage and it is not surprising that ROS generation by NPs can also lead to oxidative DNA damage. In addition, the impact of NPs on anaerobic microorganisms is an important area to explore. So far no reports have yet found that address the effects of manmade NPs on anerobes. Therefore, more research is required to unveil the toxicity mechanisms associated with different types of NPs and sizes on aerobes and anaerobes.

A new field of nanosciences has now been emerged as a diversion to NPs which is focused on the synthesis of engineered NPs by modifying or coating with different functional groups for various applications, for example, quantum dots that have tremendous optical properties. This has raised new concerns about human and environmental health. The toxic effect of engineered NPs on microorganisms has yet to be studied in greater detail. Recently numerous engineered NPs have been industrially manufactured without the knowledge of their impact on living microorganisms. However, many of these engineered NPs have been used as fluorescent labels/markers to trace or locate the cancer or tumor cells in mice and are suspected to have cytotoxic effects, though it is still unclear. Nevertheless, it is also important to test these engineered NP's toxicity to microorganisms. Hence, there is a growing concern regarding the regulations on the synthesis and production of novel nanostructures because of their potential toxicity on microorganisms and other living systems. There seems to be lack of a model to predict toxicity on living organisms based on the physicochemical characteristics and microbial toxicity of new nanomaterials that can be used for risk assessment or for safe product design.

References

Adams LK, Lyon DY and Alvarez PJ (2006a) Comparative ecotoxicity of nanoscale TiO_2, SiO_2 and ZiO_2 Water suspensions. Water Res 40: 3527–3532

Adams LK, Lyon DY, McIntosh A and Alvarez PJJ (2006b) Comparative toxicity of nano scale TiO2, SiO2 and Zno water suspensions. Water Sci Technol 54: 327–334

Astruc D, Blais JC, Daniel MC, Gatard S, Nlate S and Ruiz J (2003) Metallodendrimers and dendronized gold colloids as nanocatalysts, nanosensors and nanomaterials for molecular electronics. C R Chim 6: 1117–1127

Baker C, Pradhan A, Pakstis L, Pochan DJ and Ismat SS (2005) Synthesis and antibacterial properties of silver nanoparticles. J Nanosci Nanotechnol 5: 244–249

Balkwill DL, Maratea D and Blakemore RP (1980) Ultrastructure of a magnetotactic spirillum. J Bacteriol 141: 1399–1408

Beveridge TJ, Hughes MN, Lee H, Leung KT, Poole RK, Savvaidis I, Silver S and Trevors JT (1997) Metal-microbe interactions: contemporary approaches. Adv Microb Physiol 38: 177–243

Beydoun D, Amal R, Low G and McEvoy S (1999) Role of nanoparticles in photocatalysis. J Nanoparticle Res 1: 439–458

Bielski BH, Arudi RL and Sutherland MW (1983) A study of the reactivity of HO2/O2- with unsaturated fatty acids. J Biol Chem 258: 4759–4761

Borm PJ and Kreyling W (2004) Toxicological hazards of inhaled nanoparticles – potential implications for drug delivery. J Nanosci Nanotechnol 4: 521–531

Borm PJ, Robbins D, Haubold S, Kuhlbusch T, Fissan H, Donaldson K, Schins R, Stone V, Kreyling W, Lademann J, Krutmann J, Warheit D and Oberdorster E (2006) The potential risks of nanomaterials: a review carried out for ECETOC. Part Fibre Toxicol 3: 11

Brayner R, Ferrari-Iliou R, Brivois N, Djediat S, Benedetti MF and Fievet F (2006) Toxicological impact studies based on Escherichia coli bacteria in ultrafine ZnO nanoparticles colloidal medium. Nano Lett 6: 866–870

Brown DM, Wilson MR, MacNee W, Stone V and Donaldson K (2001) Size-dependent proinflammatory effects of ultrafine polystyrene particles: a role for surface area and oxidative stress in the enhanced activity of ultrafines. Toxicol Appl Pharmacol 175: 191–199

Brunner TJ, Wick P, Manser P, Spohn P, Grass RN, Limbach LK, Bruinink A and Stark WJ (2006) In vitro cytotoxicity of oxide nanoparticles: comparison to asbestos, silica, and the effect of particle solubility. Environ Sci Technol 40: 4374–4381

Caminade AM and Majoral JP (2004) Nanomaterials based on phosphorus dendrimers. Acc Chem Res 37: 341–348

Chatellier X, Bottero JY and Le Petit J (2001) Adsorption of a cationic polyelectrolyte on Escherichia coli bacteria: 1. Adsorption of the polymer. Langmuir 17: 2782–2790

Chen WJ, Tsai PJ and Chen YC (2008) Functional Fe(3)O(4)/TiO(2) Core/Shell Magnetic Nanoparticles as Photokilling Agents for Pathogenic Bacteria. Small 4: 485–491

Colvin VL (2003) The potential environmental impact of engineered nanomaterials. Nat Biotechnol 21: 1166–1170

De Windt W, Boon N, Van den Bulcke J, Rubberecht L, Prata F, Mast J, Hennebel T and Verstraete W (2006) Biological control of the size and reactivity of catalytic Pd(0) produced by Shewanella oneidensis. Antonie Van Leeuwenhoek 90: 377–389

Derfus AM, Chan WCW and Bhatia SN (2004) Probing the cytotoxicity of semiconductor quantum dots. Nano Lett 4: 11–18

Fang J, Lyon DY, Wiesner MR, Dong J and Alvarez PJ (2007) Effect of a fullerene water suspension on bacterial phospholipids and membrane phase behavior. Environ Sci Technol 41: 2636–2642

Fasim F, Ahmed N, Parsons R and Gadd GM (2002) Solubilization of zinc salts by a bacterium isolated from the air environment of a tannery. FEMS Microbiol Lett 213: 1–6

Fischer HC and Chan WC (2007) Nanotoxicity: the growing need for in vivo study. Curr Opin Biotechnol 18: 565–571

Fortner JD, Lyon DY, Sayes CM, Boyd AM, Falkner JC, Hotze EM, Alemany LB, Tao YJ, Guo W, Ausman KD, Colvin VL and Hughes JB (2005) C60 in water: nanocrystal formation and microbial response. Environ Sci Technol 39: 4307–4316

Frankel RB (1987) Microbial metabolism-anaerobes pumping iron. Nature 330: 208–209

Fredrickson JK, Zachara JM, Kukkadapu RK, Gorby YA, Smith SC and Brown CF (2001) Biotransformation of Ni-substituted hydrous ferric oxide by an Fe(III)-reducing bacterium. Environ Sci Technol 35: 703–712

Fu J, Ji J, Fan D and Shen J (2006) Construction of antibacterial multilayer films containing nanosilver via layer-by-layer assembly of heparin and chitosan-silver ions complex. J Biomed Mater Res A 79: 665–674

Ghafari P, St-Denis CH, Power ME, Jin X, Tsou V, Mandal HS, Bols NC and Tang XS (2008) Impact of carbon nanotubes on the ingestion and digestion of bacteria by ciliated protozoa. Nat Nanotechnol 3: 347–351

Gogoi SK, Gopinath P, Paul A, Ramesh A, Ghosh SS and Chattopadhyay A (2006) Green fluorescent protein-expressing Escherichia coli as a model system for investigating the antimicrobial activities of silver nanoparticles. Langmuir 22: 9322–9328

Gojova A, Guo B, Kota RS, Rutledge JC, Kennedy IM and Barakat AI (2007) Induction of inflammation in vascular endothelial cells by metal oxide nanoparticles: effect of particle composition. Environ Health Perspect 115: 403–409

Gorby YA, Beveridge TJ and Blakemore RP (1988) Characterization of the bacterial magnetosome membrane. J Bacteriol 170: 834–841

Heinlaan M, Ivask A, Blinova I, Dubourguier HC and Kahru A (2008) Toxicity of nanosized and bulk ZnO, CuO and TiO(2) to bacteria Vibrio fischeri and crustaceans Daphnia magna and Thamnocephalus platyurus. Chemosphere 71: 1308–1316

Hoffmann M, Martin S, Choi W and Bahnemann D (1995) Environmental applications of semiconductor photocatalysis. Chem Rev 95: 69–96

Huang Z, Zheng X, Yan D, Yin G, Liao X, Kang Y, Yao Y, Huang D and Hao B (2008) Toxicological EFFECT of ZnO nanoparticles based on bacteria. Langmuir 24: 4140–4144

Hwang ET, Lee JH, Chae YJ, Kim YS, Kim BC, Sang BI and B. GM (2008) Analysis of the toxic mode of action by silver nano-particles using stress-specific bioluminescent bacteria. Small 4: 746–750

Imlay JA (2003) Pathways of oxidative damage. Annu Rev Microbiol 57: 395–418

Jeng HA and Swanson J (2006) Toxicity of metal oxide nanoparticles in mammalian cells. J Environ Sci Health A Tox Hazard Subst Environ Eng 41: 2699–2711

Jia G, Wang H, Yan L, Wang X, Pei R, Yan T, Zhao Y and Guo X (2005) Cytotoxicity of carbon nanomaterials: single-wall nanotube, multi-wall nanotube, and fullerene. Environ Sci Technol 39: 1378–1383

Joo SH, Feitz AJ and Waite TD (2004) Oxidative degradation of the carbothioate herbicide, molinate, using nanoscale zerovalent iron. Environ Sci Technol 38: 2242–2247

Kai Y, Komazawa Y, Miyajima A, Miyata N and Yamakoshi Y (2003) [60]Fullerene as a novel photoinduced antibiotic. Fuller Nanotub Car N 11: 79–87

Kang S, Pinault M, Pfefferle LD and Elimelech M (2007) Single-walled carbon nanotubes exhibit strong antimicrobial activity. Langmuir 23: 8670–8673

Kashiwada S (2006) Distribution of nanoparticles in the see-through medaka (Oryzias latipes). Environ Health Perspect 114: 1697–1702

Ke PC and Qiao R (2007) Carbon nanomaterials in biological systems. J Phys Condens Matter 19 373101 (25 pp)

Kohen R and Nyska A (2002) Oxidation of biological systems: oxidative stress phenomena, antioxidants, redox reactions, and methods for their quantification. Toxicol Pathol 30: 620–650

Lam CW, James JT, McCluskey R and Hunter RL (2004) Pulmonary toxicity of single-wall carbon nanotubes in mice 7 and 90 days after intratracheal instillation. Toxicol Sci 77: 126–134

Lengke MF, Ravel B, Fleet ME, Wanger G, Gordon RA and Southam G (2006) Mechanisms of gold bioaccumulation by filamentous cyanobacteria from gold(III)-chloride complex. Environ Sci Technol 40: 6304–6309

Li N, Sioutas C, Cho A, Schmitz D, Misra C, Sempf J, Wang M, Oberley T, Froines J and Nel A (2003) Ultrafine particulate pollutants induce oxidative stress and mitochondrial damage. Environ Health Perspect 111: 455–460

Limbach LK, Li Y, Grass RN, Brunner TJ, Hintermann MA, Muller M, Gunther D and Stark WJ (2005) Oxide nanoparticle uptake in human lung fibroblasts: effects of particle size, agglomeration, and diffusion at low concentrations. Environ Sci Technol 39: 9370–9376

Lin Y, Taylor S, Li HP, Fernando KAS, Qu LW, Wang W, Gu LR, Zhou B and Sun YP (2004) Advances toward bioapplications of carbon nanotubes. J Mater Chem 14: 527–541

Liu S, Zhou J, Zhang CC, Cole DR, Gajdarziska M and Phelps TJ (1997) Thermophilic Fe(III)-reducing bacteria from the deep subsurface: the evolutionary implication. Science 277: 1106–1109

Lok CN, Ho CM, Chen R, He QY, Yu WY, Sun H, Tam PK, Chiu JF and Che CM (2006) Proteomic analysis of the mode of antibacterial action of silver nanoparticles. J Proteome Res 5: 916–924

Lovely DR, Stolz JF, Nord Jr. GL and Phillips EJP (1987) Anaerobic production of magnetite by a dissimilatory iron-reducing microorganism. Naure 330: 252–254

Lovley DR (1995) Bioremediation of organic and metal contaminants with dissimilatory metal reduction. J Ind Microbiol 14: 85–93

Lu Q, Moore JM, Huang G, Mount AS, Rao AM, Larcom LL and Ke P (2004) RNA polymer translocation with single-walled carbon nanotubes. Nano Lett 4: 2473–2477

Lyon DY, Adams LK, Falkner JC and Alvarezt PJ (2006) Antibacterial activity of fullerene water suspensions: effects of preparation method and particle size. Environ Sci Technol 40: 4360–4366

Lyon DY, Fortner JD, Sayes CM, Colvin VL and Hughe JB (2005) Bacterial cell association and antimicrobial activity of a C60 water suspension. Environ Toxicol Chem 24: 2757–2762

Maenosono S, Suzuki T and Saita S (2007) Mutagenicity of water-soluble FePt nanoparticles in Ames test. J Toxicol Sci 32: 575–579

Magrez A, Kasas S, Salicio V, Pasquier N, Seo JW, Celio M, Catsicas S, Schwaller B and Forro L (2006) Cellular toxicity of carbon-based nanomaterials. Nano Lett 6: 1121–1125

Mandal D, Bolander ME, Mukhopadhyay D, Sarkar G and Mukherjee P (2006) The use of microorganisms for the formation of metal nanoparticles and their application. Appl Microbiol Biotechnol 69: 485–492

Moore MN (2006) Do nanoparticles present ecotoxicological risks for the health of the aquatic environment? Environ Int 32: 967–976

Morones JR, Elechiguerra JL, Camacho A, Holt K, Kouri JB, Ramirez JT and Yacaman MJ (2005) The bactericidal effect of silver nanoparticles. Nanotechnology 16: 2346–2353

Narayan RJ, Berry CJ and Brigmon RL (2005) Structural and biological properties of carbon nanotube composite films. Mat Sci Eng B-Solid 123: 123–129

Nel A, Xia T, Madler L and Li N (2006) Toxic potential of materials at the nanolevel. Science 311: 622–627

Nies DH (2003) Efflux-mediated heavy metal resistance in prokaryotes. FEMS Microbiol Rev 27: 313–339

Oberdorster E (2004) Manufactured nanomaterials (fullerenes, C60) induce oxidative stress in the brain of juvenile largemouth bass. Environ Health Perspect 112: 1058–1062

Oberdorster E, Zhu S, Blickley TM, McClellan-Green P and Haasch ML (2006) Ecotoxicology of carbon-based engineered nanoparticles: effects of fullerene (C$_{60}$) on aquatic organisms. Carbon 44: 1112–1120

Oberdorster G, Maynard A, Donaldson K, Castranova V, Fitzpatrick J, Ausman K, Carter J, Karn B, Kreyling W, Lai D, Olin S, Monteiro-Riviere N, Warheit D and Yang H (2005a) Principles for characterizing the potential human health effects from exposure to nanomaterials: elements of a screening strategy. Part Fibre Toxicol 2: 8

Oberdorster G, Oberdorster E and Oberdorster J (2005b) Nanotoxicology: an emerging discipline evolving from studies of ultrafine particles. Environ Health Perspect 113: 823–839

Oberdorster G, Stone V and Donaklson K (2007) Toxicology of nanoparticles: a historical perspective. Nanotoxicology 1: 2–25

Pal S, Tak YK and Song JM (2007) Does the antibacterial activity of silver nanoparticles depend on the shape of the nanoparticle? A study of the Gram-negative bacterium Escherichia coli. Appl Environ Microbiol 73: 1712–1720

Partha R, Lackey M, Hirsch A, Casscells SW and Conyers JL (2007) Self assembly of amphiphilic C60 fullerene derivatives into nanoscale supramolecular structures. J Nanobiotechnol 5: 6

Porter AE, Muller K, Skepper J, Midgley P and Welland M (2006) Uptake of C60 by human monocyte macrophages, its localization and implications for toxicity: studied by high resolution electron microscopy and electron tomography. Acta Biomater 2: 409–419

Rasanen LA, Elvang AM, Jansson J and Lindstrom K (2001) Effect of heat stress on cell activity and cell morphology of the tropical rhizobium, Sinorhizobium arboris. FEMS Microbiol Ecol 34: 267–278

Reddy KM, Feris K, Bell J, Wingett DG, Hanley C and Punnoose A (2007) Selective toxicity of zinc oxide nanoparticles to prokaryotic and eukaryotic systems. Appl Phys Lett 90: 213902

Rengifo-Herrera JA, Sanabria J, Machuca F, Dierolf CF, Pulgarin C and Orellana G (2007) A comparison of solar photocatalytic inactivation of waterborne E. coli using tris(2,2′-bipyridine)ruthenium(II), Rose Bengal, and TiO$_2$. J. Solar Energy Engg. 129: 135–140

Rikans LE and Hornbrook KR (1997) Lipid peroxidation, antioxidant protection and aging. Biochim Biophys Acta 1362: 116–127

Ritz M, Tholozan JL, Federighi M and Pilet MF (2001) Morphological and physiological characterization of Listeria monocytogenes subjected to high hydrostatic pressure. Appl Environ Microbiol 67: 2240–2247

Ruparelia JP, Chatterjee AK, Duttagupta SP and Mukherji S (2007) Strain specificity in antimicrobial activity of silver and copper nanoparticles. Acta Biomater 4: 707–716

SCENIHR (2006) The appropriateness of existing methodologies to assess the potential risks associated with engineered and adventitious products of nanotechnologies. Available at: http://ec.europa.eu/health/opinions2/en/nanotechnologies/l-3/9-conclusion.htm

Seetharam RN and Sridhar KR (2007) Nanotoxicity: Threat posed by nanoparticles. Curr Scie 93: 769–770

Sera N, Tokiwa H and Miyata N (1996) Mutagenicity of the fullerene C60-generated singlet oxygen dependent formation of lipid peroxides. Carcinogenesis 17: 2163–2169

Silver S (1996) Bacterial resistance to toxic metal ions-a review. Gene 179: 9–19

Sondi I and Salopek-Sondi B (2004) Silver nanoparticles as antimicrobial agent: a case study on E. coli as a model for Gram-negative bacteria. J Colloid Interface Sci 275: 177–182

Stoimenov PK, Klinger RL, Marchin GL and Klabunde KJ (2002) Metal oxide nanoparticles as bactericidal agents. Langmuir 18: 6679–6686

Subramanian V, Wolf EE and Kamat PV (2003) Influence of metal/metal ion concentration on the photocatalytic activity of TiO2-Au composite nanoparticles. Langmuir 19: 469–474

Tang YJ, Ashcroft JM, Chen D, Min G, Kim CH, Murkhejee B, Larabell C, Keasling JD and Chen FF (2007) Charge-associated effects of fullurene derivatives on microbial structural integrity and central metabolism. Nano Lett 7: 754–760

Thill A, Zeyons O, Spalla O, Chauvat F, Rose J, Auffan M and Flank AM (2006) Cytotoxicity of CeO2 nanoparticles for Escherichia coli. Physico-chemical insight of the cytotoxicity mechanism. Environ Sci Technol 40: 6151–6156

Tsuang YH, Sun JS, Huang YC, Lu CH, Chang WH and Wang CC (2008) Studies of photokilling of bacteria using titanium dioxide nanoparticles. Artif Organs 32: 167–174

Veeranagouda Y, Karegoudar TB, Neumann G and Heipieper HJ (2006) Enterobacter sp. VKGH12 growing with n-butanol as the sole carbon source and cells to which the alcohol is added as pure toxin show considerable differences in their adaptive responses. FEMS Microbiol Lett 254: 48–54

Wei W, Sethuraman A, Jin C, Monteiro-Riviere NA and Narayan RJ (2007) Biological properties of carbon nanotubes. J Nanosci Nanotechnol 7: 1284–1297

Yamakoshi Y, Umezawa N, Ryu A, Arakane K, Miyata N, Goda Y, Masumizu T and Nagano T (2003) Active oxygen species generated from photoexcited fullerene (C60) as potential medicines: O2-* versus 1O2. J Am Chem Soc 125: 12803–12809

Zhang L, Jiang Y, Ding Y, Povey M and York D (2007) Investigation into the antibacterial behaviour of suspensions of ZnO nanoparticles (ZnO nanofluids). J Nanopart Res 9: 479–489

Zhu S, Oberdorster E and Haasch ML (2006) Toxicity of an engineered nanoparticle (fullerene, C60) in two aquatic species, Daphnia and fathead minnow. Mar Environ Res 62 Suppl: S5–S9

Environmental Monitoring by Use of Genomics and Metabolomics Technologies

Tetsuji Higashi, Yoshihide Tanaka, Randeep Rakwal, Junko Shibato, Shin-ichi Wakida and Hitoshi Iwahashi

Abstract The yeast *Saccharomyces cerevisiae* is one of the most characterized eucaryotes and its complete genome sequence was published in 1986. Thus, this organism is a good candidate for biological environmental monitoring. Omics (genomics, metabolomics) technology is being applied to environmental monitoring using yeast cells.

For genomics studies, commercially available DNA microarrays are used and selected the highly induced genes by the cadmium treatment. The functional characterization of induced genes by cadmium stress suggests the accumulation of glutathione, as the almost all genes contributing sulfur amino acid biosynthesis are significantly induced.

For metabolomics studies, capillary electrophoresis-mass spectrometry (CE-MS) is used for the separation and identification of metabolites. The metabolomics results suggest the accumulation of glutathione. This strongly agrees with the results obtained by genomics.

However, the combined results found the negative biosynthesis of glycine. This proves that the accumulation of glutathione is not the reason to activated sulfur amino acid biosynthesis. By combining the genomic analysis with the metabolomic analysis, raise the possibility of discovering new mechanism that nobody has noticed until now.

Keywords *Saccharomyces cerevisiae*

T. Higashi (✉)
Health Technology Research Center (HTRC), National Institute of Advanced Industrial Science and Technology (AIST), 1-8-31 Midorigaoka, Ikeda, Osaka 563-8577, Japan and, 16-1 Onogawa, Tsukuba, Ibaraki 305-8569, Japan
e-mail: t.higashi@aist.go.jp

1 Introduction

Presently, more than 10,000 of chemicals are produced industrially, and there are accumulating in the environment every year. In spite of the fact that these industrial chemicals make live our life rich, they have great influences on environment and ecosystem.

Bioassay is the method used to estimate chemical toxicities and assessment of environmental pollutant by using cells or microorganisms instead of whole organisms. One of the cytotoxicity program, MEIC (Multicenter Evaluation of In vitro Cytotoxicity) was organized by the Scandinavian Society for Cell Toxicology (Celemedson et al. 1996) to collect and analyze the parameters of a bioassay. The investigators compared LD50 (50% lethal dose) data obtained from whole organisms in vivo and IC50 (50% inhibition constant) obtained from a bioassay in vitro. MEIC found a correlation between the data in vivo the LD50 and in vitro the IC50. This suggests that an in vivo assay can be placed by in vitro assay. MEIC suggested that basal cytotoxicity, which can be defined as the toxicity affecting cellular components, functions, and biosynthesis, is found in all cell lines (Ekwall et al. 1998).

On the other hand, the Ames test is well-known as one of the most powerful methods for monitoring the mutagenicity of environmental samples (Reifferscheid and Heil 1996). In this system, mutants of *Salmonella typhimurium* are grown in a minimum medium and mutagenicity is estimated according to the frequency of back mutation. As the frequency of back mutation is dependent on DNA damage, we can estimate the mutagenicity of chemicals or environmental stress. An extensive literature exists on bioassay systems that

Y.J. Kim et al. (eds.), *Atmospheric and Biological Environmental Monitoring*,
DOI 10.1007/978-1-4020-9674-7_13, © Springer Science+Business Media B.V. 2009

include tests by Ames, Microtox, Umu, and others (Celemedson et al. 1996) however, the information that can be estimated is limited to the degree of toxicity or mutagenicity. Furthermore, bioassay systems sometimes mistakenly identify natural products as the toxic substance. Therefore bioassay systems are required that can be used for predicting the mechanism of environmental stress.

In recent years, DNA microarray technology has developed rapidly and been widely adopted as a tool for understanding biological systems at the genomic level (Momose and Iwahashi 2001). DNA microarray technology provides a great opportunity for bioassays of chemical and environmental toxicity because it can provide an overview of thousands of genes at the same time and shed light on how to investigate toxicological problems. In addition, DNA microarray technology can be combined with proteomics and metabolomics technology.

Metabolomics is an emerging new omics science analogous to genomics, transcriptomics and proteomics, and can be regarded as the end point of the "omics" cascade (Dettmer and Hammock 2004, Rochfort 2005). The advantage of metabolomic analysis is that the biochemical consequences of mutations and stress response mechanisms can be observed directly. As the metabolome represents a wide variety of chemical compounds, it is logical that numerous high-throughput analytical techniques are being used for metabolomics. Being non-destructive, nuclear magnetic resonance (NMR) spectroscopy is highly beneficial as a metabolomics technique (Rochfort 2005, Viant et al. 2003, Robertson 2005), but it also possesses one major disadvantage, which is that it is relatively insensitive compared to mass spectrometry (MS) (Krishnan et al. 2005). In the rapidly growing field of metabolomics, MS coupled to a chromatographic separation technique is a useful method used to profile low molecular weight compounds (Rochfort 2005, Dunn and Ellis 2005, Dunn et al. 2005, Ramautar et al. 2006, Babu et al. 2006,). Capillary electrophoresis-mass spectrometry (CE-MS) has been considered a highly promising technique for comprehensive metabolomics analysis because most of the metabolites are polar and ionic compounds. It gives high-resolution separations of cationic metabolites, anionic metabolites and nucleotides/CoA in a reasonable time, and requires a minimum amount of samples (Soga et al. 2002, 2003).

It will be obtained new findings by combining genomic analysis with metabolomic analysis. In this paper, we would like to discuss the development of a combined omics approach for environmental monitoring of chemicals using yeast system. Cadmium was selected as the model chemicals. Cadmium is one of the most widely characterized chemicals. In Japan, cadmium is well known as a chemical that may cause Itai-Itai (which means 'ouch-ouch') disease, which leads to nephrotoxicity (Shibasaki et al. 1993), hepatotoxicity (Hussain et al. 1987), serious damage to the nervous system (Figueiredo-Pereira et al. 1998), and high frequency of chromatid aberrations in Japan (Shiraishi 1975).

2 Materials and Methods

2.1 Strain and Growth Conditions

Saccharomyces cerevisiae strain S288C (α*SUC2 mal mel gal2 CUP1*) was grown in YPD medium (1% Bacto Yeast Extract, 2% polypeptone, 2% glucose) at 25°C or 30°C according to the procedure outlined by Kitagawa et al. (Kitagawa et al. 2002).

2.2 Cadmium Stress Conditions

Yeast cells grown in 200 mL of a YPD medium at 25°C or 30°C to an optical density (OD_{660}) of 1.0; addition of 0.3 mM cadmium chloride, incubated for 2 hr at 25°C or 30°C, and then harvested (Momose and Iwahashi 2001).

2.3 DNA Microarray Analysis

Figure 1 shows the schematic illustration of DNA microarray analysis procedure. Total RNA was extracted by the hot-phenol method (Kohrer and Domdey 1990). Poly (A)+RNA was purified from total RNA with an Oligotex-dT30 mRNA purification kit (Takara, Otsu, Shiga, Japan). Fluorescently labeled cDNA was synthesized by oligo dT-primed

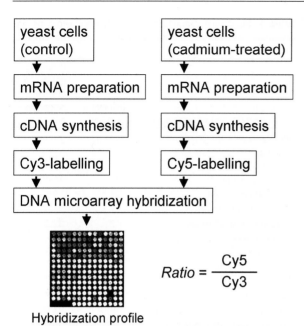

$$Ratio = \frac{Cy5}{Cy3}$$

Hybridization profile

Fig. 1 The schematic illustration of DNA microarray analysis procedure. (See Section 2.3)

polymerization using PowerScript™ reverse transcriptase (Clontech, Palo Alto, CA, USA). The pool of nucleotides in the labeling reaction included 0.5 mM dGTP, dATP and dCTP; and 0.2 mM dTTP and 0.5 mM Cy3-UTP (Amersham-Pharmacia Biotech, Little Chalfont, Buckinghamshire, England) or Cy5-UTP. cDNA made from the poly(A)+ RNA of the control was fluorescently labeled with Cy3 (represented as green), and that of the cadmium-treated sample was labeled with Cy5 (represented as red). For each label, 2–4 µg of poly(A)+ RNA was used, and the same amount of each poly(A)+ RNA was used on one slide. The two labeled cDNA pools were mixed and hybridized simultaneously to a microarray. DNA chips of a yeast genome probe were hybridized with labeled cDNA probes under cover slides for 24–36 hr at 65°C. Each set of hybridizations was performed with an independent RNA preparation. After hybridization, the labeled microarrays were washed and dried. Subsequently, the labeled microarrays were scanned using a confocal laser ScanArray 4000 (GSI Lumonics, Billerica, MA, USA) system. The resulting image data were quantitated using the QuantArray Quantitative Microarray Analysis application program (GSI Lumonics, Billerica, MA, USA). The fluorescent intensity of each spot on images was subtracted from each background,

the ratios of intensity Cy5/Cy3 were calculated, normalized with no-probe spots as the negative control to exclude nonspecific hybridization, and normalized with ACT1 (YFL039C actin gene) as the positive control. So far, ACT1 gene is widely accepted as the positive control of stress response in yeast using northern blot (Godon et al. 1998, Zahringer et al. 1998) and RT-PCR (Delaunay et al. 2000, Vido et al. 2001). All of those calculation and normalization above were performed using GeneSpring (Silicon Genetics, Redwood City, CA, USA).

2.4 Capillary Electrophoresis-Mass Spectrometry (CE-MS) (Tanaka et al. 2007)

Figure 2 shows photograph of CE-MS instrument. CE-MS experiments were performed using a P/ACE MDQ capillary electrophoresis system (Beckman Coulter, Tokyo, Japan). An electrospray ionization (ESI) interface (Agilent Technologies Japan, Tokyo, Japan) was used to couple the CE instrument to an Esquire 3000 plus ion trap mass spectrometer (Bruker Daltonics, Ibaraki, Japan).

Sheath liquid consisting of 5 mM formic acid in 50% (v/v) 2-propanol was applied coaxially at a rate of 10 µL/min by using a syringe pump (model 74900, Cole-Parmer, Vernon Hills, IL, USA). Yeast extracts were prepared after the stress conditions by filtration of yeast cells (0.45 µm membrane filter), the cells were frozen in liquid nitrogen, and then crushed to a very fine powder using a chilled mortar and pestle. Yeast metabolites were extracted from the freeze-dried yeast cells (ca. 100 mg) in 700 µL of cold chloroform and 500 µL of cold methanol followed by the addition

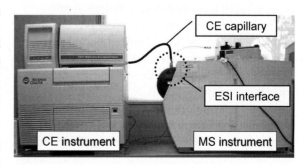

Fig. 2 CE-MS instrument photograph. (See Section 2.4)

of 500 µL of cold Milli-Q water containing a known amount of methionine sulfone as an internal standard. The resulting mixture was centrifuged at 15,000 rpm for 15 sec. The supernatant was then filtered through an ultrafiltration membrane (Ultrafree-MC filter, 5000 NMWL, Millipore, Bedford, MA, USA). The filtered solution was stored in the deep freeze at −80°C until analyzed. Analytical conditions were established based on a previously reported method (Sato et al. 2004, Soga et al. 2003).

3 Results and Discussion

3.1 Analyses of the Gene Response to the Cadmium Stress (Momose and Iwahashi 2001)

Performance of DNA microarray analyses of the global transcriptional response of the yeast genome after exposure of cells to cadmium (0.3 mM for 2 hr at 30°C) and repeated four independent experiment sets (control and cadmium treated), obtaining four sets of microarray data. The genes with ratios of hybridization values (treated: control) more than 2.0 at least 3 of 4 experiments are considered to be up-regulated by cadmium treatment. Moreover, the genes with ratios of hybridization values under 0.5 at least 3 of 4 experiments are considered to be down-regulated by cadmium treatment. By these criteria, 310 genes respond with increased mRNA levels and 322 genes respond with decreased mRNA levels. Those ratios averaged (Cy5/Cy3) of hybridization intensity for analysis.

Table 1 (Momose and Iwahashi 2001) shows the biological characteristics and number of up-regulated genes grouped according to their biological role.

These genes were annotated using functional categories assigned by the Munich Information Center for Protein Sequences (MIPS http://www.mips. biochem.mpg.de/). The highly up-regulated categories are cell rescue, defense, cell death and ageing (13%), transport facilitation (9%), energy (9%), metabolism (8%), and ionic homeostasis (7%). Among the subcategories, nitrogen and sulfur transport (50%), amino acid transporters (23%), nitrogen and sulfur utilization (22%), nitrogen and sulfur metabolism (23%), allatoin

Table 1 Functional characterization of induced genes by the cadmium treatment (Momose and Iwahashi, Environ. Toxicol. Chem., 20, 2353–2360)

Gene categories and subcategories	Number of genes*	Percentage of total genes of each categories (%)
Cell rescue, defense, cell death and ageing	47	13.0
detoxification	21	17.6
stress response	27	16.0
Transport facilitation	28	9.1
amino acid transporters	7	24.0
allantoin and allantoate transporters	2	22.2
ion transporters	8	10.3
metal ion transporters	3	20.0
Energy	22	8.9
Metabolism	82	7.8
nitrogen and sulfur metabolism	17	23.0
nitrogen and sulfur transport	4	50.0
nitrogen and sulfur utilization	8	21.6
regulation of nitrogen and sulfur utilization	5	17.2
amino acid metabolism	36	17.6
C-compound and carbohydrate metabolism	26	10.4
Ionic homeostasis	9	7.4
homeostasis of metal ions	5	8.6
Protein destination	41	7.2
other protein-destination activity	2	28.6
protein folding and stabilization	6	10.3
Cellular communication/signal transduction	8	6.1%
osmosensing	2	11.8
nutritional response pathway	2	8.3
Cellular organization	109	4.9
vacuolar and lysozomal organization	7	12.7
Cellular biogenesis	9	4.5
Cellular transport and transportmechanism	19	3.9
Cell growth, cell division and DNA synthesis	29	3.5
Transcription	27	3.5
Protein synthesis	5	1.4
Not clear	5	3.5
Unclassified proteins	100	4.1
Total number of up-regulated genes	310	4.6

The ratios (treated/control) of the up-regulated genes were more than 2.0.

* There are 230 genes overlapped in more than two categories.

Categories and subcategories are acorrding to MIPS.

The subcategories which induced more than 8% were shown in italic.

and allatonate transporters (22%), and other protein-destination activity (29%) are highly induced.

On the other hand, suppressed genes are particularly found in the protein synthesis category, especially ribosomal protein (30%), and others are widely involved in all categories (3–7%) (Table 2 (Momose and Iwahashi 2001)). A report on yeast indicated that the transcripts encoding ribosomal proteins are suppressed accompanied with the adaptation to abrupt transfer from a fermentable to nonfermentable carbon source in 10 min (Kuhn et al. 2001). This may suggest the existence of some kind of adaptation to the resistance to cadmium toxicity.

Table 2 Categories and subcategories of the genes reppressed by cadmium treatment (Momose and Iwahashi, Environ. Toxicol. Chem., 20, 2353–2360)

Gene Categories and subcategories	Number of genes*	Percentage of total genes of each categories (%)
Protein synthesis	73	21.0
ribosomal protein	61	29.6
translation	8	12.9
other protein-synthesis activities	2	12.5
Cellular organization	170	7.6
Cellular biogenesis	12	6.0
Ionic homeostasis	7	5.7
Metabolism	60	5.7
Transcription	39	5.0
Protein destination	27	4.7
Cell rescue, defense, cell death and ageing	15	4.2
Transport facilitation	12	3.9
Cellular transport and transportmechanism	18	3.7
Cell growth, cell division and DNA synthesis	30	3.7
Energy	9	3.6
Cellular communication/signal transduction	4	3.1
Not clear	7	4.9
Unclassified proteins	91	3.7
Total number of up-regulated genes	322	4.8

The ratios (treated/control) of the down-regulated genes were less than 0.5.
* There are 252 genes overlapped in more than two categories.
Categories and subcategories are acorrding to MIPS.
The subcategories which induced more than 10% were shown in italic.

3.2 Glutathione Biosynthesis is Activated by the Cadmium Treatment

Focusing on up-regulated genes, in particular, almost all the genes involved in sulfur amino acid metabolism, including the transporters of sulfate and methionine and the sulfur salvage pathway, of which the final product is glutathione, were particularly induced (*MET14, MET17, SUL1*, Table 3 (Momose and Iwahashi 2001)). Table 3 (Momose and Iwahashi 2001) shows the list of highly 46 up-regulated genes (ratio > 4) with ratio and the standard deviation.

The gene that have the highest ratio is YLL057C, which is reported to be an Fe(II)-dependent sulfonate/α-ketoglutarate dioxygenase (Hogan et al. 1999) and could have been involved in the detoxification of cytochrome P450 according to MIPS. Several common stress genes, such as *HSP26, HSP12, GRE1*, and *ATX1*, and other 8 genes of transport facilitation are highly up-regulated. 13 Genes are involved in sulfur amino acid metabolism, the transporters of sulfate and methionine, and the sulfur salvage pathway. These results indicate that glutathione synthesis is included via whole sulfur amino acid synthesis. Recently it is also reported the strong induction of 9 enzymes of the sulfur amino acid biosynthetic pathway by cadmium (Vido et al. 2001).

Yeast cells have two specific proteins to bind with cadmium: *YCF1p* and *CUP1p*. *YCF1p* is the part of the defense system of yeast and transport cadmium that uses glutathione as an SH-conjugate into a vacuole to decrease the concentration of cadmium in cytosol (Li et al. 1997). *Cup1p* is a metallothioneine that has many cysteine residues and is known to bind to copper and cadmium (Tohoyama et al. 1992). After the addition of cadmium, the yeast cells used glutathione for a defense system; therefore, there is a depletion of glutathione and cysteine because cysteines are components of both glutathione and metallothioneine. Consequently, cells need to activate the sulfur salvage pathway via sulfur amino acid metabolism to have de novo synthesis of glutathione molecules. Almost all genes are up-regulated in both the sulfur amino acid biosynthesis and the sulfur salvage pathway (Fig. 3 (Momose and Iwahashi 2001)).

This result coincides with the recent report that cadmium induced the yeast *GSH1* gene using a functional sulfur-amino acid regulatory network (Dormer

Table 3 The genes highly up-regulated* by cadmium (Momose and Iwahashi, Environ. Toxicol. Chem., 20, 2353–2360)

ORF	ratio	±	SD	Common name	Description
YLL057C	59.9	±	7.8		similarity to E.coli dioxygenase
YKL001C	21.4	±	3.8	MET14	adenylylsulfate kinase
YPL223C	20.0	±	5.6	GRE1	Induced by osmotic stress
YLL055W	19.0	±	5.7		similarity to Dal5p
YBR072W	17.4	±	6.1	HSP26	heat shock protein 26
YBR294W	14.4	±	4.5	SUL1	Putative sulfate permease
YIL166C	12.5	±	8.1		similarity to allantoate permease Dal5p
YNL277W	10.3	±	5.0	MET2	homoserine O-trans-acetylase
YLR303W	9.8	±	4.1	MET17/ MET25/ MET15	O-Acetylhomoserine-O-Acetylserine Sulfhydralase
YLR136C	9.0	±	4.7	TIS11	tRNA-specific adenosine deaminase 3
YIR017C	8.6	±	6.4	MET28	Transcriptional activator of sulfur amino acid metabolism
YDL124W	8.5	±	3.5		similarity to aldose reductases
YLR364W	8.0	±	4.0		hypothetical protein
YOR382W	7.6	±	5.0		hypothetical protein
YAL067C	7.5	±	2.5	SEO1	Suppressor of Sulfoxyde Ethionine resistance
YJR010W	7.3	±	3.8	MET3	ATP sulfurylase
YCL040W	7.1	±	5.1	GLK1	aldohexose specific glucokinase
YOL162W	6.9	±	4.7		strong similarity to hypothetical protein YIL166c
YHR008C	6.9	±	4.0	SOD2	mitochondrial superoxide dismutase (Mn) precursor
YNL015W	6.7	±	2.7	PBI2	proteinase B inhibitor 2
YFL055W	6.7	±	5.5	AGP3	Amino acid permease
YDR070C	6.4	±	2.9		hypothetical protein
YDR253C	6.4	±	3.1	MET32	Transcriptional regulator of sulfur amino acid metabolism
YFL057C	5.9	±	3.5	AAD6	strong similarity to aryl-alcohol dehydrogenases
YJR137C	5.8	±	2.1	ECM17 (MET5)	ExtraCellular Mutant involved in cell wall biogenesis and architecture
YPR167C	5.8	±	1.7	MET16	3'phosphoadenylylsulfate reductase
YGR055W	5.7	±	3.8	MUP1	high affinity methionine permease
YLR092W	5.6	±	3.2	SUL2	high affinity sulfate permease
YFR030W	5.5	±	3.4	MET10	subunit of assimilatory sulfite reductase
YPL054W	5.3	±	3.9	LEE1	protein of unknown function
YMR058W	5.1	±	3.0	FET3	cell surface ferroxidase
YFL014W	5.1	±	0.7	HSP12	12 kDa heat shock protein
YDL059C	5.1	±	4.1	RAD59	recombination and DNA repair protein
YGL121C	5.0	±	2.7		hypothetical protein
YHL040C	4.9	±	2.5	ARN1	ferrichrome-type siderophore transporter
YIL127C	4.8	±	3.5		weak similarity to Smy2p
YGR043C	4.7	±	3.1		strong similarity to transaldolase
YDR223W	4.6	±	2.3		similarity to Ifh1p
YCL025C	4.6	±	3.7	AGP1	Amino acid permease
YNL259C	4.5	±	3.8	ATX1	Antioxidant protein and metal homeostasis factor
YOL165C	4.5	±	4.2	AAD15	putative aryl alcohol dehydrogenase
YBR253W	4.4	±	3.2	SRB6	transcription factor, part of Srb\/Mediator complex
YOL164W	4.4	±	3.3		similarity to Pseudomonas alkyl sulfatase
YGR155W	4.4	±	2.5	CYS4	Cystathionine beta-synthase
YBR296C	4.3	±	2.3	PHO89	Na+-coupled phosphate transport protein
YKR049C	4.2	±	2.1		hypothetical protein
YLL060C	4.2	±	1.9	GTT2	glutathione S-transferase

*The genes with ratios of hybridization values (treated: control) beyond 2.0 at least 3 of 4 experiments and the averages of ratios were up-regulated more than 4-fold or more by cadmium. SD means standard deviation

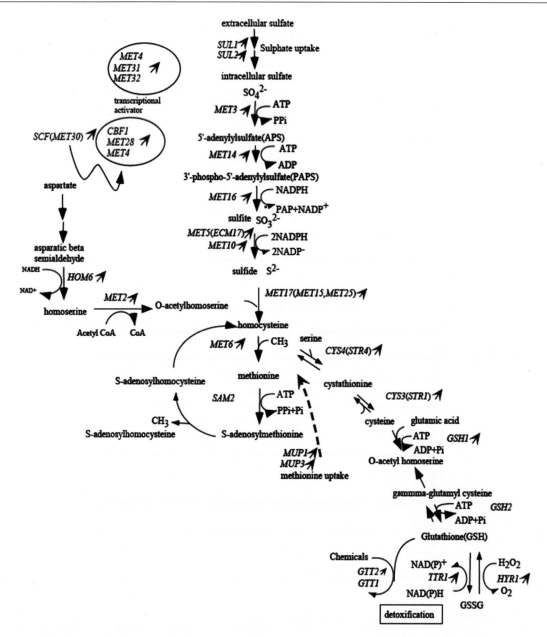

Fig. 3 Schematic representation of the biological steps involved in the sulfur assimilation and biosynthesis of methionine, SAM, cysteine and the sulfur salvage pathway of which the final product is glutathione, and the glutathione recycling cycle in *Saccharomyces cerevisiae* and the genes encode the enzymes and transcription factors. The genes up-regulated by cadmium are indicated by *arrows* (Momose and Iwahashi, *Environ. Toxicol. Chem.*, 20, 2353–2360)

et al. 2000). The validated centrometric sequences (CDE1 motif, G/TCACGTG (Mellor et al. 1991)) are found in the promoters of some methionine-regulated genes and have been shown to be bound by the regulatory protein *Cbf1* (Dormer et al. 2000), in 26 sites of the genes up-regulated by cadmium (Momose and Iwahashi 2001). These genes may have been induced by cadmium using the same transcriptional factors as methionine biosynthesis.

Cadmium has been known to cause oxidative stresses (Figueiredo-Pereira et al. 1998, Hatcher et al. 1995, Brenann and Schiestl 1996, Nigam et al. 1995, Jianghai et al. 2005) and damage to chromatin (Shiraishi 1975). After cadmium treatment, some, but not

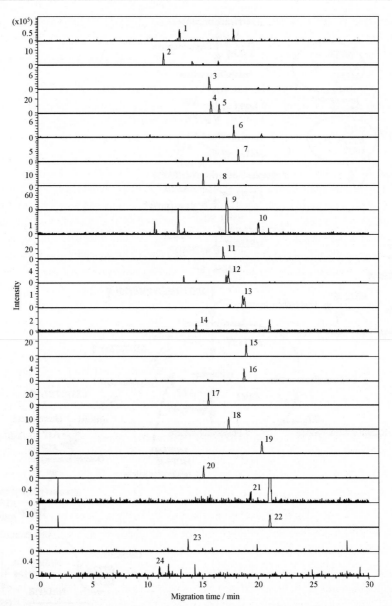

Fig. 4 The typical results of CE-MS metabolomics analysis. Selected ion electropherograms obtained from a yeast extract spiked with 23 potential sulfur-related metabolites. Spiked concentrations: 50 μg/mL for each target metabolite, 150 μg/mL for glutamic acid and reduced glutathione. 1: glycine (m/z = 76), 2: alanine (m/z = 90), 3: serine (m/z = 106), 4: homoserine (m/z = 120), 5: threonine (m/z = 120), 6: cysteine (m/z = 122), 7: aspartic acid (m/z = 134), 8: homocysteine (m/z = 136), 9: glutamic acid (m/z = 148), 10: O-acetyl-l-serine (m/z = 148), 11: methio-nine (m/z = 150), 12: O-acetyl-L-homoserine (m/z = 162), 13: methionine S-oxide (m/z = 166), 14: cysteinylglycine (m/z = 179), 15: methionine sulfone (internal standard) (m/z = 182), 16: O-succinyl-L-homoserine (m/z = 220), 17: cystathionine (m/z = 223), 18: cystine (m/z = 241), 19: L-γ-glutamylcysteine (m/z = 251), 20: homocystine (m/z = 269), 21: oxidized glutathione (m/z = 308), 22: reduced glutathione (m/z = 308), 23: S-adenosyl-L-homocysteine (m/z = 385) and 24: S-adenosyl-L-methionine (m/z = 399)

all, genes are responsive to oxidative stress (Momose and Iwahashi 2001).

3.3 Determination of Sulfur-Related Metabolites in Response to Cadmium Stress (Tanaka et al. 2007)

Genomics are mainly evaluation systems for induced functions but not the products of induced functions. In contrast, metabolomics is a system for evaluating substances as the products of induced functions. Metabolomics can yield direct evidence of cellular stress. CE-MS is suitable for analysis of low molecular weight and ionic substances. The majority of metabolites are considered as ionic and small substances.

Figure 4 shows the typical results of CE-MS metabolomics analysis (Tanaka et al. 2007).

Three replicate sample preparations under the same conditions (initial, control incubation for 1–3 hr, stress incubation for 1–3 hr) and two replicate injections of each sample were performed. MS electropherograms of representative yeast metabolites in cadmium stress-induced and control yeast extracts (2 hr) are shown in Fig. 5 (Tanaka et al. 2007).

Several metabolites showed remarkable differences between control and stress-treated yeast cells. The peak area ratios of the analyte/ internal standard were calculated from their corresponding selected ion

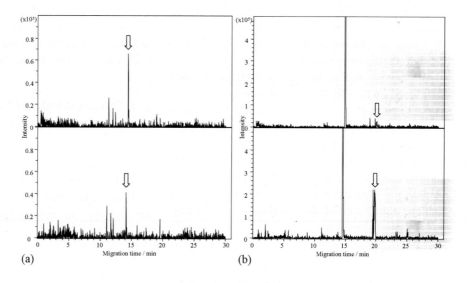

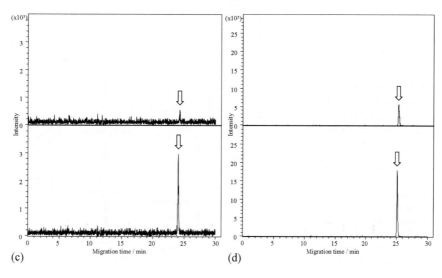

Fig. 5 Representative electropherograms showing differences in cadmium stress response. (**a**) glycine (m/z = 76), (**b**) O-acetyl-L-homoserine (m/z = 162), (**c**) L-γ-glutamylcysteine (m/z = 251), (**d**) glutathione (m/z = 308). Upper electropherograms: Cadmium stress treatment for 2 hr, lower electropherograms: control for 2 hr. The *arrows* indicate the peaks of the target analytes. (Tanaka et al., *J. Pharm. Biomed. Anal.* 44, 608–613)

Table 4 Differences in sulfur-related metabolites during Cd stress response in yeast cells (Tanaka et al., *J. Pharm. Biomed. Anal.* 44, 608–613.)

Component	Stress treatment with 300 μM cadmium chloride		
	1 h	2 h	3 h
Glycine	0.653±0.072	0.411±0.074	0.375±0.042
Serine	0.276±0.062	0.237±0.016	0.527±0.057
Homoserine	0.712±0.079	0.921±0.087	1.024±0.129
Threonine	0.489±0.062	0.343±0.004	0.310±0.021
Aspartic acid	1.540±0.135	1.222±0.070	1.214±0.163
Glutamic acid	1.043±0.050	1.241±0.053	1.316±0.098
Methionine	0.273±0.037	0.326±0.002	0.386±0.057
O-Acetyl-L-homoserine	5.562±2.788	14.776±6.964	16.288±6.784
O-Succinyl-L-homoserine	1.333±0.371	1.668±0.196	1.326±0.108
Cystathionine	0.802±0.137	0.588±0.020	0.487±0.068
Glutathione, reduced	2.089±0.411	2.660±0.569	2.584±0.517

The average (mean ± SD) values obtained from the difference in stress/control concentration ratio of the metabolites at the same time point.

electropherograms of the target metabolites. In Table 4 (Tanaka et al. 2007), the indicative values of the cadmium stress response were obtained by dividing the peak area ratio for the control condition to each stress condition at the same time point.

We briefly discuss the sulfur metabolism pathway based on the CE-MS measurements as shown in Fig. 6 (Tanaka et al. 2007). In this study, five cationic metabolites (i.e. cysteine, homocysteine, oxidized glutathione, *S*-adenosyl-L-homocysteine and *S*-adenosyl-L-methionine) were below detection limits. The results showed that a common feature of the cadmium stress response is obtained from the accumulation of glutathione and *O*-acetyl-L-homoserine. A very interest-

ing observation of this investigation was that cadmium stress induced the depletion of glycine and the strong accumulation of L-γ-glutamylcysteine. This observation has lead to the speculation that glycine may be rate-limiting for the production of glutathione.

3.4 Cross-Linking of Genomics and Metabolomics

Figure 6 shows the result of combining the genomic analysis with the metabolomic analysis.

Most importantly, "the biosyntheses of glutathione are actively" is considering in both genomic and

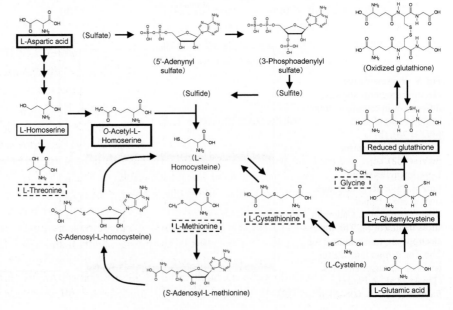

Fig. 6 An overview of sulfur metabolism pathway and the differences in cadmium stress response in yeast. The frames indicated in *bold, narrow and dash lines* on the compound name show the changes in stress response as increase, unchanged and decrease metabolites, respectively. Undetected metabolites are indicated in parentheses. (Tanaka et al., *J. Pharm. Biomed. Anal.* 44, 608–613)

metabolomic analysis. However, the conclusion with an active biosyntheses of gultathione cannot be led by the combined results by Genomics and metabolomics. As the biosynthesis of glycine is not promoted, we cannot accept the enhanced accumulation of glutathione. Homoserine becomes *O*-acetyl-L-homoserine when it is acetylated and carbon skeleton of cysteine is supplied. On the other hand, homoserine becomes *O*-phospho-L-homoserine when it is phosphorylated and it is metabolized to glycine. A gene expression is not observed in the direction of phosphorylation whereas, the direction of acetylation is observed. The result of the metabolomic analysis also shows the accumulation of *O*-acetyl-L-homoserine and a depletion of glycine, and this result is in good agreement with genomic analysis. Positive biosyntheses of gultathione are impossible as long as biosyntheses of glycine are not activated. As for glycine, the induction of each gene is not admitted from a result of genomic analytical though there is a possibility that it is synthesized by other metabolic pathway. It is interesting that the consideration led from genomics in parallel with metabolomics differ from result of obtaining each genomics and metabolomics alone. It is shown the possibility of discover that new mechanism nobody has noticed until now according to combining the both technologies.

4 Conclusion

The information concerning the toxicity of cadmium in a simple experiment makes use of the yeast microarray system. All genes corresponding to sulfur metabolism are up-regulated by cadmium (Fig. 3). This result indicates that cadmium caused the depletion of SH-compounds in the cells and suggests that cadmium is binding to antioxidant molecules. In the cells, oxidative stress and free radicals must increase during the cadmium treatment because SH-compounds act as oxidative and free radical compounds scavengers and their decrease leads to the induction of genes related to oxidative stress and free radical stress. These stresses can be the cause of nephrotoxicity (Shibasaki et al. 1993), hepatotoxicity (Hussain et al. 1987), damage of the nervous system (Figueiredo-Pereira et al. 1998), and chromatid aberrations (Shiraishi 1975). Therefore, the multifunctional

effect of cadmium is determined by only one kind of method, DNA microarrays.

On the other hand, the CE-MS method was applied to the investigation of qualitative and quantitative difference of yeast metabolites in response to cadmium stress. The depletion of glycine and the strong accumulation of L-γ-glutamylcysteine as well as the accumulation of glutathione and *O*-acetyl-L-homoserine in yeast cells was a novel finding of this study.

"The biosyntheses of glutathione are actively" is considering in both Genomic and Metabolomic analysis. However, the conclusion with an active biosyntheses of glutathione cannot be led by the combined results by genomics and metabolomics. The biosyntheses of glycine, which is one of the three components of glutathione, are not promoted. It is interesting that the consideration led from genomics in parallel with metabolomics differ from result of obtaining each genomics and metabolomics alone. It is shown the possibility of discover that new mechanism nobody has noticed until now according to combining the genomic analysis with the metabolomic analysis.

References

Babu C.V.S., Song E.J., Babar S.M.E., Wi M.H. and Yoo Y.S. (2006), Capillary electrophoresis at the omics level: Towards systems biology. *Electrophoresis*, 27, 97–110.

Brenann R.J. and Schiestl R.H. (1996), Cadmium is an inducer of oxidative stress in yeast. *Mutat Res*, 356, 171–178.

Celemedson C., et al. (44Scientists) (1996), MEIC evaluation of acute systemic toxicity. *ATLA*, 24, 251–272.

Delaunay A., Isnard A.D. and Toledano M.B. (2000), H_2O_2 sensing through oxidation of the Yap1 transcription factor. *EMBO J*, 19, 5157–5166.

Dettmer K. and Hammock B.D. (2004), Metabolomics – A new exciting field within the "omics" sciences. *Environ Health Perspect*, 112, A396–A397.

Dormer U.H., Westwater J., McLaren N.F., Kent N.A., Mellor J. and Jamieson D.J. (2000) Cadmium-inducible expression of the yeast *GSH1* gene requires a functional sulfur-amino acid regulatory network. *J Biol Chem*, 275, 32611–32616.

Dunn W.B. and Ellis D.I. (2005), Metabolomics: Current analytical platforms and methodologies. *Trends Anal Chem*, 24, 285–294.

Dunn W.B., Bailey N.J.C. and Johnson H.E. (2005), Measuring the metabolome: current analytical technologies. *Analyst*, 130, 606–625.

Ekwall B., Clemedson C., Crafoord B., Ekwall B., Hallander S., Walum E., and Bondesson I. (1998), MEIC evaluation of acute systemic toxicity. *ATLA*, 26, 571–616.

Figueiredo-Pereira M.E., Yakushin S. and Cohen G. (1998), Disruption of the intracellular sulfhydryl homeostasis by cadmium-induced oxidative stress leads to protein thiolation and ubiquitination in neuronal cells. *J Biol Chem*, 273, 12703–12709.

Godon C. et al. (1998), The H_2O_2 stimulon in *Saccharomyces cerevisiae*. *J Biol Chem*, 273, 22480–22489.

Hatcher E.L., Chen Y. and Kang Y.J. (1995), Cadmium resistance in A549 cells correlates with elevated glutathione content but not antioxidant enzymatic-activities. *Free Radic Biol Med*, 19, 805–812.

Hogan D.A., Auchtung T.A. and Hausinger R.P. (1999), Cloning and characterization of a sulfonate/α-ketoglutarate dioxygenase from *Saccharomyces* cerev. *J Bacteriol*, 181, 5876–5879.

Hussain T., Shukla G.S. and Chandra S.V. (1987), Effects of cadmium on superoxide dismutase and lipid peroxidation in liver and kidney of growing rats – in vivo and in vitro studies. *Pharmacol Toxicol*. 60, 355–358.

Jianghai L., Yingmei Z., Dejun H. and Gang S. (2005), Cadmium induced MTs synthesis via oxidative stress in yeast *Saccharomyces cerevisiae*. *Mol Cell Biochem*, 280, 139–145.

Kitagawa E., Takahashi J., Momose Y. and Iwahashi H. (2002), The effects of the pesticide thiuram: Genome-wide screening of indicator genes by yeast DNA microarray. *Environ Sci Technol*, 36, 3908–3915.

Kohrer K. and Domdey H. (1990), Guide to yeast genetics and molecular biology. In Guthrie C., Funk G.R., editor. *Methods Enzymol.*, Academic Press, San Diego, 194, 398–401.

Krishnan P., Kruger N.J. and Ratcliffe R.G. (2005), Metabolite fingerprinting and profiling in plants using NMR. *J Exp Bot*, 56, 255–265.

Kuhn K.M., DeRisi J.L., Brown P.O. and Sarnow P. (2001), Global and specific translational regulation in the genomic response of *Saccharomyces cerevisiae* to a rapid transfer from a fermentable to a nonfermentable carbon source. *Mol Cell Biol*, 21, 916–927.

Li Z.S., Lu Y.P., Zhen R.G., Szczypka M., Thiele D.J. and Rea P.A. (1997), A new pathway for vacuolar cadmium sequestration in *Saccharomyces cerevisiae*: *YCF1*-catalyzed transport of bis(glutathionato) cadmium. *Proc Natl Acad Sci USA*, 94, 42–47.

Mellor J., Rathjen J., Yian W., Branes C.A. and Dowell S.J. (1991), DNA-binding of CPF1 is required for optimal centromere function but not for maintaining methionine prototrophy in yeast. *Nucleic Acid Res*, 19, 2961–2969.

Momose Y. and Iwahashi H. (2001), Bioassay of cadmium using a DNA microarray: Genome-wide expression patterns of *Saccharomyces cerevisiae* response to cadmium. *Environ ToxicolChem*, 20, 2353–2360.

Nigam D., Shukla G.S. and Agarwal A.K. (1999), Glutathione depletion and oxidative damage in mitochondria following exposure to cadmium in rat liver and kidney. *Toxicol Lett*, 106, 151–157.

Ramautar R., Demirci A. and de Jong G.J. (2006), Capillary electrophoresis in metabolomics. *Trends Anal Chem*, 25, 455–466.

Reifferscheid G. and Heil J. (1996), Validation of the SOS/umu test using test results of 486 chemicals and comparison with the Ames test and carcinogenicity data. *Mutat Res*, 369, 129–145.

Robertson D.G. (2005), Metabonomics in toxicology: A review. *Toxicol Sci*, 85, 809–822.

Rochfort S. (2005), Metabolomics reviewed: A new "Omics" platform technology for systems biology and implications for natural products research. *J Nat Prod*, 68, 1813–1820.

Sato S., Soga T., Nishioka T. and Tomita M. (2004), Simultaneous determination of the main metabolites in rice leaves using capillary electrophoresis mass spectrometry and capillary electrophoresis diode array detection. *Plant J*, 40, 151–163.

Shibasaki T., Ohno I., Ishimoto F. and Sakai O. (1993), Characteristics of cadmium-induced nephrotoxicity in Syrian hamsters. *Nippon Jinzo Gakkai Shi*, 8, 913–917.

Shiraishi Y. (1975), Cytogenetic studies in 12 patients with itai-itai disease. *Human Genetics*, 27, 31–44.

Soga T., Ohashi Y., Ueno Y., Naraoka H., Tomita M. and Nishioka T. (2003), Quantitative metabolome analysis using capillary electrophoresis mass spectrometry. *J Proteome Res*, 2, 488–494.

Soga T., Ohashi Y., Ueno Y., Naraoka H., Tomita M. and Nishioka T. (2003), Quantitative metabolome analysis using capillary electrophoresis mass spectrometry. *J Proteome Res*, 2, 488–494.

Soga T., Ueno Y., Naraoka H., Ohashi Y., Tomita M. and Nishioka T. (2002), Simultaneous determination of anionic intermediates for Bacillus subtilis metabolic pathways by capillary electrophoresis electrospray ionization mass spectrometry. *Anal Chem*, 74, 2233–2239.

Tanaka Y., Higashi T., Rakwal R., Wakida S., Iwahashi, H. (2007), Quantitative analysis of sulfur-related metabolites during cadmium stress response in yeast by capillary electrophoresis-mass spectrometry. *J Pharm Biomed Anal*, 44, 608–613.

Tohoyama H., Tomoyasu T., Inoue M., Joho M. and Murayama T. (1992), The gene for cadmium metallothionein from cadmium-resistant yeast appears to be identical to *CUP1* in copper-resistant strain. *Curr Genet*, 21, 275–280.

Viant M.R., Rosenblum E.S. and Tjeerdema R.S. (2003), NMR-based metabolomics: A powerful approach for characterizing the effects of environmental stressors on organism health. *Environ Sci Technol*, 37, 4982–4989.

Vido K., Spector D., Lagniel G., Lopez S. and Toledano M. (2001), A proteome analysis of the cadmium response in *Saccharomyces cerevisiae*. *J Biol Chem*, 276, 8469–8474.

Zahringer H., Holzer H. and Nwaka S. (1998), Stability of neutral trehalase during heat stress in *Saccharomyces cerevisiae* is dependent on the activity of the catalytic subunits of cAMP-dependent protein kinase, Tpk1 and Tpk2. *Eur J Biochem*, 255, 544–551.

A Gene Expression Profiling Approach to Study the Influence of Ultrafine Particles on Rat Lungs

Katsuhide Fujita, Yasuo Morimoto, Akira Ogami, Isamu Tanaka, Shigehisa Endoh, Kunio Uchida, Hiroaki Tao, Mikio Akasaka, Masaharu Inada, Kazuhiro Yamamoto, Hiroko Fukui, Mieko Hayakawa, Masanori Horie, Yoshiro Saito, Yasukazu Yoshida, Hitoshi Iwahashi, Etsuo Niki and Junko Nakanishi

Abstract In recent years, industrial or commercial products incorporating nanomaterials have triggered concerns for human health. Manufactured nanomaterials are unstable and tend to form secondary particles by agglomeration. Little is known about the cytotoxicity induced by nano-sized secondary particles dispersed in aqueous solution. In the present study, we have attempted to disperse ultrafine nickel oxide particles in water (Uf-NiO), characterize the physicochemical properties, and instill into the trachea of rat lungs. Analysis by inductively couple plasma mass spectrometry (ICP-MS) and transmission electron microscope (TEM) revealed that the Uf-NiO particles began to be cleared from the lungs immediately after treatment, and that low levels of the particles were present at 6 months post-instillation. Genome-wide expression analysis using DNA microarray revealed that intratracheal instillation of Uf-NiO particles led to a rapid increase in the expression of chemokines and genes involved in inflammation. These changes were most pronounced at 1 week post-instillation with Uf-NiO. The expression of Mmp12 mRNA, encoding macrophage metalloelastase 12, was strongly induced immediately following intratracheal instillation. However, expression returned to control levels by 6 months post-instillation expression of various other genes categorized into the detection of chemical stimulus were increased at this time point, at which time the inflammatory response had diminished. These results suggest that residual Uf-NiO in the lungs subacutely initiated distinct cellular events through signal transduction after resolution of the inflammatory response. We conclude that gene expression analysis using DNA microarrays can be extremely useful in assessing the influence of utrafine particles on biological systems.

Keywords DNA microarray · Nano · Ultrafine nickel oxide · Intratracheal instillation · Lung · Rat

1 Introduction

In recent years, concern over the influence of ultrafine (nano) particles on human health and the environment has risen due to advances in the development of nanotechnology. Especially, based on the clinical assessments of dusts, asbestos, or these substitutions exposure, there has been more interest in the influences of ultrafine particles on pulmonary inflammation, fibrosis, and cancer etc. The toxicological effects of ultrafine particles delivered to the lungs of rodents by inhalation have been reported (Warheit et al. 2004; Grassian et al. 2007; Sayes et al. 2007; Warheit et al. 2007). Gene expression profiling using DNA microarrays have been performed to assess the effects of pulmonary fibrosis or lung injury on gene expression in the lung (Katsuma et al. 2001; McDowell et al. 2003; Kaminski and Rosas 2006; Studer and Kaminski 2007). The aim of such studies is to identify clusters of genes involved in the progression of pulmonary diseases. In addition, the toxicological effects of nanoparticles on gene expression have been evaluated using DNA microarrays (Chen et al. 2006;

K. Fujita (✉)
Health Technology Research Center (HTRC), National Institute of Advanced Industrial Science and Technology (AIST), Onogawa 16-1, Tsukuba, Ibaraki, 305-8569, Japan
e-mail: ka-fujita@ aist.go.jp

Y.J. Kim et al. (eds.), *Atmospheric and Biological Environmental Monitoring*,
DOI 10.1007/978-1-4020-9674-7_14, © Springer Science+Business Media B.V. 2009

Chou et al. 2008). An approach can provide a clear analysis of the molecular events induced by nanoparticle exposure.

Nickel excels in terms of durability, and it has therefore been used extensively in industrial applications such as plating, catalysis, and as a raw material for the production of stainless steels and coins etc. Ultrafine nickel oxide has been widely used in the production of electronic materials such as mixed oxide ferrites, varistors and thermistors, and has also been used for the production of high-quality pigments for glass and ceramics.

A number of reports have described the pulmonary toxicity assessments of the nickel oxide exposure. The carcinogenicity of nickel compounds is inversely linked to their solubility in aqueous solution. Insoluble nickel compounds, such as nickel oxide, are more carcinogenic than soluble compounds, such as Ni (II) acetate, chloride, or sulfate (Kasprzak et al. 2003). Animal studies have linked nickel oxide to the development of both lung cancer and acute lung injury and inflammation (Morimoto et al. 1995; Kawanishi et al. 2002). Since a lot of information can be referred, we reasoned that Uf-NiO would be a suitable agent to use to evaluate how manufactured nanomaterials influence the rat lung.

Ultrafine particles are unstable and tend to agglomerate in aqueous solution. To date, little is known of the cytotoxicity or biological responses induced by ultrafine particles dispersed in aqueous solution and well-characterized. The preparation of nanoparticles and the characterization of such particles by robust physicochemical techniques are required to resolve this problem (Oberdörster et al. 2005).

In the present study, we developed an aqueous system of dispersed, nano-sized ultrafine nickel oxide (Uf-NiO) particles, and characterized the physicochemical properties of these particles. We further examined the distribution of nickel in rat organ samples after intratracheal instillation of these Uf-NiO particles. DNA microarray analysis was used to evaluate changes in gene expression in the rat lung in response to Uf-NiO exposure. Our results have demonstrated the utility of gene expression profiling by DNA microarray analysis for the evaluation of the influence of well-characterized ultrafine particles on rat lungs.

2 Material and Methods

2.1 Source and Characterization of Nano-Sized Nickel Oxide

The nano-sized nickel oxide particles used in this study were purchased from Nanostructured & Amorphous Materials Inc. (Houston, TX). The manufacturer's specifications indicated an average primary particle size of 10–20 nm, and a specific surface area of 50–80 m^2/g. The nano-sized nickel oxide particles dissolved in distilled water (0.5% w/w) were dispersed by an ultrasonic homogenizer (450 W, 90 min at 10 min intervals), and centrifuged at $8,883 \times g$ for 20 min. The supernatant was 1.0 μm membrane-filtered in order to remove the loose and coarse flocs. The size of ultrafine nickel oxide particles was characterized by a transmission electron microscope (TEM) at 200 kV (Leo, Germany), and a laser light diffraction methods using the Microtrac® UPA150 (Nikkiso, Tokyo, Japan).

2.2 Animals

Nine-week-old male Wistar rats (Clea Japan, Tokyo, Japan) treated with intratracheal instillation were housed for 6 months with a chow diet *ad libitum* in a room controlled with a temperature at 22°C. One week, 1 month, 3 months, and 6 months after the intratracheal instillation of 0.4 mL distilled water or Uf-NiO (100 μg/0.4 mL distilled water or 200 μg/0.4 mL distilled water). Prior to initiating animal exposure tests, we have observed pulmonary granuloma formation by approximately 250 μg Uf-NiO. Thus, Uf-NiO was treated with at dosages of 100 μg and 200 μg. The lungs of anesthetized rats were perfused with physiological saline, excised and then stored at −80 °C until used for lung weight measurements ($n = 5$), nickel distribution in organs ($n = 5$), and microarray analysis ($n = 2$–4). Animal procedures were approved by University of Occupational and Environmental Health, Japan and National Institute of Advanced Industrial Science and Technology, Japan Animal Care and Use Committees.

2.3 Nickel Distribution in Organs

A small portion (ca. 50 mg) of homogenized samples of lung, brain, and liver was digested with a mixture of 2 mL HNO_3, 0.5 ml HCl and 4 mL H_2O_2 using a microwave digester (Multiwave 3000, Anton Paar GmbH, Austria). The digested samples were diluted to 50 mL with distilled-deionized water and analyzed with an inductively coupled plasma mass spectrometer (ICP-MS, Agilent Technologies, Santa Clara, CA). The detection limit of nickel in original organ was 0.1 μg/g. The nickel distribution was observed by a transmission electron microscope (TEM) at 200 kV (Leo, EM-922, Oberkochen, Germany).

2.4 Microarrays Analysis

The right lungs were disrupted and homogenized with a Tissue Ruptor (Qiagen, Tokyo, Japan). Total RNA from the homogenates was extracted using the RNeasy Midi kit (Qiagen, Tokyo, Japan) following the manufacturer's instructions. RNA quality and concentration were measured using an Agilent 2100 bioanalyzer (Agilent Technologies, Santa Clara, CA) and a NanoDrop ND-1000 (NanoDrop Technologies, Wilmington, DE). cRNA synthesized from cDNA were amplified in the presence of fluorescent Cyanine 3-CTP. Labeled cRNA was used for hybridization onto 4 × 44 K Whole Rat Genome Oligo Multiplex Microarray slides (Agilent Technologies, Santa Clara, CA) containing approximately 41,000 oligonucleotide probes at 65 °C for 17 hours. Hybridized microarray slides were washed according to the Agilent user's manual instructions, and were scanned with an Agilent DNA Microarray Scanner (Agilent Technologies, Santa Clara, CA). The scanned images were analyzed numerically using the Agilent Feature Extraction Software version 9.5.3.1. Statistical analysis was performed using the commercial software GeneSpring GX (Agilent Technologies, Santa Clara, CA). Expressed genes were categorized into various gene functional categories on the basis of Gene Ontology (GO) database (http://www.geneontology.org/) and Fatigo database (http://fatigo.bioinfo.cipf.es/).

3 Results and Discussion

3.1 Characterization of Uf-NiO

Using the laser light diffraction method performed with the Microtrac® device, we estimated the particle size and specific surface area of the nano-sized nickel oxide particles dispersed in distilled water to be 26.0 nm and 104.6 m^2/g respectively. We confirmed that these physicochemical properties were maintained for 3 weeks. TEM revealed that the homogeneously dispersed particles formed crystalline structures, and that their diameter ranged approximately from 3 to 15 nm (Fig. 1).

Based on these results, we defined the nano-sized particles as ultrafine nickel oxide (Uf-NiO), which was used for animal studies according to the study design (Fig. 2).

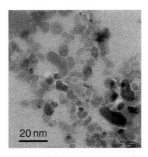

Fig. 1 TEM image of Uf-NiO dispersed in distilled water

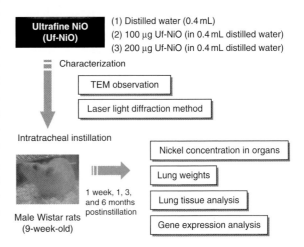

Fig. 2 Study design

3.2 Nickel Distribution in Lung, Liver and Brain

Table 1 shows time-dependent changes in nickel concentrations in rat lung after intratracheal instillation. ICP-MS analysis indicated that the concentrations of nickel in lung tissues exhibited a time-dependent decrease immediately after intratracheal instillation. Approximately 40% of the instilled dosage was detected at 1 week post-instillation, and approximately 10–20% remained at 3 months post-instillation. We detected a small amount of nickel oxide particles in the lungs at 6 months post-instillation. In liver and brain, the nickel concentrations were below the detection limit.

To validate the results of the ICP-MS analysis, we attempted to observe the nickel oxide particles in the lungs by TEM studies (Fig. 3).

Zero-loss images demonstrated that some particles remained in the lung at 1 month post-instillation with 100 μg or 200 μg Uf-NiO. The particles were identified as nickel oxide particles by electron energy-loss spectroscopy (EELS) analysis. We observed some traces in lung tissue samples at 6 months post-instillation with 100 μg or 200 μg Uf-NiO by zero-loss images. However, energy-dispersive X-ray spectroscopy (EDS) and EELS analyses revealed that these traces were not nickel oxide particles (data not shown). These results differed from the ICP-MS analysis. The discrepancy between these results needs further examination, however, it is clear that only a very small amount of nickel oxide particles remained in the lungs at 6 months post-instillation.

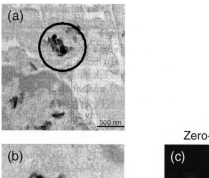

Zero-loss image

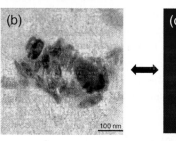
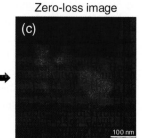

Fig. 3 Nickel oxide particles in rat lungs exposed to 200 μg Uf-NiO at 1 month post-instillation; (**a**) and (**b**) nano-sized particles observed by zero-loss images, (**c**) the nickel mapping image obtained by electron energy-loss spectroscopy (EELS) analysis

Uf-NiO (data not shown). Histopathological analyses revealed significant inflammatory lesions in lung tissue at 3 days and 1 month post-instillation (Fig. 4). Persistent inflammatory lesions in lung tissue were observed in 100 μg Uf-NiO exposed rats at 1 month post-instillation, and 200 μg Uf-NiO exposed rats at 6 month post-instillation (data not shown). However, no histopathological abnormalities were detected in liver, kidney, spleen, cerebrum, cerebellum, testis, or nasal cavity tissues. These results suggested that rat lungs reacted to Uf-NiO immediately after instillation.

3.3 Changes in Lung Weight and Histopathological Finding

We observed significant increases in lung weights 3 days, 1 week, 1 and 3 months post-instillation in animals that were exposed to either 100 μg or 200 μg

3.4 Gene Expression Analysis

We compared gene expression patterns at different times after treatment with the Uf-NiO particles. At 1 week post-instillation, the number of genes whose expression was increased over 2-fold by exposure to either 100 μg or 200 μg Uf-NiO was 177. In con-

Table 1 Detection of nickel in the lungs over time following intratracheal instillation of Uf-NiO

| Exposure group | Nickel in μg/lung (mean ± SE) by posttreatment period | | | |
	1 wk	1 mo	3 mo	6 mo
100 μg Uf-NiO	45.8 ± 3.6	34.7 ± 0.9[b]	16.8 ± 1.3[b]	2.7 ± 0.9[b]
200 μg Uf-NiO	81.1 ± 4.2[a]	59.3 ± 6.0[a,b]	30.9 ± 5.7[b]	17.9 ± 3.0[a,b]

[a] $P \leq 0.05$ compound to 100 μg Uf-NiO exposure group on the same posttreatment day.
[b] $P \leq 0.05$ compound to days 1 wk within the exposure group.

Post-instillation exposure

3 days | 1 month

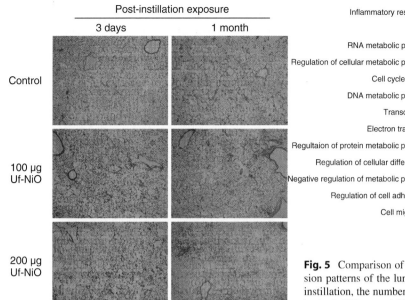

Control

100 μg
Uf-NiO

200 μg
Uf-NiO

Fig. 4 Inflammatory lesions in lung tissue observed in Uf-NiO exposed rats at 3 days and 1 month post-instillation

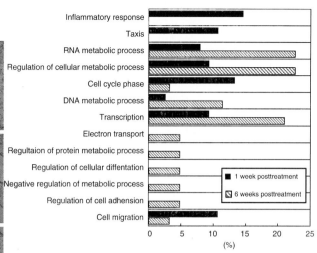

Inflammatory response
Taxis
RNA metabolic process
Regulation of cellular metabolic process
Cell cycle phase
DNA metabolic process
Transcription
Electron transport
Regultaion of protein metabolic process
Regulation of cellular diffentation
Negative regulation of metabolic process
Regulation of cell adhesion
Cell migration

■ 1 week posttreatment
▨ 6 weeks posttreatment

0   5   10   15   20   25
(%)

Fig. 5 Comparison of time-dependent changes in gene expression patterns of the lungs exposed to Uf-NiO. At 1 week post-instillation, the number of genes whose expression was induced by both 100 μg and 200 μg Uf-NiO to more than 2-fold baseline was 177. At 6 months post-instillation, the number of genes whose expression was induced by both 100 μg and 200 μg Uf-NiO to more than 2-fold was 111. *Bold bars* show the ratio of gene induced at 1 week post-instillation, and hatched bars for these at 6 month post-instillation

trast, at 6 months post-instillation, the number of genes whose expression was increased over 2-fold by exposure to either 100 μg or 200 μg Uf-NiO was 111. The genes whose expression increased were categorized into various functional categories on the basis of the Fatigo database (Biological process. Level: 5). The results revealed a distinct pattern of expression of different gene ontology categories at 1 week and 6 months post-instillation. Uf-NiO exposure increased the expression of genes involved in the inflammatory response (GO: 0006954) 1 week post-instillation. In contrast, the genes whose expression was increased 6 months post-instillation included those involved in RNA metabolic process (GO: 0016070), the regulation of cellular metabolic process (GO: 0031323), and transcription (GO: 0006350) (Fig. 5).

Table 2 depicts the relative expression of individual genes whose expression was altered by Uf-NiO and which are involved in the inflammatory response (GO: 0006954). These included various genes representative of those involved in the inflammatory response, such as Ccl2 (MCP-1), encoding proteins involved in a monocyte chemoattractant protein, Cxcl10 (IP-10), C-X-C motif chemokine ligand 10, encoding proteins involved in the induction of DNA synthesis, and cell proliferation, Cxcl2 (Mip-2), a chemokine involved in

the pulmonary inflammatory response, Cxcl1 (CINC-1), which acts as a neutrophil chemoattractant and may play a role in the acute phase inflammatory response, Cxcl9 (Mig), C-X-C motif chemokine ligand 9, which plays a role in the recruitment of mononuclear cells. The changes in expression of these genes was most remarkable at 1 week post-instillation with either 100 μg or 200 μg Uf-NiO, suggesting that Uf-NiO strongly induced genes coding for proinflammatory chemokines. However, increased expression of these genes was not observed by 3 months post-instillation.

Matrix metalloproteinases (MMPs) comprise a family of proteinases, that mediate the degradation of extracellular matrix components and that also cleave non-matrix proteins, including growth factors, chemoattractants and cell surface receptors (Lagente et al. 2005). We observed a significant increase in the expression of Mmp12, encoding a macrophage metalloelastase/matrix metalloproteinase, immediately after intratracheal instillation of Uf-NiO, which subsequently gradually decreased over a 6 months period (Table 3). On the other hand, the expression of the genes coding for antiproteases, such as Timp1, encoding tissue inhibitor of metalloproteinases 1, which acts

Table 2 Expressed genes associated with the inflammatory response in response to Uf-NiO

Genbank	Gene name	Description	100μg Uf-NiO				200 μg Uf-NiO			
			1 wk	1 mo	3 mo	6 mo	1 wk	1 mo	3 mo	6 mo
XM_213425	Ccl12(predicted)	chemokine (C-C motif) ligand 12 (predicted)	13.3	1.9	n.d.	n.d.	28.8	5.9	4.2	0.5
NM_031530	Ccl2/MCP-1	chemokine (C-C motif) ligand 2	11.0	1.6	1.0	0.5	15.4	4.1	10.1	1.0
NM_139089	Cxcl10/IP-10	chemokine (C-X-C motif) ligand 10	10.3	0.4	1.6	n.d.	13.5	10.3	7.3	1.1
NM_012881	Spp1/OSP	secreted phosphoprotein 1	8.4	1.7	0.7	0.3	9.8	10.4	n.d.	1.3
NM_053647	Cxcl2/Mip-2	chemokine (C-X-C motif) ligand 2	4.3	2.6	1.0	0.9	4.9	2.1	2.7	0.9
NM_030845	Cxcl1/CINC-1	chemokine (C-X-C motif) ligand 1	4.2	3.0	1.2	0.9	3.9	1.7	0.5	1.7
NM_138522	Cinc-2	gene model 1960, (NCBI)	3.5	1.7	0.7	1.9	3.7	1.2	2.4	1.3
NM_016994	C3	complement component 3	3.3	1.5	1.2	0.7	3.5	1.5	0.2	1.1
NM_012696	Kng1	kininogen 1	3.1	1.4	2.0	0.6	3.4	1.1	1.5	1.1
NM_138879	Sele	selectin, endothelial cell	2.5	0.8	0.6	0.7	2.7	1.9	2.7	0.9
NM_012696	Kng1	kininogen 1	2.1	1.1	2.0	0.3	2.4	1.4	1.9	1.4
XM_342824	Ccl19 (predicted)	chemokine (C-C motif) ligand 19 (predicted)	2.1	1.3	0.8	1.1	2.1	1.6	7.9	1.1
NM_053858	Ccl4/Mip1-b	chemokine (C-C motif) ligand 4	1.9	1.6	1.4	0.8	2.1	0.8	0.8	0.7
NM_145672	Cxcl9/Mig	chemokine (C-X-C motif) ligand 9	1.9	n.d.	1.0	0.6	2.0	1.0	3.7	1.4

n.d., not determined

as an inhibitor of metalloprotease, were not increased (data not shown). Taken together, the lung weight changes, nickel distribution in the lungs, and the histopathological findings, suggest that proinflammatory mediators were elicited by Uf-NiO exposure, and that matrix metallopeptidases might degrade the extracellular matrix components, elastin in the walls of alveoli or vascular endothelial cell.

We also observed a changes in the expression of Hmox1/Ho1 in response to exposure to Uf-NiO (Table 4). Hmox1/Ho1, encoding heme oxygenase 1, which catalyzes the oxidative cleavage of heme to biliverdin is involved in the oxidative stress response

Table 3 Enhanced expression of Mmp-12 gene in response to Uf-NiO exposure

Exposure group	Fold change			
	1 wk	1 mo	3 mo	6 mo
100 μg Uf-NiO	21.1	23.9	14.1	n.d.
200 μg Uf-NiO	34.7	13.5	4.9	1.9

n.d., not determined

(GO: 0006979). Hmox1/Ho1 gene expression in rat lungs was increased during the acute and chronic phases of crystalline silica or chrysotile asbestos exposure (Nagatomo et al. 2006, 2007). The gene may be similarly regulated as an anti-oxidative defense factor in lungs injury by Uf-NiO. However, no significant increasing of expression of other genes involved in the oxidative stress was observed.

Among genes involved in apoptosis (GO: 0006915), Lcn2, encoding lipocalin 2, a member of the lipocalin superfamily with diverse functions such as the regulation of inflammatory responses, control of cell growth and development, tissue involution, and apoptosis (Sunil et al. 2007), was up-regulated by Uf-NiO (Table 5). Interestingly, increased expression of Lcn2 by 200 μg Uf-NiO was maintained during the 6 months evaluation period.

Numerous genes encoding proteins known to be involved in the detection of chemical stimulus (GO: 0009593) were found to be induced after the

Table 4 Expressed genes associated with the oxidative stress response in response to Uf-NiO

Genbank	Gene name	Description	100μg Uf-NiO				200μg Uf-NiO			
			1 wk	1 mo	3 mo	6 mo	1 wk	1 mo	3 mo	6 mo
NM_012580	Hmox1/Ho1	heme oxygenase (decycling) 1	2.4	3.2	1.9	0.7	2.7	2.3	4.1	1.6
NM_017051	Sod2	superoxide dismutase 2, mitochondrial	2.0	1.5	1.2	0.7	2.2	0.9	1.8	1.3
NM_031614	Txnrd1	thioredoxin reductase 1	1.1	1.2	1.4	0.6	1.6	0.7	1.5	1.0
NM_031510	Idh1	isocitrate dehydrogenase 1	0.9	1.4	1.5	0.3	1.5	0.6	0.3	0.9
NM_031614	Txnrd1	thioredoxin reductase 1	0.7	1.2	1.4	0.3	1.5	0.7	0.2	0.8
NM_017232	Ptgs2/Cox2	prostaglandin-endoperoxide synthase 2	1.2	1.2	0.7	n.d.	1.4	1.1	n.d.	n.d.
NM_183403	Gpx2	glutathione peroxidase 2	0.9	0.9	1.1	0.4	1.4	0.8	0.3	0.8
NM_057114	Prdx1	peroxiredoxin 1	1.1	1.7	1.0	0.7	1.3	1.1	1.0	1.1
NM_012520	Cat	catalase	0.4	0.8	1.8	0.1	1.3	0.7	n.d.	0.5

n.d., not determined

inflammatory response had subsided at 6 months post-instillation. Table 6 indicates top 20 up-regulated genes by 100 μg Uf-NiO at 6 months post-instillation.

Also, a large number of genes were up-regulated by 200 μg Uf-NiO at 6 months post-instillation. The molecular function of these genes was assigned to olfactory receptors predicted. Olfactory receptors

share a seven transmembrane domain structure with many neurotransmitter and hormone receptors, and are responsible for the recognition of, and G-protein mediated transduction of, odorant signals (Buck and Axel 1991). Chemokine receptors are known to be members of the G-protein-coupled receptor (GPCR) superfamily (Onuffer and Horuk 2002). We detected

Table 5 Expressed genes associated with apoptosis in response to Uf-NiO

Genbank	Gene name	Description	100μg Uf-NiO				200μg Uf-NiO			
			1 wk	1 mo	3 mo	6 mo	1 wk	1 mo	3 mo	6 mo
NM_012829	Cck	cholecystokinin	2.3	0.4	1.0	2.1	0.6	0.3	3.8	3.9
NM_130741	Lcn2	lipocalin 2	9.5	5.0	2.7	0.7	9.9	4.2	9.0	3.0
NM_053388	Gjb6/Cx30	gap junction membrane channel protein beta 6	1.3	2.4	1.5	1.2	3.5	1.2	1.6	2.9
XM_230377	Traf6 (predicted)	Tnf receptor-associated factor 6 (predicted)	0.9	0.6	1.3	0.8	1.0	0.1	0.6	2.4
XM_343413	Pdcd7 (predicted)	programmed cell death protein 7 (predicted)	1.9	0.9	0.4	4.9	0.6	0.4	4.2	2.2
XM_240178	Acin1	apoptotic chromatin condensation inducer 1	1.6	0.9	0.4	3.2	0.6	0.4	1.5	2.1
NM_001012179	Fxr1 h	fragile X mental retardation gene 1	1.9	0.9	0.6	3.5	0.8	0.4	5.1	2.1
NM_130406	Faf1	Fas-associated factor 1	1.4	1.2	1.1	2.2	0.9	0.6	4.5	2.0
NM_021837	Mycs	myc-like oncogene	1.5	1.5	0.6	2.8	1.3	1.9	2.3	2.0

n.d., not determined

Table 6 Expressed genes associated with detection of chemical stimulus in response to Uf-NiO

Genbank	Gene name	Description	100 µg Uf-NiO				200 µg Uf-NiO			
			1 wk	1 mo	3 mo	6 mo	1 wk	1 mo	3 mo	6 mo
NM_001000897	Olr837	olfactory receptor 837	n.d.	n.d.	n.d.	4.7	1.2	1.4	1.1	4.7
NM_001000167	Olr153 (predicted)	olfactory receptor 153 (predicted)	n.d.	n.d.	n.d.	4.7	n.d.	n.d.	3.6	2.5
NM_001000533	Olr1581 (predicted)	olfactory receptor 1581 (predicted)	2.1	0.9	0.4	4.5	n.d.	n.d.	n.d.	1.9
NM_001000615	Olr749 (predicted)	olfactory receptor 749 (predicted)	n.d.	n.d.	n.d.	4.3	n.d.	n.d.	1.9	1.8
NM_001000523	Olr1381 (predicted)	olfactory receptor 1381 (predicted)	1.2	n.d.	0.1	4.3	n.d.	1.3	n.d.	1.8
NM_001000118	Olr20 (predicted)	olfactory receptor 20 (predicted)	n.d.	n.d.	n.d.	4.2	n.d.	n.d.	2.2	1.8
NM_001001008	Olr53 (predicted)	olfactory receptor 53 (predicted)	n.d.	n.d.	n.d.	3.8	n.d.	n.d.	n.d.	1.7
NM_001000771	Olr1455 (predicted)	olfactory receptor 1455 (predicted)	1.1	1.0	0.7	3.8	n.d.	n.d.	n.d.	1.7
NM_001001035	Olr232 (predicted)	olfactory receptor 232 (predicted)	n.d.	n.d.	0.4	3.8	n.d.	n.d.	1.5	1.7
NM_173333	Olr1361	olfactory receptor 1361	n.d.	n.d.	0.3	3.7	n.d.	n.d.	4.0	1.7

n.d., not determined

a small amount of nickel oxide particles in the lungs at 6 months post-instillation by ICP-MS analysis (Table 1). In contrast, high expression levels of genes associated with inflammatory response, oxidative stress were not determined (Tables 2 and 4). Thus, Uf-NiO exposure might induce signal transduction by chemosensory pathways mediated by currently unidentified chemokine receptor systems following the acute inflammatory responses.

4 Conclusion

Rats were intratracheally-instilled with utrafine nickel oxide (Uf-NiO) particles dispersed in an aqueous solution. The detection of nickel in the lungs decreased over the time course of our study. Microarray analyses revealed that Uf-NiO induced high expression of genes involved in the inflammatory response, chemokines, and Mmp12, coding for the macrophage elastase. The increased expression of these genes was most marked at 1 week post-instillation with Uf-NiO. Taken together, the changes in lung weight, nickel distribution, and histopathological findings, suggested that proinflammatory mediators were elicited by Uf-NiO exposure, and that matrix metallopeptidase might degrade the extracellular matrix com-

ponent, elastin in the walls of alveoli or vascular endothelial cell. In addition, Uf-NiO exposure might stimulate a chemosensory pathway mediated by unidentified chemokine receptors following the acute inflammatory responses. We conclude that gene expression analysis using DNA microarrays can be extremely useful in assessing the influence of nanosized particles on biological systems.

Acknowledgments The authors thank Ms. Kitagawa, E. of Roche Diagnostics, Japan for helping with microarray data analysis and invaluable comments. This research was funded by New Energy and Industrial Technology Development Organization of Japan (NEDO) Grant "Evaluating risks associated with manufactured nanomaterials (P06041)".

References

Buck L, Axel R (1991) A novel multigene family may encode odorant receptors: a molecular basis for odor recognition. Cell 65:175–187

Chen HW, Su SF, Chien CT et al. (2006) Titanium dioxide nanoparticles induce emphysema-like lung injury in mice. FASEB J 20:2393–2395

Chou CC, Hsiao HY, Hong QS et al. (2008) Single-walled carbon nanotubes can induce pulmonary injury in mouse model. Nano Lett 8:437–445

Grassian VH, O'shaughnessy PT, Adamcakova-Dodd A et al. (2007) Inhalation exposure study of titanium dioxide nanoparticles with a primary particle size of 2 to 5 nm. Environ Health Perspect 115:397–402

Kaminski N, Rosas IO (2006) Gene expression profiling as a window into idiopathic pulmonary fibrosis pathogenesis: Can We identify the right target genes? Proc Am Thorac Soc 3:339–344

Kasprzak KS, Sunderman FW Jr, Salnikow K (2003) Nickel carcinogenesis. Mutat Res 533:67–97

Katsuma S, Nishi K, Tanigawara K et al. (2001) Molecular monitoring of bleomycin-induced pulmonary fibrosis by cDNA microarray-based gene expression profiling. Biochem Biophys Res Commun 288:747–751

Kawanishi S, Oikawa S, Inoue S et al. (2002) Distinct mechanisms of oxidative DNA damage induced by carcinogenic nickel subsulfide and nickel oxides. Environ Health Perspect 110:789–791

Lagente V, Manoury B, Nénan S et al. (2005) Role of matrix metalloproteinases in the development of airway inflammation and remodeling. Braz J Med Biol Res 38:1521–1530

McDowell SA, Gammon K, Zingarelli B et al. (2003) Inhibition of nitric oxide restores surfactant gene expression following nickel-induced acute lung injury. Am J Respir Cell Mol Biol 28:188–198

Morimoto Y, Nambu Z, Tanaka I et al. (1995) Effects of nickel oxide on the production of tumor necrosis factor by alveolar macro-phages of rats. Biol Trace Elem Res 48: 287–296

Nagatomo H, Morimoto Y, Ogami A et al. (2007) Change of heme oxygenase-1 expression in lung injury induced by chrysotile asbestos in vivo and in vitro. Inhal Toxicol 19:317–323

Nagatomo H, Morimoto Y, Oyabu T et al. (2006) Expression of heme oxygenase-1 in the lungs of rats exposed to crystalline silica. J Occup Health 48:124–128

Oberdörster G, Oberdörster E, Oberdörster J (2005) Nanotoxicology: an emerging discipline evolving from studies of ultrafine particles. Environ Health Perspect 113:823–839

Onuffer JJ, Horuk R (2002) Chemokines, chemokine receptors and small-molecule antagonists: recent developments. Trends Pharmacol Sci 23:459–467

Sayes CM, Marchione AA, Reed KL et al. (2007) Comparative pulmonary toxicity assessments of C60 water suspensions in rats: few differences in fullerene toxicity in vivo in contrast to in vitro profiles. Nano Lett 7:2399–2406

Studer SM, Kaminski N (2007) Towards systems biology of human pulmonary fibrosis. Proc Am Thorac Soc 4:85–91

Sunil VR, Patel KJ, Nilsen-Hamilton M, et al. (2007) Acute endotoxemia is associated with upregulation of lipocalin 24p3/Lcn2 in lung and liver. Exp Mol Pathol 83: 177–187

Warheit DB, Laurence BR, Reed KL et al. (2004) Comparative pulmonary toxicity assessment of single-wall carbon nanotubes in rats. Toxicol Sci 77:117–125

Warheit DB, Webb TR, Reed KL et al. (2007) Pulmonary toxicity study in rats with three forms of ultrafine-TiO2 particles: differential responses related to surface properties. Toxicology 230:90–104

Effects of Endocrine Disruptors on Nervous System Related Gene Expression: Comprehensive Analysis of Medaka Fish

Emiko Kitagawa, Katsuyuki Kishi, Tomotaka Ippongi, Hiroshi Kawauchi, Keisuke Nakazono, Katsunori Suzuki, Hiroyoshi Ohba, Yasuyuki Hayashi, Hitoshi Iwahashi and Yoshinori Masuo

Abstract A number of environmental compounds have been reported to interact with the endocrine system and are thus referred to as endocrine disrupting chemicals. Although it is well known that estrogenic endocrine disrupting compounds stimulate the abnormal expression of the vitellogenin gene in the livers of various aquatic male vertebrates, leading to hepatic production of the vitellogenin protein, there are only few reports investigating the effects of estrogenic pollutants on function related to the nervous system. In the present study, male medaka was exposed to five estrogenic compounds: 17β-Estradiol, 17α-ethinylestradiol, 4-nonylphenol, 4-tert-octylphenol and bisphenol A. Gene profiles were then obtained using a medaka DNA microarray where probes consisted of genes derived from EST libraries prepared from the liver, ovary and brain. To evaluate the effects of these compounds on gene expression related to the nervous system, we estimated comprehensive impact values using the Pearson correlation coefficient by comparing the gene expression profiles of untreated males versus treated males. Results indicated that 17β-Estradiol and 17α-ethinylestradiol exhibited a strong impact on gene expression derived from both the liver and ovary library, whereas some of the 17α-ethinylestradiol treatments disrupted genes derived from the brain library and also nervous system specific gene expression. It is suggested that neuronal damage could occur in medaka following exposure to endocrine disrupting compounds.

Y. Masuo (✉)
Human Stress Signal Research Center, National Institute of Advanced Industrial Science and Technology (AIST), Tsukuba 305-8569, Japan
e-mail: y-masuo@aist.go.jp

Keywords Endocrine disrupting chemical · Nervous system related gene · Medaka · Comprehensive impact value

1 Introduction

The endocrine system is influenced by both natural and synthetic estrogenic compounds, such as 17β-estradiol (E2) and ethinyl estradiol (EE2), respectively. These chemicals mimic endogenous hormones and are called endocrine disrupting compounds (EDCs). Recently, a number of compounds, such as some pesticides, herbicides, detergents, paints and plastics, were reported to act as EDCs. Estrogenic EDCs abnormally stimulate vitellogenin gene expression and protein production in the liver of various aquatic male vertebrates, via an estrogen receptor-mediated mechanism. Exogenous EDCs show dramatic effects on the sexual development of various vertebrates (Balch et al. 2004, Fitch and Denenberg 1998, Willoughby et al. 2005). Based on these studies, it has been proposed that EDCs might influence neuronal function. Indeed, estrogen exerts certain effects on the central nervous system (CNS) mediated by G-protein-coupled receptors (Kelly et al. 2002). The expression of G-protein-coupled receptor genes has been studied in rats administered with alkylphenolic EDCs, 4-nonylphenol (NP), 4-tert-octylphenol (OP) and bisphenol A (BpA) (Ishido et al. 2005). These chemicals may contribute to etiology of hyperkinetic disorders (Ishido et al. 2004, Masuo et al. 2004).

A Japanese medaka (*Oryzias latipes*) has been widely used for the study of the effects of EDCs. It was demonstrated in a male medaka that estrogenic chemicals significantly induce female gene products such as vitellogenin (Arukwe et al. 2002) and choriogenin H (Murata et al. 1997). Exposure to endogenous or exogenous estrogenic compounds also cause inter-sex (testis-ova; oocyte formation in the testes) or sex reversal in the gonad of a mature male medaka (Balch et al. 2004, Yamamoto and Matsuda 1963, Kang et al. 2002a, b). Interestingly, if androgens or estrogens are applied during a sensitive period of embryonic or juvenile medaka development, the sex can be functionally reversed (Kuhl et al. 2005, Kuhl and Brouwer 2006). This may imply that hormonally-active chemicals would alter sex-related brain function in a medaka. In the present study, we assessed changes in mRNA expression induced by estrogenic chemicals with a medaka DNA microarray (Kishi et al. 2006). We selected E2, EE2 and three suspected EDCs, NP, OP and BpA that have been known to exhibit estrogenic activity in vivo (Laws et al. 2000, An et al. 2002). Adult male medaka was exposed to these estrogenic chemicals and gene expression was examined in the whole body. The results suggest that the gene expression profiling in medaka is a powerful tool for screening EDCs that may affect the CNS.

2 Materials and Methods

2.1 Animals

A Japanese medaka *(Oryzias latipes*, orange-red variety or "Himedaka") was originally purchased from Tsuchiura Goldfish Fishery (Tsuchiura, Ibaraki, Japan) and maintained at the Japan Pulp & Paper Research Institute, Inc. for several generations. The brood stock were maintained at $24 \pm 1°C$ in UV-disinfected, dechlorinated, carbon-treated tap water with a 16h light-8h dark photoperiod. The fishes were fed *Artemia nauplii* (<24h after hatching) twice daily. A medaka selected for this study was approximately 6 months post-hatch and were fully mature (around 3 cm in body length). Growth conditions followed guidelines recommended by the international toxicity test protocol (OECD 1992; http://www.env.go.jp/chemi/kagaku/). For microarray analysis, female and male fishes were collected and were flash-frozen in liquid nitrogen and stored at $-80°C$ for RNA extraction.

2.2 Treatment and Sample Preparation

17β-Estradiol (E2), 17α-ethinylestradiol (EE2), 4-nonylphenol (NP), 4-*tert*-octylphenol (OP) and bisphenol A (BpA) were purchased from Wako Pure Chemical Industries, Ltd. (Osaka, Japan). A mature male medaka was exposed to the estrogenic chemicals for 21 days in 10-litre glass chambers containing 6 litre test solutions at concentrations of 100 ng/l E2 and EE2, 100 μg/l NP and OP, and 2 mg/l BpA. Test solutions for each estrogenic chemical were prepared daily by adding 150 μl of the chemical stock solutions to 30 litres of dechlorinated tap water in 40-litre glass aquaria. The test solutions were delivered to each test chamber by a peristaltic pump (Cole Parmer Instrument Co., IL, USA) at a flow rate of 1.25 l/hr. The test solution in each chamber was, consequently, renewed 5 times daily. Two test chambers were used for each treatment group, and 7 fishes were placed in each chamber. Fishes were maintained under a 16h:8h light:dark photoperiod and fed *A. nauplii* (<24 h after hatching) twice a day. The test chambers were cleaned once weekly, and residual bait and fences in the test chambers were removed daily. Throughout the exposure period, the DO concentration (mean $\pm$ SD) was 8.1 ± 0.1, and pH was 7.5 ± 0.1. The water temperature in all test chambers was $24 \pm 0.6°C$. On the last day of exposure, three fishes in each treatment group were flash-frozen in liquid nitrogen and stored at $-80°C$ for RNA extraction.

2.3 Construction of a Medaka Microarray

60-mer oligonucleotide probes were designed for each sequence on the TIGR (The Institute for Genomic Research) *Oryzias laptipes* Gene Index. Probe design was based on genomic resources updated on May 17th, 2004 (Release 5.0). The total number of sequences available at this time was 26,689, including 12,849 TCs (Tentative Consensus), 13,669 singleton ESTs (Expressed Sequence Tags), and 171 singleton ETs (Expressed Transcripts). Seven probes were designed for each sequence. Hereafter, the word "gene" is used to express each sequence even though some sequences might not be in its complete form or might only represent alternatively spliced forms. The medaka microarray was synthesized using MAS Technology (NimbleGen Systems, Inc., Madison, WI, USA) (Kuhl and Brouwer 2006).

TIGR database was constructed with many EST library derived from many laboratories. For example, the brain library of 8DH, the liver libraries of 19D, 8I0, 9OG and F2M. and ovary of 8HV and 9S8 are, One EST library (8DH) in the TIGR database was derived from the brain. The biological source of 8DH was a mixture of adult female/male tissues, and consisted of 1532 TCs and 1673 singleton ESTs. These genes are specifically related to neural function. Libraries derived from the liver and ovary were also extracted from the TIGR, since their function is completely different from the nervous system and they are likely to play a key role in the feminization reaction. Liver libraries consisted of 459 TCs and 161 singleton ESTs and were referred to as 19D, 8I0, 9OG and F2M. The libraries derived from ovary, referred to as 8HV and 9S8, consisted of a total of 551TCs and 268 singleton ESTs. Consequently, our medaka microarray consisted of at least 3205, 620 and 819 probes derived from the brain, liver or ovary libraries, respectively.

2.4 Microarray Analysis

Total RNA was isolated from whole frozen medaka using an EASYPrep RNA kit (TaKaRa Bio Inc., Shiga, Japan) according to the manufacturers protocols. The RNA was converted to double-stranded cDNA, followed by the synthesis of biotin-labeled cRNA using *in vitro* transcription as described elsewhere (Eberwine et al. 1992, Nuwaysir et al. 2002). cRNA was then purified and fragmented to an average size of 50 to 200 bp, and hybridized with a microarray. Hybridization and scanning were carried out with standard procedures (NimbleGen Systems, Inc.). The expression level of each gene was calculated by averaging the signal intensity from seven different probes. Signals between each array were normalized using RMA (Robust Multi-chip Analysis) normalization (Irizarry et al. 2003). GeneSpring 7.3 (Agilent Technologies, CA, USA) was used for further analysis. In order to exclude unreliable data, gene expression with low intensity (10th percentile in each array) was eliminated from each array. Genes that were differently expressed, with statistical signif-

icance, were selected from reliable genes by one-way ANOVA in GeneSpring.

3 Results

3.1 Gene Expression Profiles in the Brain of Untreated Male and Female Medaka

As we focus on neural gene expression during feminization by chemical exposure, we first compared gene expression between normal males and females. The brain was approximately 2% (w/w) of the total body weight in medaka. Since our gene expression data were obtained from the whole body, it was important to consider whether gene expression intensity derived from each library was sufficient for further gene expression analysis. Following normalization, gene expression intensity varied from 1 to 5 orders of magnitude. Median intensity of whole body data was about 200 for both males and females (Fig. 1A). The mean intensity of gene expression from the liver libraries, derived from the largest organ in medaka, gave the highest value of gene expression, approximately 550 (Fig. 1B). Although genes shown in Fig. 1C were derived from ovarian libraries, we found that some of these genes were highly expressed in males, suggesting that the ovarian library contained some universal function genes and hence may have cross-hybridized with probes. As shown in Fig. 1D-F, approximately 70% of genes derived from the brain library gave values higher than 100, indicating that the function of specific organs, such as neural or brain function, can successfully be investigated using whole body expression data.

The correlation coefficient values (CCs) amongst male and female individuals were 0.958 and 0.987 respectively (Fig. 1D and E), indicating that overall gene expression in the brain library was highly reproducible amongst individuals. As shown in Fig. 1F, the difference between male and females in terms of gene expression in the brain library was very small (CC = 0.963). On the other hand, male/female differences were much more pronounced for genes derived from the liver and ovary libraries (CC = 0.765 and 0.643, respectively) (Fig. 1B, C). These data suggest that the gene expression profiles in males and females are quite similar in the brain.

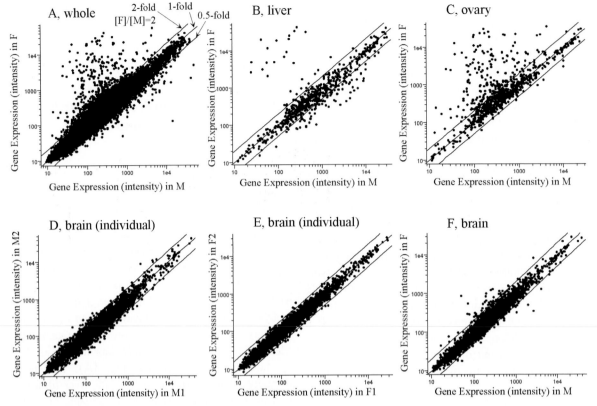

Fig. 1 Scatter plots of medaka microarray analyses of mRNA expression. Normalized fluorescent signal intensities were plotted for each gene group; **A**, all genes on the microarray (26,689 genes); **B**, the liver library derived genes (620); **C**, the ovary library derived genes (819); **D**, **E** and **F**, the brain library derived genes (3,205). In **A**, **B**, **C** and **F**, the x and y axes indicate the mean intensities of 2 males (M) and 2 females (F), respectively. In **D** and **E**, the x and y axes indicate the intensities of individual male 1 (M1) and male 2 (M2), and female 1 (F1) and female 2 (F2), respectively. The three lines in each graph indicate the 2-fold, 1-fold, and 0.5-fold borders in [y]/[x]

3.2 Alterations in Gene Expression After EDC Treatments

Conditions of exposure with EDCs were selected in accordance with previous studies and involved: 100 ng/L E2 and EE2, 100 μg/L NP and OP, and 2 mg/L BpA. Under these conditions, induction of vitellogenin production and testis-ova formation were observed during 21 days of exposure in adult male medaka (Kang et al. 2002a, b, 2003, Gronen et al. 1999, Seki et al. 2002). No histological abnormality was observed in the gonads of control males. High incidences (100%, 8/8) of testis-ova were observed in males treated with E2 and EE2. Testis-ova formation was observed in 4 out of 8 males (50%) receiving BpA treatment (data not shown) although no histological abnormality was observed in control male gonads. In contrast, only one of the 8 males (12.5%) exhibited testis-ova whilst receiving NP or OP treatment (data not shown).

To investigate individual differences, gene expression profiles were obtained separately from 3 individuals receiving each treatment. The impact of estrogenic compounds on gene expression patterns was evaluated by assessing similarity in gene expression between treated and untreated males. Impact values were estimated using the Pearson correlation coefficient (Pearson CC). The intensity of gene expression was converted to a natural logarithm scale, and comparisons were made between individual samples and the means of untreated samples. The Pearson CC values were obtained with gene expression profile from whole data (Fig. 2A) and with those from each EST derived genes (Fig. 2B–D). For visualization and to aid interpretation, Fig. 2 presents data as 1.0 minus Pearson

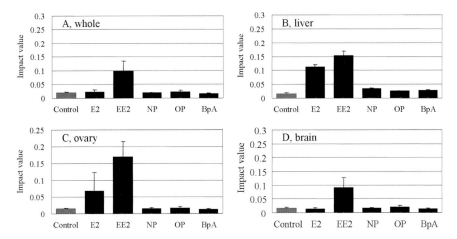

Fig. 2 Comprehensive effects of EDCs on gene expression. Gene expression patterns after treatment with EDCs were obtained from 3 individual fishes with independent hybridizations. Comprehensive impact values were evaluated with the similarity of gene expression patterns between the means of untreated-males and individual males using the Pearson CC (1.0-Pearson CC). Zero indicates no difference from untreated males. The values are expressed as mean ± SEM (n=3) for each gene group; **A**, all genes on the microarray (26,689 genes); **B**, liver library derived genes (620); **C**, ovary library derived genes (819); **D**, brain library derived genes (3,205)

CCs. This is because a value of 1.0 in the Pearson CC indicates complete identity between the two series of data sets whilst 0 indicates no correlation. Consequently a 0 in Fig. 2 means that gene expression patterns of the treated male and untreated controls are identical.

As shown in Fig. 2A and D, E2, NP, OP and BpA caused minor disruption to gene expression at the whole genomic level and with brain-related genes. E2 and EE2 treatments caused significant alteration in gene expression in the liver with only minor differences between individuals (Fig. 2B). Ovary-related gene expression however, did vary amongst individuals (Fig. 2C). Under the conditions tested in the present study, the largest impact value on brain-related genes was observed during EE2 treatment (Fig. 2D), suggesting that EE2 was the chemical most able to disrupt gene expression in the nervous system. In addition, one of the fishes treated with OP exhibited slightly higher impact on brain-related genes than other chemicals, except fishes treated with EE2.

3.3 Nervous System Related Genes

Treatments with E2, EE2, NP, OP and BpA caused significant changes ($P < 0.01$ by ANOVA) in the expression of 9, 25, 3, 30 and 14 genes, respectively, in the brain library derived genes. Since plural chemicals affected an identical gene in some cases, a total of 73 genes appeared to respond to EDC treatments. However, many of these genes were not annotated or were functionally unknown (data not shown). This gene set also included general function genes such as ribosomal protein genes, even though the library was derived from mRNAs expressed in the brain. Thus, it was deemed improper to attempt to elucidate the mechanism of how such genes exert effect on the neuronal system, by using simple annotation of these genes.

Table 1 lists genes that had been allocated particular keywords. We searched the appropriate tentative annotation on TIGR (http://www.tigr.org/tigr-scripts/tgi/T_reports.cgi?species=o_latipes). We found 346 genes on the microarray and applied these genes in subsequent analysis of nervous system related genes. Gene expression profiles were similar between male and female fishes exhibiting adequate expression signals (Fig. 3A). Treatment with EE2 exhibited significant impact on gene expression (Fig. 3B) with the same tendency as Fig. 2D. Other treatments did not significantly disrupt the gene expression of nervous system related genes. E2, EE2, NP, OP, and BpA caused significant alterations in 6, 27, 3, 20 and 7 genes, respectively. Indeed, 55 genes showed changes in the expression (Fig. 4). It was clear that genes being

Table 1 Number of genes allocated particular keyword

Category	
Keywords for TIGR search	Number of genes found
brain and brain parts	
brain-specific	9
brain, cerebro, encephalic, encephalo	48
cerebellar, cerebellum, cerebelli	14
cerebral, cerebrum, cerebri	2
pituitary gland, hypophysis, hypophyses, pituitary	5
medulla oblongata, medullary, bulbar	1
thalamus, thalami, thalamo, thalamic, hypothalamus, hypothalami, hypothalamic, subthalamic	1
hippocampus, hippocampi, hippocampal	1
brainstem, brain stem, diencephalon, diencephalic, hindbrain, metencephalon, midbrain, mesencephalon, mesencephalic, myelencephalon, myelencephalic, olfactory bulb, bulbus olfactorius, optic tectum, optic tecta, tectum, tectal, pineal gland, pineal body, pineal body, telencephalon, telencephalic, spinal cord, spinal, septum, dissepiment, septal, septal area, septal region	0
neuron and glia	
Neuron	6
motor neuron, motoneuron, motoneuronal, motoneurone, motor nerve, motor neurone	7
axon, axonal	4
synapse, synaptic	8
dendrite, dendritic	2
Neuropilin	4
Glia	2
astrocyte, astrocytic, astroglia, stellate	2
nerve fiber, nerve fibre, oligodendrocyte, oligodendrocytic, oligodendroglia, ependymal cell, ependymal	0
Myelin	16
nurotransmitter	
Neurotransmitter	2
Glutaminase	6
glutamate receptor	9
N-methyl-D-aspartate receptor, NMDA receptor	2
kainate receptor, KA receptor	2
mGluR	1
GABA, gamma-aminobutyric-acid, gamma aminobutyric acid [receptor]	8
GABA, gamma-aminobutyric-acid, gamma aminobutyric acid [transporter]	5
glycine receptor	2
acetylcholine receptor, Ach receptor	6
5-HT, serotonin, 5-hydroxytryptamine [receptor]	4

Table 1 (continued)

Category	
Keywords for TIGR search	Number of genes found
Dopamine	1
AMPA receptor, catecholamine, noradrenalin, norepinephrine, epinephrine, adrenalin	0
G protein coupled receptor	21
Hormone	
prolactin, PRL	2
adrenocorticotropic hormone, adrenocorticotropin, corticotropin, ACTH	1
growth hormone, somatotropin, somatotrophic hormone, GH	5
gonadotropin, GTH, follicle-stimulating hormone, FSH, luteinizing hormone, LH	1
melanocyte-stimulating hormone, melanotropin	1
thyroid-stimulating hormone, TSH, melanophore-stimulating hormone, somatolactin, SL	0
gonadotropin-releasing hormone, GnRH	6
thyrotropin-releasing hormone, TRH	3
somatostatin	1
corticoropin-releasing hormone	1
arginine vasopressin, AVP	1
neuropeptide Y, NPY	1
B-type natriuretic peptide	2
C-type natriuretic peptide	7
adenylyl cyclase, adenylate cyclase, adenylylcyclase, adenylate cyclase-activating	8
guanylyl cyclase, guanyl cyclase	17
growth hormone-releasing hormone, GHRH, prolactin-releasing peptide, arginine vasotosin, AVT, melanin-concentrating hormone, MCH, neurotensin, endomorphin, nociceptin, orphanin	0
orexin	2
endorphin	1
enkephalin	3
dynorphin	1
substance P	1
sense	
olfaction, olfactory	13
vision, visual, optic, ocular	23
taste, gustation, gustatory	1
tactile sense, tactile, audio, lateral line nerves	0
others	
neuro	19
nerve	1
nervous	2
neural	23
neurofilament	3
neuroblastoma	3

Table 1 (continued)

Category	
Keywords for TIGR search	Number of genes found
neurofibromatosis	1
neuropathy	2
neurovirulence	1
neuroglobin	1
neurolin	2
neuroglycan	1
neurogranin	1
neurocan	1
neurogenin, neuroD	4
neurotrypsin	1
histamine	1
acetylcholine	7
amyloid beta, beta-amyloid	6
neurotrophin	1
neurotrophic	2
neurobeachin	2
neuroendocrine	2

up-regulated particularly with the EE2 treatment were neurotransmitter related (Fig. 4). EE2 was shown to enhance the expression of glutamate (Glu) receptors, such as metabotropic Glu receptor 8, Glu receptor KA1 subunit, Glu receptor 6 precursor, and NMDA receptor NR1 subunit precursor. Increased expression was also observed in 5-hydroxytryptamine 3 receptor B subunit precursor, γ-aminobutyric acid (GABA) receptor α-4 subunit precursor, sodium- and chloride-dependent GABA transporter 1, and mitochondrial glutaminase precursor. It is particularly worth noting that gonadotropin-releasing hormone, one of the key

hormones for feminization, was also highly induced by the EE2 treatment.

4 Discussion

Although EDCs are known to impair a variety of physiological functions including neural development (Kelly et al. 2002, Ishido et al. 2005, 2004), investigations of the effect of estrogenic EDCs on sexually mature medaka have mostly been restricted to vitellogenin assays and gonad histology. The fact that the sex of medaka can be functionally reversed by the exposure to EDCs (Kuhl et al. 2005, Kuhl and Brouwer 2006) prompted us to study the effects of EDCs on the CNS. We have developed and used a medaka DNA microarray to evaluate the effects of various EDCs on the nervous system. In contrast to rodents such as mice and rats, the species of fish used in this study, medaka, have certain limitations in terms of dissecting the brain into several regions. However, gene expression profiling in medaka could be a highly useful tool to investigate the comprehensive effects of environmental chemicals, such as exogenous estrogens.

In the present study, we first tried to extract the local reaction to EDCs from whole body gene expression without organ dissection. Although the expression levels of the genes derived from the brain library were low, we successfully confirmed that the levels of these genes were sufficient for detection, even in the whole body preparation. The genes derived from the brain library contained many organ non-specific

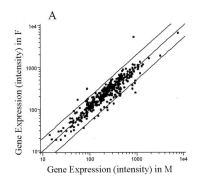

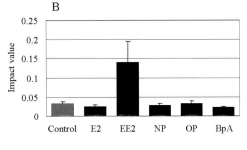

Fig. 3 Nervous system related gene expression. **A**: Scatter plots of neural mRNA expression in normal males and females. Normalized fluorescent signal intensities of 346 nervous system related genes were plotted; x and y axes indicate the mean intensities of 2 males (M) and 2 females (F), respectively. The three lines in the graph indicate the 2-fold, 1-fold and 0.5-fold borders in [y]/[x]. **B**: Comprehensive effects of EDCs on nervous system related gene expression. Comprehensive impact values are represented as Fig. 2, and expressed as mean ± SEM (n=3)

Neurotransmission

glutamate receptor

AU168925 similar to 1935043, metabotropic glutamate receptor 8, partial (13%)
AU180369 similar to 475546, glutamate receptor KA1 subunit, partial (2%)
AU167356 similar to 96128, glutamate receptor 6 precursor, partial (25%)
TC42702 similar to O93338, N-methyl-D-aspartate receptor NR1 subunit precursor, partial (10%)

serotonin receptor

AU242419 weakly similar to 8132427, 5-hydroxytryptamine 3 receptor B subunit precursor, partial (27%)

acetylcholine receptor

BJ492200 similar to P09628, acetylcholine receptor protein delta chain precursor, partial (32%)

GABA receptor, transporter

BJ012342 similar to P20237, gamma-aminobutyric-acid receptor alpha-4 subunit precursor, partial (3%)
TC39788 similar to P31648, sodium- and chloride-dependent GABA transporter 1, partial (26%)

glutaminase

AU171728 similar to A41009 glutaminase precursor mitochondrial, partial (24%)

G protein coupled receptor

TC30547 similar to Q86SQ4, G protein-coupled receptor 126, partial (7%)
AU171027 weakly similar to S40454, G protein-coupled receptor GPCR21, partial (18%)
TC34404 similar to Q8BKF8, G protein-coupled receptor 37, partial (12%)
AU168581 weakly similar to 12711485, G-protein coupled receptor GPR86, partial (28%)
TC39293 weakly similar to Q86SQ4, G protein-coupled receptor 126, partial (11%)
NP849699 G protein coupled progestin receptor alpha [Oryzias latipes]
BJ008313 weakly similar to 9651839, G-protein coupled receptor EDG-7, partial (42%)
TC32580 homologue to P24799, adenylatecyclase-stimulating G alpha protein, partial (84%)
TC34722 weakly similar to Q9EQL6, G-protein-coupled receptor induced protein GIG2 (Fragment), partial (58%)

voltage-gated channel

TC41209 similar to Q9BXT2, neuronal voltage-gated calcium channel gamma-6 subunit, partial (24%)

Brain and brain parts

pituitary gland

TC38690 weakly similar to AAF73770, pituitary tumor-transforming gene protein binding factor, partial (34%)
TC33120 weakly similar to P53801, pituitary tumor-transforming gene 1 protein-interacting protein, partial (36%)

cerebellum

NP822059 zinc finger of the cerebellum 4 [Oryzias latipes]

brain-specific

TC33485 similar to Q8BL11, brain-specific angiogenesis inhibitor 1 precursor homolog, partial (13%)
TC40444 similar to Q80ZF8, brain-specific angiogenesis inhibitor 3, partial (15%)
TC38309 similar to AAR08137, brain-specific alpha actinin 1 isoform, partial (31%)

brain

BJ027071 similar to A40138, glycogen phosphorylase brain, partial (14%)
BJ526100 weakly similar to S32404, myosin heavy chain I brain, partial (16%)
TC34236 weakly similar to P17439, D-glucosyl-N-acylsphingosine glucohydrolase, partial (22%)
BJ001997 homologue to Q01082, spectrin beta chain brain 1 (spectrin non-erythroid beta chain 1), partial (6%)

Hormone, neuro peptide

TC36096 similar to Q91V16, growth hormone-inducible soluble protein, partial (94%)
NP800397 prolactin [Oryzias latipes]
TC39671 homologue to AAS18432, prodynorphin, partial (27%)
BJ496973 homologue to P37892, carboxypeptidase (Enkephalin convertase), partial (30%)
AU170811 homologue to P37892, carboxypeptidase (Enkephalin convertase), partial (45%)
TC32957 Q9DD49, progonadoliberin III prec. (salmon-type gonadotropin-releasing hormone) {O. latipes}, complete
TC37801 Q8AYR6, C-type natriuretic peptide-1, complete
AU169827 homologue to 22830877, C-type natriuretic peptide-2 {O. latipes}, complete
TC33364 similar to P01170, somatostatin II precursor, partial (58%)
BJ517679 similar to 11878215, thyrotropin-releasing hormone receptor 2, partial (20%)
TC39620 Q98UH9, membrane guanylyl cyclase OlGC8, complete
NP423659 guanylate cyclase activating protein 2 [Oryzias latipes]

Neuron and glia

AU172296 weakly similar to P28685, contactin 2 precursor (axonin-1), partial (5%)
TC38399 similar to Q9W6S8, survival motor neuron protein, partial (56%)
TC32019 similar to Q28296, myelin and lymphocyte protein, partial (55%)
BJ003259 weakly similar to 12652591, sphingomyelin phosphodiesterase 2, partial (39%)
TC33133 weakly similar to O13098, middle molecular weight neurofilament protein NF-M(1), partial (4%)
TC35570 homologue to P24505, synaptotagmin A (synaptic vesicle protein O-p65-A), partial (7%)

Others

BJ002453 homologue to 2582522, neural src interacting protein, partial (24%)
TC38771 weakly similar to I55454, neuroglycan C precursor, partial (18%)
NP418754 Y1 olfactory receptor [Oryzias latipes]
AU180635 homologue to 4062850, neural plakophilin-related arm-repeat protein (NPRAP), partial (1%)
BJ489630 homologue to 20809566, beta-amyloid binding protein precursor, partial (31%)
AU167801 weakly similar to T30532, neural cell adhesion molecule L1 homolog, partial (7%)
TC33410 weakly similar to Q64322, neural proliferation differentiation and control protein-1 prec., partial (7%)
TC32986 weakly similar to P97785, TGF-beta related neurotrophic factor receptor1, partial (12%)

E2 EE2 NP OP BpA

Expression Ratio

Trust

0.1 0.5 0.7 1.5 2.0 4.0 5.0

Fig. 4 The effects of EDC treatment on nervous system related gene expression. Fifty-five genes shown here were either up- or down-regulated by at least one of the EDCs. These genes were selected by ANOVA ($P < 0.05$). Colors indicate the gene expression ratios of the treated male compared with those of the untreated male; from *green* (suppressed) to *red* (induced). Each individual data point is illustrated separately. Hit coverage in the similarity search was shown in parentheses (%)

genes: 104 and 108 of the brain derived genes overlapped with the liver and ovary libraries, respectively. Nonetheless, the brain library included a number of brain-specific genes and it was thus possible to elucidate effects on brain function using brain library-derived gene expression, to some extent.

We estimated the magnitude of the effect imposed by EDCs on gene expression with comprehensive impact values. The huge amount of data generated from microarray analysis sometimes makes interpretation of data difficult. Comprehensive impact values can be easily calculated with any type of data set to allow comparison with other data sets. In this report, we used the genes derived from the brain library and the genes related to the nervous system to compute comprehensive impact values. As expected from histological observations, as the incidence of testis-ova was 100% in both the E2 and EE2 treated males, the comprehensive impacts of E2 and EE2 to liver- and ovary-derived gene expression were also very high. However, the degree of impact was significantly different for each individual fish (Fig. 2). In fact, impacts on ovary-derived gene expression were very small for two individuals in the E2 treatment group (Fig. 2C), implying that the testes of these two individuals were not severely damaged or only few testis-ova might have been formed. This might be due to the progression of feminization and/or other toxic effects. Since the altered ovary library-derived genes were part of downstream reactions in the feminization process, the impact on liver might appear relatively earlier, compared with genes derived from the ovary.

Treatment of male medaka with EE2 resulted in a significant increase in the gene expression of Glu receptors such as metabotropic Glu receptor 8, Glu receptor KA1 subunit, Glu receptor 6 precursor and NMDA receptor NR1 subunit precursor (Fig. 4). Glutamate is the most widespread excitatory neurotransmitter in the mammalian brain. Two classes of glutamate receptor have been described: the ionotropic receptors (ligand-gated ion channels) named after selective agonists, i.e., N-methyl-D-aspartate (NMDA) and kainate receptors, and the metabotropic receptors (G protein-coupled receptors) (Dingledine and Conn 2000). Thus, the present results suggest strong stimulation of Glu neurotransmission via its receptors by EE2 treatment. On the other hand, other EDCs did not modify the expression of neurotransmitter-related genes under the experimental conditions used in this study.

EE2 seems to be more potent than the other EDCs tested here, although dose-responses should be further investigated. It is known that Glu-containing neurons play important roles in the CNS of vertebrates (Meldrum 2000). It is possible that EE2 stimulates neurotransmission in both the peripheral and central nervous system. The enhanced Glu neurotransmission may be involved in excitotoxic neuronal damage (Vajda 2002). Indeed, the stimulation of Glu receptors causes calcium-dependent neuronal cell death, such as apoptosis after transient ischemia (Arias et al. 1998). Therefore, it is possible that neuronal damage could occur in medaka following exposure to EE2.

We also observed a significant increase in the expression of the serotonin (5-HT) 3 receptor B subunit precursor gene. The 5-HT systems play important roles in the CNS and are suggested to be involved in feeding, aggression, depression and stress behavior (Larson and Summers 2001). Concerning GABA elements, EE2 enhanced the expression of the GABA receptor alpha-4 subunit precursor gene and the sodium- and chloride-dependent GABA transporter 1 genes. Increased expression of GABA receptors and GABA transporters indicate the likely up-regulation of GABA neurotransmission that are the major inhibitory neurons in the CNS. Several lines of evidence suggest that estrogen interacts with dopaminergic neurons in the brain (Kuppers et al. 2000, Agrati et al. 1997). Moreover, we recently demonstrated in the rat that neonatal treatment with BpA, NP and OP caused a deficit in the development of dopaminergic neurons, which resulted in motor hyperactivity (Ishido et al. 2005, 2004). In the present study however, we did not detect any change in the gene expression of dopaminergic elements. This may be due to different sensitivities to endocrine disruptors and/or different organization of the nervous system in medaka.

We also observed a significant increase in the expression of the gonadotropin-releasing hormone (GnRH) gene as a result of EE2 treatment. This finding supports previous study showing a stimulation of GnRH by EE2 in *Xenopus laevis* (Urbatzka et al. 2006). Hypothalamic GnRH controls hypophyseal gonadotropins, follicle-stimulating hormone and luteinizing hormone (Thornton and Geschwind 1974). These hormones play important roles in gonadal development and the feminization reaction.

The magnitude of disruption caused by EDC exposure on nervous system related gene expression was smaller than that on genes derived from the liver and ovary. However, some of the important genes related to nervous system related function were significantly up/down-regulated as a direct result of EDC exposure. The gene expression profiling with DNA microarrays in medaka may be a powerful tool for screening pre-behavioral marker genes that reflect alteration in the CNS.

Acknowledgments This work was supported by the Japan Science and technology Agency (JST).

References

Agrati P, Ma ZQ, Patrone C, Picotti GB, Pellicciari C, Bondiolotti G, Bottone MG, Maggi A (1997) Dopaminergic phenotype induced by oestrogens in a human neuroblastoma cell line. Eur. J. Neurosci., 9, 1008–1016.

An BS, Kang SK, Shin JH, Jeung EB (2002) Stimulation of calbindin-D(9k) mRNA expression in the rat uterus by octylphenol, nonylphenol and bisphenol. Mol. Cell. Endocrinol., 191, 177–186.

Arias C, Becerra-Garcia F, Tapia R (1998) Glutamic acid and Alzheimer's disease. Neurobiology, 6, 33–43.

Arukwe A, Kullman SW, Berg K, Goksoyr A, Hinton DE (2002) Molecular cloning of rainbow trout (Oncorhynchus mykiss) eggshell zona radiata protein complementary DNA: mRNA expression in 17beta-estradiol- and nonylphenol-treated fish. Comp. Biochem. Physiol. B Biochem. Mol. Biol., 132, 315–326.

Balch GC, Mackenzie CA, Metcalfe CD (2004) Alterations to gonadal development and reproductive success in Japanese medaka (Oryzias latipes) exposed to 17alpha-ethinylestradiol. Environ. Toxicol. Chem., 23, 782–791.

Dingledine R, Conn PJ (2000) Peripheral glutamate receptors: molecular biology and role in taste sensation. J. Nutr., 130, 1039S–1042S.

Eberwine J, Spencer C, Miyashiro K, Mackler S, Finnell R (1992) Complementary DNA synthesis in situ: methods and applications. Methods Enzymol., 216, 80–100.

Fitch RH, Denenberg VH (1998) A role for ovarian hormones in sexual differentiation of the brain. Behav. Brain. Sci., 21, 311–327; discussion 327–352.

Gronen S, Denslow N, Manning S, Barnes S, Barnes D, Brouwer M (1999) Serum vitellogenin levels and reproductive impairment of male Japanese Medaka (Oryzias latipes) exposed to 4-tert-octylphenol. Environ. Health Perspect., 107, 385–390.

Irizarry RA, Hobbs B, Collin F, Beazer-Barclay YD, Antonellis KJ, Scherf U, Speed TP (2003) Exploration, normalization, and summaries of high density oligonucleotide array probe level data. Biostatistics, 4, 249–264.

Ishido M, Masuo Y, Sayato-Suzuki J, Oka S, Niki E, Morita M (2004) Dicyclohexylphthalate causes hyperactivity in the rat concomitantly with impairment of tyrosine hydroxylase immunoreactivity. J. Neurochem., 91, 69–76.

Ishido M, Morita M, Oka S, Masuo Y (2005) Alteration of gene expression of G protein-coupled receptors in endocrine disruptors-caused hyperactive rats. Regul. Pept., 126, 145–153.

Kang IJ, Yokota H, Oshima Y, Tsuruda Y, Hano T, Maeda M, Imada N, Tadokoro H, Honjo T (2003) Effects of 4-nonylphenol on reproduction of Japanese medaka, Oryzias latipes. Environ. Toxicol. Chem., 22, 2438–2445.

Kang IJ, Yokota H, Oshima Y, Tsuruda Y, Oe T, Imada N, Tadokoro H, Honjo T (2002b) Effects of bisphenol a on the reproduction of Japanese medaka (Oryzias latipes). Environ. Toxicol. Chem., 21, 2394–2400.

Kang IJ, Yokota H, Oshima Y, Tsuruda Y, Yamaguchi T, Maeda M, Imada N, Tadokoro H, Honjo T (2002a) Effect of 17beta-estradiol on the reproduction of Japanese medaka (Oryzias latipes). Chemosphere, 47, 71–80.

Kelly MJ, Qiu J, Wagner EJ, Ronnekleiv OK (2002) Rapid effects of estrogen on G protein-coupled receptor activation of potassium channels in the central nervous system (CNS). J. Steroid Biochem. Mol. Biol., 83, 187–193.

Kishi K, Kitagawa E, Onikura N, Nakamura A, Iwahashi H (2006) Expression analysis of sex-specific and 17beta-estradiol-responsive genes in the Japanese medaka, Oryzias latipes, using oligonucleotide microarrays. Genomics, 88, 241–251.

Kuhl AJ, Brouwer M (2006) Antiestrogens inhibit xenoestrogen-induced brain aromatase activity but do not prevent xenoestrogen-induced feminization in Japanese medaka (Oryzias latipes). Environ. Health Perspect., 114, 500–506.

Kuhl AJ, Manning S, Brouwer M (2005) Brain aromatase in Japanese medaka (Oryzias latipes), Molecular characterization and role in xenoestrogen-induced sex reversal. J. Steroid Biochem. Mol. Biol., 96, 67–77.

Kuppers E, Ivanova T, Karolczak M, Beyer C (2000) Estrogen: a multifunctional messenger to nigrostriatal dopaminergic neurons. J. Neurocytol., 29, 375–385.

Larson ET, Summers CH (2001) Serotonin reverses dominant social status. Behav. Brain Res., 121, 95–102.

Laws SC, Carey SA, Ferrell JM, Bodman GJ, Cooper RL (2000) Estrogenic activity of octylphenol, nonylphenol, bisphenol A and methoxychlor in rats. Toxicol. Sci., 54, 154–167.

Masuo Y, Ishido M, Morita M, Oka S (2004) Effects of neonatal treatment with 6-hydroxydopamine and endocrine disruptors on motor activity and gene expression in rats. Neural Plast., 11, 59–76.

Meldrum BS (2000) Glutamate as a neurotransmitter in the brain: review of physiology and pathology. J. Nutr., 130, 1007S–1015S.

Murata K, Sugiyama H, Yasumasu S, Iuchi I, Yasumasu I, Yamagami K (1997) Cloning of cDNA and estrogen-induced hepatic gene expression for choriogenin H, a precursor protein of the fish egg envelope (chorion). Proc. Natl. Acad. Sci. USA, 94, 2050–2055.

Nuwaysir EF, Huang W, Albert TJ, Singh J, Nuwaysir K, Pitas A, Richmond T, Gorski T, Berg JP, Ballin J, McCormick M, Norton J, Pollock T, Sumwalt T, Butcher L, Porter D, Molla M, Hall C, Blattner F, Sussman MR, Wallace RL, Cerrina F, Green RD (2002) Gene expression analysis using oligonucleotide arrays produced by maskless photolithography. Genome Res., 12, 1749–1755.

Seki M, Yokota H, Matsubara H, Tsuruda Y, Maeda M, Tadokoro H, Kobayashi K (2002) Effect of ethinylestradiol on the reproduction and induction of vitellogenin and testis-ova in medaka (Oryzias latipes). Environ. Toxicol. Chem., 21, 1692–1698.

Thornton VF, Geschwind, II (1974) Hypothalamic control of gonadotropin release in amphibia: evidence from studies of gonadotropin release in vitro and in vivo. Gen. Comp. Endocrinol., 23, 294–301.

Urbatzka R, Lutz I, Opitz R, Kloas W (2006) Luteinizing hormone, follicle stimulating hormone, and gonadotropin releasing hormone mRNA expression of Xenopus laevis in response to endocrine disrupting compounds affecting reproductive biology. Gen. Comp. Endocrinol., 146, 119–125.

Vajda FJ (2002) Neuroprotection and neurodegenerative disease. J. Clin. Neurosci., 9, 4–8.

Willoughby KN, Sarkar AJ, Boyadjieva NI, Sarkar DK (2005) Neonatally administered tert-octylphenol affects onset of puberty and reproductive development in female rats. Endocrine., 26, 161–168.

Yamamoto TO, Matsuda N (1963) Effects of estradiol, stilbestrol and some alkyl-carbonyl androstanes upon sex differentiation in the medaka. Orvzias latipes. Gen. Comp. Endocrinol., 3, 101–110.

Assessment of River Health by Combined Microscale Toxicity Testing and Chemical Analysis

Sagi Magrisso and Shimshon Belkin

Abstract A battery of commercially available toxicity bioassays was applied to assess the quality status of two polluted coastal rivers in Israel: the Yarkon, polluted with treated domestic wastewater, and the Kishon, that at the time of sampling served as a conduit of industrial wastewaters. Samples from the latter displayed much higher toxicity; in both cases, sediment toxicity was considerably higher than that of the water. Highest sensitivities were exhibited by assays based on the micro-crustaceans *Daphnia pulex* and *Thamnocephalus platyurus*. A new index (total relative toxicity) is presented for river health assessment, which integrates and normalizes the toxicity of all identified pollutants. Calculated total relative toxicity values were in excellent agreement with the biological toxicity tests. We highlight the significance of toxicity bioassays as an essential component of any river monitoring program, both for the assessment of river health and for following the progress of remediation schemes. For this purpose, it is proposed that a river-specific panel of toxicity bioassays is selected, representing different trophic levels and taxonomic complexity.

Keywords Toxicity testing · Bioassays · River pollution · Sediments · *Vibrio fischeri* · *Tetrahymena thermophila* · *Daphnia pulex* · *Brachionus plicatilis* · *Artemia salina*

1 Introduction

Two complementary approaches may be adopted for monitoring for the presence of toxic pollutants: chemical analysis or toxicity bioassays. Analytical chemical methodologies, often characterized by high sensitivity, high reliability, and low detection thresholds (Kohler et al., 2000) are essential for regulatory purposes but may lack the ability to assess toxicity, bioavailability and potential antagonistic/synergistic effects of pollutants (Vangheluwe et al., 1999). The limitations of chemical analysis are particularly apparent when the chemical nature of pollutants is unknown, in which case an extensive array of instrumentation needs to be used, often in a time consuming and costly manner. Toxicity bioassays quantify the negative effects of the sample on a population of test organisms, with no attempt to analyze the exact composition of the sample (Quershi et al., 1998). Over the years, toxicity bioassays have been described and put to routine use that are based on a variety of organisms, from well-established fish and Daphnia toxicity tests to cell based systems (Blaise et al., 1998). In the last two decades, while traditional methodologies have not been abandoned, there is an increasing tendency to adopt microscale bioassays that are relatively simple, require very low logistic support and minimal special training (Blaise et al., 1998).

The observed toxicity in any given assay system is influenced by numerous environmental factors and is strongly dependant on the test organism (Bancon-Montigny et al., 2001). Consequently, increasing evidence suggests that a single bioassay can not provide a full picture of the toxicity of an environmental sample (Bierkens et al., 1998); a more accurate assessment of substance toxicity or of environmental

S. Belkin (✉)
Department of Plant and Environmental Sciences, Institute of Life Sciences, The Hebrew University of Jerusalem, Jerusalem 91904, Israel
e-mail: shimshon@vms.huji.ac.il

Y.J. Kim et al. (eds.), *Atmospheric and Biological Environmental Monitoring*,
DOI 10.1007/978-1-4020-9674-7_16, © Springer Science+Business Media B.V. 2009

"health" calls for a test battery composed of different bioassays, based on different test organisms representing different complexities and trophic levels (Castillo et al., 2000). Members of the test panel should be selected from among species of ecological relevance to the study area, and combine a broad chemicals, detection spectrum and a sufficiently high sensitivity.

In response to these needs there is a growing availability of microscale toxicity bioassays, many of them marketed commercially (Blaise et al., 1998). Many of the available kits include the test organism in a "dormant" state (immobilized algae, lyophilized bacteria, crustaceans' eggs etc.), allowing relatively long shelf lives and alleviating the need for growing the organism in preparation for conducting the assay (Jansen et al., 2000).

River pollution is a world-wide phenomenon (Demuth et al., 1993; Ronco et al., 1995), with severe environmental and health-related implications (Castillo et al., 2000; Ritter et al., 2002). The most important contamination sources are domestic and industrial wastewaters, urban and agricultural run-off, and wet or dry atmospheric deposition (Ritter et al., 2002). While in most cases the primary receptacle of the pollutant inputs is the flowing water stream, most of the foreign chemicals are eventually trapped in the sediments, which act as a sink (Vangheluwe et al., 1999). River water quality may improve dramatically soon after removal of the pollution source, but this is not the case for sediments, which are often characterized by pollutants, concentrations much higher than that of the water above them (Chapman, 1986; (Liu et al., 1996). Sediment-associated pollutants may be soluble in the interstitial pore water (Vangheluwe et al., 1999) or adsorbed onto sediment particles, plants and organisms (Pardos and Blaise, 1999) As sediment perturbations as well as changes in chemical conditions may lead to the release of pollutants back into the water, it is essential that sediment toxicity is assessed along with that of the water above them as a means to evaluate aquatic system health Winger et al., 1998).

This concept was tested in the present study using two Israeli coastal plain rivers, both currently undergoing rehabilitation programs following many years of pollution and neglect (Bar-or, 2000). One of the Rivers (Kishon) has been a conduit for industrial wastewaters for several decades (Oren et al., 2006; Avishai et al., 2002, 2004), while the water flowing in the other (Yarkon) has mostly been municipal wastewater, ranging in quality from raw sewage to high quality effluents (Bar-or, 2000).

2 Materials and Methods

2.1 Study Sites

Both studied rivers run east to west in the Israeli coastal plain, the Kishon (70 Km long) emptying into the Mediterranean north of the city of Haifa while the Yarkon (27 Km) enters the sea through the city of Tel Aviv. In both cases, the upper parts of the river are relatively pollution-free. The last 7 Km of the Kishon are characterized by significant industrial pollution. Most of the water flowing from the central section of the Yarkon westwards is actually wastewater effluents, while the section comprising the outlet into the sea is characterized by a high degree of seawater infiltration.

2.2 Sampling

River water (RW) and sediment samples were collected during 2000–2002 at five sites along each of the rivers, selected to include areas representing three different water quality and contamination levels (Table 1). In both cases the first and easternmost sample (Y1 and K1) was upstream of the introduction of the first waste stream into the river; the next three samples (Y2-Y4, K2-K4), all polluted, were located in a westward progression and the terminal westernmost sample (Y5, K5) was in the estuarine area were river waters were mixed with coastal Mediterranean seawater.

RW samples were collected by 250 ml sterile polypropylene bottles (Falcon 3024, Becton Dickinson, France), stored on ice, filtered (0.45 μm pore size filters, Millipore Corporation, USA) to remove particulate matter, and stored at 4°C. Sediments were collected by pulling out 20 cm long cores using a specially constructed 70 cm diameter plastic corer. The cores were immediately divided into 10 cm length segments and placed in sterile 50 ml test tubes, which were transferred on ice to the lab. Interstitial water (IS) was extracted by centrifugation

Table 1 Sampling sites and geographical coordinates, arranged east to west

Yarkon river			Kishon river		
Sampling site	Position	General quality	Name	Position	General quality
Y1	32°06′18.1″/34°55′40.11″	Clean	K1	32°41′51.1″/35°06′10.8″	Clean
Y2	32°07′31.86″/34°53′43.12″	Contaminated	K2	32°47′34.3″/35°03′02.1″	Contaminated
Y3	32°07′12.1″/34°52′02.5″	Contaminated	K3	32°47′14.1″/35°03′23.2″	Contaminated
Y4	32°06′23.33″/34°50′57.86″	Contaminated	K4	32°481′06.9″/35°02′43.4″	Contaminated
Y5	32°05′50.35″/34°47′37.8″	Estuarine	K5	32°48′27.8″/35°01′43.6″	Estuarine

(10 min, 3700 rpm, 4°C; Eppendorf 5810R, Eppendorf, Germany), filtered (0.45 μm) to remove residual particulates and kept at 4°C.

2.3 Sample Preparation

Liquid samples were adjusted to the pH required by each test protocol, and serially diluted in either artificial seawater or a freshwater buffer solution, supplied in the toxicity bioassay kits for the marine and freshwater assays, respectively.

2.4 Organic Contaminant Analyses

Organic chemical contaminants were analyzed in extracts of both total sediment and interstitial pore water samples. Pore water was extracted twice by dichloromethane (first extraction: 500 ml sample, 100 ml Petrol ether, 100 ml dichloromethane; second extraction: 100 ml dichloromethane). Total sediment (50 g) was first extracted by 100 ml acetone for 12 hours, and then was further extracted as described above for pore water. Separation of the main organic contaminants was carried out using high pressure liquid chromatography (HP5989A column, (Hewlett Packard, USA). Compounds were identified by retention time and quantified by comparison to standards.

2.5 Total Relative Toxicity of Organic Compounds

We have defined a value termed "total relative toxicity" (TRT), which sums up the calculated mammalian toxicity of all identified pollutants in a sample, taking into account both their concentration and their toxicity. As toxicity data we have used published data (Toxnet; http://toxnet.nlm.nih.gov), either of oral rat or mouse intravenous LD50. For each sample, TRT was calculated as follows:

$$\text{TRT} = \sum Concentration\,(i) \times LD50i/LD50n$$

Concentration (i) denotes the concentration in mg/l of each substance, LD50i is the LD_{50} (mg/Kg) of the same substance and $LD_{50}(n)$ is the LD_{50} (mg/Kg) of the least toxic substance identified in the sample. For oral rat and mouse IV, these chemicals were 1,2 bis-benzenedicarboxylic acid and tetradecane, respectively.

2.6 Toxicity Bioassays

All toxicity bioassays except the *Vibrio fischeri* bioluminescence assay (Azur Environmental, Carlsbad, CA) were purchased as ready to use kits from Creasel (Belgium); assays were conducted according to the manufacturer's instructions, as briefly summarized in Table 2. Sediment toxicity was routinely assayed by testing the extracted interstitial water which was then tested as a regular water sample. In addition, a direct sediment exposure assay was carried out for Yarkon river samples using the ostracod *Heterocypris incongruens* (Table 2). This assay could not be used for the Kishon River sediments due to their high salinity.

2.7 Calculation of Toxicity

Toxicity values obtained in the course of this study are presented either as LC_{50} (sample concentration

Table 2 Organisms and test kits used

Organism	Commercial assay name*	Preparation	Sample size	Endpoint
Brachionus plicatilis (Rotifer, marine)	Rotoxkit M^{TM}	Hatching of eggs in artificial seawater, (ASW) 1 h light 3000 lux and 28 h dark	5 larva in 300 µl ASW	Death, 24 h
Artemia salina ("brine shrimp", marine)	Artoxkit M^{TM}	Hatching of eggs in artificial seawater, 1 h light 3000 lux and 24 h dark	10 larva in 1 ml ASW	Death, 24 h
Thamnocephalus platyurus	Thamnotox F^{TM}	Hatching of eggs in freshwater, 22 h light 3500 lux	10 larva in 1 ml freshwater	Death, 24 h
Daphnia pulex	Daphtox F^{TM}	Hatching of eggs in freshwater, 72–96 h light 6000 lux	5 larva in 5 ml freshwater	Death, 24 h
Heterocypris incongruens	Ostracodtoxkit F^{TM}***	Hatching of eggs in freshwater, 52 h, light 3500 lux	10 larvae, 300 µl sediment sample, 2 ml algal suspension *Rophidocelis subcapitata*	Death, 6 days
Tetrahymena thermophila	Protoxkit F^{TM}	Dilute stock to O.D$_{440}$ 0.03	40 µl protozoa suspension, 40 µl food, 2 ml sample	OD$_{440}$ 0.03, 24 h
Raphidocelis subcapitata	Algaltoxkit F^{TM}	De-immobilization from alginate beads	250 µl of algal suspension, 25 ml sample in a 10 cm cuvette	Growth inhibition, 72 h
*Vibrio fischeri***	Microtox®	20-fold dilution of resuscitated cell suspension in 2% NaCl	100 µl	EC$_{50}$ (15 min)

* Suffix: F – freshwater tests; M – marine tests

**The *Vibrio fischeri* bioluminescence inhibition assay was conducted as previously described (Ribo and Kaiser, 1987), with a slight modification. Samples were serially diluted in a 96 well microtiter plate to a volume of 100µl, and 20µl of cell suspension was added. Luminescence was measured after 15 minutes exposure.

***Sediment test

causing a 50% mortality in the test population) or EC$_{50}$ (sample concentration causing a 50% inhibition of bioluminescence in the *Vibrio fischeri* assay).

Mortality rates R for deriving LC$_{50}$ values were calculated from Ri = Ni/N$_0$, where Ri is the mortality rate at sample concentration i, Ni is the number of dead organisms at the same concentration, and N$_0$ is the number of organisms in the toxicant-free control.

For Ostracodtoxkit F^{TM}, sediment toxicity is demonstrated as the percentage of mortality rate only. *Vibrio fischeri* bioluminescence inhibition EC$_{50}$ values were calculated as previously described (Quershi et al., 1998). All toxicity results are presented as the sample concentration (%) that yields either EC$_{50}$ or LD$_{50}$ in the test organism.

3 Results

3.1 Yarkon River

Toxicity of the Yarkon samples, the less polluted out the two tested rivers was examined in five stations marked Y1 to Y5. Y1 is the non-polluted upstream sampling site, while Y5 is in close to the sea coast, composed mostly of seawater infiltrating upriver. In view of the marine nature of the last segment, there was a need to include assays using marine microorganisms.

As a preliminary screen, all water (RW) and interstitial water (IS) samples were tested by the *Vibrio fischeri* bioluminescence inhibition assay; no toxicity was detected in any of the samples (data not shown).

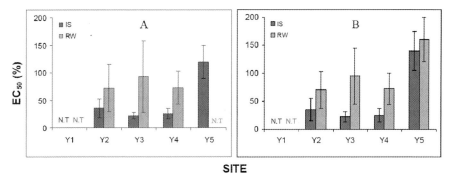

Fig. 1 ThamnotoxF^{TM} (**A**) and DaphtoxF^{TM} (**B**) toxicity of Yarkon water (RW) and interstitial water (IS). NT – no toxicity

Similarly, no toxicity was observed when the protozoan *Tetrahymena thermophila* served as the test organism (not shown). In contrast, clear toxic effects were observed when the same samples were tested using the two planktonic crustaceans *Thamnocephalus platyurus* and *Daphnia pulex* (Fig. 1) as well as by direct sediment exposure of the ostracod *Heterocypris incongruens* (Fig. 2).

From the results presented in Figs. 1 and 2, it may be clearly observed that in all cases, toxicity of the interstitial waters was higher (= lower EC$_{50}$ values) of that of the flowing water above, re-emphasizing once more the "self cleaning" nature of the flowing water and the sink characteristics of the sediment. Also apparent are the changes in observed toxicity along the river: no toxicity at the unpolluted origin (Y1), high toxicity at point Y2 where the treated wastewaters join the flow, and a gradual decrease towards the outflow. Sample Y5, representing the river section strongly influenced by seawater, displayed the lowest toxicity. This

trend was also observed for the direct sediment assays (Fig. 2); in that case, a small but reproducible degree of mortality was also observed for the Y1 sample.

3.2 Kishon River

Due to the high concentrations of industrial wastewaters in the Kishon River at the time of testing, salinity of both the river water and the interstitial water was high (1–4%). We have therefore initially used marine organisms, the rotifer *Brachionus plicatilis* and brine shrimp *Artemia salina* as the test organisms, as well as *Vibrio fischeri*. The first two organisms did not reveal any sample toxicity except for very low values at one location (K2; data not shown).

The *Vibrio fischeri* results similarly revealed no flowing water toxicity, but displayed a very clear pattern for the interstitial samples: no toxicity at K1, high toxicity at K2, and a gradual decrease towards K5 (Fig. 3).

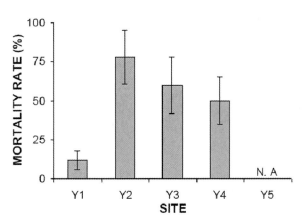

Fig. 2 Mortality rate of Ostracodtoxkit F^{TM} after direct exposure to Yarkon river sediment samples. N.A – not applicable because of high salinity. n=3

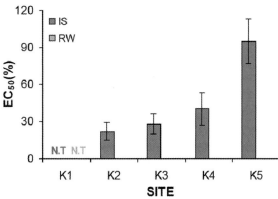

Fig. 3 Microtox (15 min) EC$_{50}$ of Kishon river samples. N.T = no toxicity

Fig. 4 ThamnotoxF™ (**A**) and DaphtoxF™ (**B**) LC$_{50}$ (24 h) of Kishon river samples

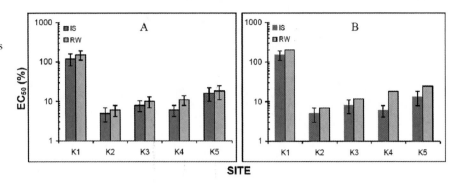

In view of the apparent low sensitivity of the marine test organisms, the two organisms found most responsive in the Yarkon system, *Thamnocephalus platyurus* and *Daphnia pulex*, were also applied to the Kishon samples. Since these organisms require a low salinity, the Kishon samples were diluted 10-fold to generate the most concentrated sample in the tested dilution series. Due to the high initial toxicity of the samples, it was possible to measure and calculate EC$_{50}$ values from the effects of the diluted samples (Fig. 4A – *T. platyurus*; Fig. 4B – *D. pulex*).

In both cases, location K1 displayed a very low but measurable toxicity (EC$_{50}$ close to 100%). The rest of the samples exhibited a much higher toxicity, which gradually decreased along the progression towards the sea, similarly to the pattern emerging from the *Vibrio fischeri assay* (Fig. 3). Toxicity decreased along the river from K2 to K5 and no toxicity was observed at K1. As for the Yarkon, interstitial water toxicity was much higher than that of the water above. Kishon toxicity values were in general much higher than those exhibited by the Yarkon samples.

3.3 Total Relative Toxicity

To try and correlate the toxicity values derived from the bioassays conducted in the course of the present study with the actual presence of pollutants found in the same samples, an HPLC analysis of extractable organics was carried out (Table 3).

Also listed in Table 3 are published mammalian toxicity values for the same compounds. Using the specific concentration and toxicity of each of the chemicals, TRT values were calculated for each of the samples. Due to a technical problem, sample K3 data were not available. The results for the other samples (Fig. 5)

indicate a high potential toxicity in all of the samples, including the "clean" K1.

Also clear from the data is that the chemicals present in the interstitial water represent only a fraction (as low as less than 1%) of the total toxic load of the sediments. Similarly to the indications provided by the crustaceans' assays, there was a gradual decrease of this load along the flow of the stream.

Past publications (Kronfeld and Navrot, 1975; Oren et al., 2006) have indicated the presence of organic and inorganic pollutants in the Yarkon and Kishon Rivers, and an in vitro cellular genotoxicity in the latter (Avishai et al., 2002, 2004), but to date no systematic investigation of the rivers' toxicity to water biota was conducted. In the present study we have attempted to assess the hazardous potential in both water and sediments, by combining two approaches: toxicity bioassays and chemical analysis. To allow quantification of the toxic potential in the chemicals identified by the latter approach, we have devised an artificial global toxicity (TRT) parameter that can serve as a semi-quantitative indicator of the potential risk inherent in the sample.

Our results clearly indicate that microscale bioassays can serve as efficient tools for monitoring river "health"; they can provide important information on the extent of pollution, as well as allow to assess the success of rehabilitation measures. For this purpose, however, it is essential that the appropriate test organisms are selected. In the two rivers tested, the following sensitivity ranking was observed: *D. pulex* = *T. platyurus* > *V. fischeri* > *A. salina* > *B. plicatilis*.

The two planktonic freshwater crustaceans were thus found to be the most useful, while the very common bacterial *V. fischeri* assay was much less sensitive. In fact, if the latter were used as a single toxicity indicator, many toxic samples would have yielded false negative results. To minimize such errors, it is

Table 3 Relative toxicity calculation in the Kishon River. For each compound, concentration data (upper value) are denoted in the interstitial pore water (IS, mg/l) and/or in the total sediment (TS, mg/l). The lower value (bold) in each cell denotes the relative toxicity.

Compound	K1 IS	K1 TS	K2 IS	K2 TS	K4 IS	K4 TS	K5 TS	LD$_{50}$ IV Mouse mg/Kg	LD$_{50}$ Oral Rat mg/Kg	Relative toxicity factor
4-Methyl Phenol		0.3 **9**	19 **1197**						207	63
Isooctyl mercaptoacetate				13 **481**					348	37
Pentanoic acid			26 **572**						600	22
Tributyl Phosphate			0.4 **4.4**	0.07 **0.8**	4 **43.6**				1200	10.9
Phenol nonyl							0.4 **3.2**		1620	8
2-Methyl Naphtalene						0.5 **4**	1.7 **13.6**		1630	8
Butanoic acid			0.4 **2.6**						2000	6.5
Benzenacetic acid			0.3 **1.7**						2250	5.8
1-Undecanol				12 **52.2**					3000	4.35
4-Hydroxy-4-Methyl-2-Pentanone	0.02 **0.07**								4000	3.3
DL-Limonene			0.2 **0.5**						5000	2.6
1-Octanamine N,N dioctyl	0.4 **0.9**		463 **1079**			1.1 **2.6**			5600	2.33
1,2 Benzenedicarboxylic acid, bis				18 **18**		18 **18**	2.4 **2.4**	13060		1
Hexadecanoic acid			0.15 **15.3**		0.2 **20.4**	1.6 **163**	2.1 **214**	57		102
Undecane			2 **22.4**		21 **235**			517		11.2
Tridecane					0.2 **1**	116 **580**	3.2 **16**	1161		5
Dodecane			1.5 **3.3**		0.3 **0.7**	5 **11**		2672		2.2
Pentadecane			0.2 **0.34**		0.1 **1.7**	2 **3.4**		3493		1.7
Tetradecane			0.7 **0.7**	10 **17**	0.1 **0.1**	4 **4**		5800		1
Sum TRT	1	19	1820	1648	301	768	250			

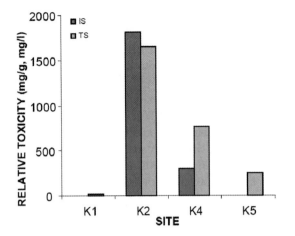

Fig. 5 Calculated total relative toxicity of Kishon river samples. IS- interstitial extracts; TS- total sediment extracts

recommended that a battery of bioassays be applied. Selection of the panel members should be based upon preliminary studies, combined with known operational and performance parameters such as simplicity, chemicals detection spectrum, exposure routes, or taxonomic diversity. Correct interpretation of toxicity data also necessitates knowledge of the physical and chemical parameters of the studied waters (Canna-Michaelidou et al., 2000).

A very clear and recurring observation in practically all locations sampled was that river sediments serve as a sink for water-borne pollutants. Furthermore, we have shown that sediment interstitial water toxicity, while significant, represents only a small fraction of the overall toxic potential of the sediment and that a significant amount of toxic pollutants is adsorbed to solid sediments constituents. In all cases, sediment toxicity was high close to the sources of pollution and gradually decreased downstream. While such adsorbed chemicals may not be directly bioavailable, they nevertheless represent a significant risk in cases of direct exposure or ingestion. This may occur in cases were sediments are disturbed and resuspended, not an unlikely scenario in shallow rivers.

This observation may have a significant impact on river rehabilitation programs. The first and obvious step in any such plan calls for improvement of the water quality by limiting the pollutant input. While such steps may indeed dramatically improve water quality, it may be expected that changes in sediment quality will be much slower. Furthermore, it is highly likely that sediment pollutants will be continuously

released into the water over long periods. Thus, dredging and removal of the sediment may be the only viable option for toxicants removal within a reasonable time period.

Our results also point out the significance of inclusion of toxicity bioassays as an important component of any river monitoring program. The data yielded by such assays can serve as an important tool for assessment of river health, as well as for following the progress of remediation schemes. For this purpose, it is proposed that a river-specific panel of toxicity bioassays is selected, representing different trophic levels and taxonomic complexity.

Acknowledgments The study was supported by The Israeli Ministry for Environmental Protection and the Yarkon River Authority. The considerable help of the management and staff of the Yarkon and Kishon River Authorities is gratefully acknowledged.

References

Avishai N, Rabinowitz C, Moiseeva E, Rinkevich B (2002) Genotoxicity of the Kishon River, Israel: the application of an in vitro cellular assay. Mutation Research 518: 21–37

Avishai N, Rabinowitz C, Rinkevich B (2004) A 2.5-year genotoxicity profile for a partially restored polluted river. Environmental Science and Technology 38: 3482–3487

Bancon-Montigny, C, Lespes G, Potin-Gautier M (2001) Optimisation of the storage of natural freshwater before organotin speciation. Water Research 35: 224–232

Bar-or, Y (2000) Restoration of the rivers in Israel's coastal plain. Water, Air and Soil Pollution 123: 311–321

Bierkens J, Klein G, Corbisier P, Van Den Heuvel R, Verschaeve L, Weltens R, Schoeters G (1998) Comparative sensitivity of 20 bioassay for soil quality. Chemosphere 37: 2935–2947

Blaise C,Wells PG, Lee K (1998) Microscale testing in aquatic toxicology: introduction, historical perspective, and context. In: Wells PG, Lee K, Blaise C (eds) Microscale Testing in Aquatic Toxicology Advances, Techniques, and Practice, pp. 1–9. CRC Press: Boca Raton

Canna-Michaelidou S, Nicolau AS, Neopfytou E, Christodoulidou M (2000) The use of a battery of microbiotests as a tool for integrated pollution control: evaluation and perspective in Cyprus. In: Persoone G, Janssen C, De Coen W (eds) New Microbiotests for Routine Toxicity Screening and Biomonitoring, pp. 39–48. Kluwer Academic, New York

Castillo GC, Vila, IC, Neild E (2000) Ecotixicity assessment of metals and wastewater using multitrophic assays. Environmental Toxicology 15: 370–375

Chapman PM (1986) Sediment quality criteria from the sediment quality triad: an example. Environmental Toxicology and Chemistry 5: 957–964

Demuth S, Casillas E, Wolfe DA, and McCain BB (1993) Toxicity of saline and organic solvent extracts of sediments from Boston Harbor, Massachusetts and the Hudson River-Raritan Bay estuary, New York using the Microtox® bioassay. Archive of Environmental Contamination and Toxicology 25: 377–386

Jansen C R, Vangheluwe M, van Sprang P (2000) A brief review and critical evaluation of the status of mictobiotests. In: Persoone G, Janssen C, De Coen W (eds) New Microbiotests for Routine Toxicity Screening and Biomonitoring, pp. 27–37. Kluwer Academic, New York

Kohler S, Belkin S, Schmid DR (2000) Reporter gene bioassays in environmental analysis. Fresenius Journal of Analytical Chemistry 366:769–779

Kronfeld J, Navrot J (1975) Aspects of the trace metal contamination in the coastal rivers of Israel. Water Air and Soil pollution 4: 127–134

Liu D, Aoyama I, Okamura H, Dutka BJ (1996) Enhancement of toxicant release from sediment by sonication and sodium ligninsulfonate. Environmental Toxicology and Water Quality 11: 195–203

Oren A, Aizenshtat Z, Chefetz B (2006) Persistent organic pollutants and sedimentary organic matter properties: a case study in the Kishon River, Israel. Environmental Pollution 141: 265–74

Pardos M, Blaise C (1999) Assessment of toxicity and genotoxicity of hydrophobic organic compounds in wastewater. Environmental Toxicology 14: 241–247

Quershi AA, Bulich AA, Isenberg, D.L (1998) Microtox toxicity test systems – where they stand today. In: Wells PG, Lee K, Blaise C (eds) Microscale testing in aquatic toxicology advances, techniques, and practice, pp. 185–200. CRC Press: Boca Raton

Ribo JM, Kaiser KLE (1987) *Phoobacterium phosphoreum* toxicity bioassay. I. Test procedures and applications. Toxicity Assessment 25: 305–323

Ritter L, Solomon K, Sibley P, Hall K, Keen P, Mattu G, Linton B (2002) Sources, pathways, and relative risks of contamination in surface water and groundwater: a perspective prepared for the Walkerton inquiry. Journal of Toxicology and Environmental Health 65: 1–142

Ronco AE, Sobrero MC, Rossini GDB, Alzuet PR, Dutka BJ (1995) Screening for sediment toxicity in the Rio Santiago basin: a baseline study. Environmental Toxicology and Water Quality 10: 35–39

Toxnet (http://toxnet.nlm.nih.gov/), United States National Library of Medicine

Vangheluwe ML, Janssen CR, Van Sprang PA (1999) Selection of bioassays for sediment toxicity screening. In: Persoone G, Janssen C, De Coen W (eds) New Microbiotests for Routine Toxicity Screening and Biomonitoring, pp. 449–458. Kluwer Academic, New York

Winger P V, Lasier PJ, Jackson BP (1998) The influence of extraction procedure on ion concentrations in sediment pore water. Archive of Environmental Contamination and Toxicology 35: 8–13

Saccharomyces cerevisiae as Biosensor for Cyto- and Genotoxic Activity

Jost Ludwig, Marcel Schmitt and Hella Lichtenberg-Fraté

Abstract Conventionally, toxicological bioassays are based on rodent models to evaluate the toxic effects of chemical compounds and to study the mechanism of action of toxicants. However, scientific developments are required to keep in line with regulatory frameworks, such as existing EU guidelines for assessment of manufactured chemicals (67/548/EEC, 93/67/EEC, and 83/571/EEC) and the EU regulatory framework for chemicals (REACH, EC1907/2006) concerning in part also existing chemicals. Scientific developments are thus directed towards rapid and reliable high-throughput assays to evaluate more accurately and more mechanistically the potential hazards of large numbers of chemicals. The yeast *Saccharomyces cerevisiae* is a promising model for such assays because it is amenable to genetic studies and because of the vast amount of genomics knowledge, resources, and manipulative tools associated with this unicellular fungus. The high degree of homology of essential cellular organization and metabolism shared by *S. cerevisiae* and higher eukaryotes has enabled the study of aspects of cellular toxicity and phenomena of relevance to human biology at the molecular level.

Keywords *Saccharomyces cerevisiae*

1 Regulatory Framework

1.1 REACH

On 1 June 2007 the European Community Regulation on chemicals and their safe use (EC 1907/2006), dealing with the Registration, Evaluation, Authorisation and Restriction of Chemical substances (REACH,) entered into force. The REACH Regulation concerns chemical substances with a yearly production or import rate exceeding 1t with the aim to improve the protection of human health and the environment by better and earlier identification of their intrinsic properties. Manufacturers and importers will be required to gather information on the properties of their chemical substances, which will allow their safe handling, and to register the information in a central database run by the European Chemicals Agency (ECHA) in Helsinki. Approximately 30.000 existing substances and 400 new substances per year are incurred for the registration whereby common standards for risk assessment and a uniform regulatory framework for new and existing substances are implemented. Problematic substances such as cancerogenic, genotoxic and teratogenic compounds or those with persistent or bioaccumulating potential require a special accreditation. REACH comprises comprehensive Chemical Safety Report Guidelines for all before September 1981 introduced existing substances exceeding 1t per year production. It is thought that for most of these substances the necessary data for risk assessment are not available, only 140 are sufficiently described.

Against this background and under consideration of the fact that the European Commission for all

H. Lichtenberg-Fraté (✉)
University of Bonn, IZMB, Molekular Bioenergetics,
Hella Lichtenberg-Fraté, IZMB, Universität Bonn,
53115 Bonn, Germany
e-mail: h.lichtenberg@uni-bonn.de

Y.J. Kim et al. (eds.), *Atmospheric and Biological Environmental Monitoring*,
DOI 10.1007/978-1-4020-9674-7_17, © Springer Science+Business Media B.V. 2009

future risk assessment stipulates the reduction of animal testing to the absolute minimum, the increasing demand for alternative and meaningful biotest procedures becomes apparent. Conventional toxicological bioassays are based on rodent models to evaluate the toxic effects of chemical compounds and to study the mechanism of action of toxicants. New scientific developments are thus required to keep in line with regulatory frameworks such as existing EU guidelines for assessment of manufactured chemicals (67/548/EEC, 93/67/EEC, and 83/571/EEC) and REACH and to meet ethical considerations like the reduction of animal testing. Current developments are thus directed towards rapid and reliable high-throughput assays to evaluate more accurately and more mechanistically the potential hazards of large numbers of chemicals. The yeast *Saccharomyces cerevisiae* – among other model organisms – may provide a most promising model for such assays because it is amenable to genetic studies and because of the vast amount of genomics knowledge, resources and manipulative tools associated with this unicellular fungus. The high degree of homology of essential cellular organization and metabolism shared by *S. cerevisiae* and higher eukaryotes has enabled the study of aspects of cellular toxicity and phenomena of relevance to human biology at the molecular level.

2 Biotest Based Analysis of Environmental Pollutants

Based on the increasing entry of chemical contaminants in the aquatic ecosystem and altered legal frameworks demands for the development of sensitive and reliable methods for assessment of the potential impact of these contaminants on aquatic organisms cumulate (Huggett et al., 1992; McCarthy und Shugart, 1990). Obtaining evidence for the existence of environmentally relevant, that means those with a negative impact on wildlife, appears to be an increasing challenge for conventional environmental analytics. The comprehensive gathering of potential hazardous substances within a complex sample is upon singular application of chemical analytical methods and procedures hardly possible. Such processes are most suited to detect known compounds at very low concentrations with highest accuracy. Unknown or complex samples

among which the single substances impact on each other (synergistic or antagonistic effects) are however often not detected, or lack a suitable set of criteria for assessment of the actual hazard. Therefore additional impact directed tests are necessary. These are in most cases based on ecotoxicological approaches and assess the summation of all effects of all components within a sample in correlation to the individual test organism. The most simple and straight forward impact tests assess sample caused growth inhibition. Since toxic compounds and substances always exhibit a negative effect on the growth of an organism that is independent of their specific mechanism, this parameter represents a reliable set of criteria for assessment of the toxic properties of a given sample.

2.1 Growth Inhibition Tests

Growth inhibition or chronic toxicity tests that comprise several generations are carried out to determine the long-term effects of pollutants on the biota. Pollutants comprise naturally occurring substances (in potentially hazardous concentrations) and xenobiotics, chemicals that are foreign to biological systems and include industrial chemicals, pesticides, herbicides, and fungicides. As many of these chemicals enter the aquatic environment, aquatic toxicity testing requires rapid and representative bioassays. Well-developed systems, representing different trophic levels, comprise different bacterial species, algae, and ciliates for toxicity evaluations because their response to chemical content in water is considered as significant, their life cycle is short, their response to environmental change is fast and they are easy to maintain. However, test results obtained with bacteria, as prokaryotes, are not representative for eukaryotic organisms as well. Eukaryotic unicellular organisms like algae are generally employed for the detection of phytotoxicity and ciliates are not commonly used as test organisms outside the USA. Given the vast variety of chemicals to be tested under REACH,, the use of short-term cytotoxicity assays involving eukaryotic cells for the initial screening of chemicals not only aids in establishing priorities for the selection of chemicals that should be tested in vivo but also decreases the time in which potential toxicants can be evaluated. Short-term toxicity testing involving the eukaryotic yeast can in addition complement the traditional bacterial test batteries in order to get a comprehensive picture of toxic effects of pollutants.

Growth inhibition tests however do not enable mechanistic studies or evaluations. Pollutants can exhibit their hazardous potential in several different manners, e.g. specific inhibition of enzymes like it is the case for cytochrom-oxidase by cyanide or on a more general cellular level by causing oxidative stress as for heavy metals. Among the heavy metals the transition metals such as copper, zinc, nickel, and iron are involved in redox processes of respiratory activity. Others, such as magnesium, nickel, and cobalt, are part of complex molecules, stabilize protein structures, and serve in maintaining the osmotic balance, like potassium, or stabilize various enzymes (Zn^{2+} fingers) and DNA through electrostatic forces, like magnesium and zinc. Although considered as essential cofactors for a variety of enzymatic reactions and for important structural and functional roles in cell metabolism, metals at high concentrations are potent toxic pollutants (Eide, 2000, 2001). Toxic effects can include blocking of functional groups on important biomolecules as well as denaturation of enzymes (Blackwell et al., 1998) and DNA damage as for copper, chromium, zinc, and nickel. Another mechanism of metal toxicity (copper, nickel, cadmium) involves the transition-metal-catalyzed generation of reactive oxygen species (ROS) in the Fenton reactions and subsequent lipid peroxidation (Stohs and Bagchi, 1995). Furthermore, metal toxicity (cadmium, cobalt) may be also caused by depletion of glutathione (GSH), considered as a major antioxidant in eukaryotic cells (Fortuniak et al., 1996).

Thus besides basic impact tests like growth inhibition the implementation of specific endpoint analysis for identification of differentiated mechanisms is necessary. Such endpoints may be the determination of the genotoxic potential of a substance or the effect on key enzymes or functions like regulatory factors.

2.2 Gentoxicity Tests

Genotoxicity is a collective term that describes the general genetic material injury which can be due to a variety of impairment mechanisms. Among the number of genotoxic detection systems true mutagenicity tests

and indicator tests are to be differentiated. In addition, biomarkers have been formulated to differentiate into those of exposure (substance or its metabolites in biological fluids, mutagenicity, protein and DNA adducts) and of effects (chromosome aberrations (CAs), sister chromatid exchanges (SCEs), micronuclei (Mn), COMET assay, HPRT mutants) measuring the alterations of important genetic targets.

Standardized prokaryotic mutagenicity procedures like the Ames-Test (Ames et al., 1973; Gee et al., 1996) that detects the back mutation of a histidin-synthesis-defective *Salmonella thyphimurium* strain in the presence of a test substance enable the direct verification of single mutational events. The DNA-tail test (Comet assay; Östling und Johanson, 1984) is based on electrophoretic separation to visualize strand breaks of DNA molecules of single cells like those of protozoans, green algae, calm cells, primary fish hepatocytes, fish or mammalian cell lines.

Hampsey (1991) reported a system for the direct detection of mutations of each of the six base-pair substitutions in isogenic *S. cerevisiae* strains by selection for an essential cysteine in codon 22 of the *CYC1* gene (iso-1-cytochrome c). Through reverse mutation of the substitutions the previously lost enzymatic function can be restored and selected for the CYC+-phenotype. Scoring enables the determination of the genotoxic potential of the given samples according to the number of revertants. Additionally known tests are the alkali filter elution test (clams and green algae, Kohn and Grimek-Ewig, 1973), the chromosomal aberration (CA) frequency (Müller et al., 1982) and the micronucleus-test (Countryman and Heddle, 1976) in human lymphocytes.

Indicator test systems do not differentiate between point-, strand break or chromosomal mutations but, as biological indicators of genotoxic risk in environmental or occupational exposure, rely on DNA-repair signalling-reporter responses as in indirect measure for genotoxic damage. Monitoring of toxicant-induced gene expression changes can potentially provide a powerful alternative or supplement to traditional approaches to toxicological screening. The SOS chromotest (Fish et al., 1987; Quillardet et al., 1982, 1985; Otha et al., 1984) involves Lac+ reversion of a F0 plasmid borne lacZ gene and reports induction of a SOS repair gene (*sulA*) in a genetically engineered *Escherichia coli* strain as indirect measure for DNA damage. In Germany, the governmental survey of wastewater (monitoring of plant effluents) comprises the umu-test (ISO 13829; Reifferscheidt et al., 1991; Oda et al., 1985) as the relevant biological parameter. This assay utilizes prokaryotic *Salmonella typhimurium* TA 1535 pSK1002 cells; the detected genetic damage (transcriptional activation of the umuC repair gene and its lacZ coupled calorimetric detection) represents DNA damage from point mutations. More recently, *Saccharomyces cerevisiae* tester strains as simple eukaryotic model organisms have been utilized for detection of genotoxic potential with yeast-optimized green fluorescent protein fused to the RAD54 (radiation repair; Schmitt et al., 2002, Lichtenberg-Fraté, 2003) or RNR2 (ribonucleoside-diphosphate reductase) promoter (Walmsley et al., 1997; Afanassiev et al., 2000). In principle, simple eukaryotic test systems may well be added to the know test batteries due to the potential extrapolation and transfer of results to higher organisms.

2.3 Resistances

The influence of general mechanisms that confer resistance to a great variety of compounds has so far received little attention in the context of genotoxicity testing. Multidrug resistance (MDR) mechanisms include enzymatic degradation or metabolic inactivation of drugs (Davies, 1994), the alteration of drug targets (Spratt, 1994), intracellular sequestration and active export across the plasma membrane. Active efflux by specific or MDR transporters prevents the access of toxic compounds to their intracellular targets and is believed to contribute significantly to acquired bacterial resistance (especially in important pathogens) because of the very broad variety of recognized substrates and co-operation with other mechanisms of resistance (Kolaczkowski and Goffeau, 1997).

The multidrug efflux systems encountered in eukaryotic cells are very similar to those of prokaryotes (van Bambeke et al., 2000). Multidrug transporters are membrane proteins found in nearly every cell, these proteins transport a great variety of structurally and chemically dissimilar compounds and remove them from the cell, a process of high significance for toxicological assessment. In the yeast *S. cerevisiae,* a so called pleiotropic drug resistance (PDR) network (Balzi and Goffeau, 1995) comprises the major deter-

minants of multiple drug resistance (Kolaczkowska and Goffeau, 1999). Induction by a drug or toxin involves the upregulation of expression of well characterized genes that encode transporters involved in drug extrusion and thus multidrug resistance. The majority of these proteins have broad overlapping specificities for compounds with very different chemical structures and cellular targets: transported compounds include a variety of anticancer drugs, antibiotics, antifungals, detergents and ionophores (Kolaczkowski and Goffeau, 1997; Kolaczkowski et al., 1998). Potentially, the activities of these ABC-type extrusion systems confound attempts to develop bioassays for cyto- and genotoxins, since continuous removal of the relevant compound from the cell compromises the sensitivity of any assay.

2.4 The Yeast Indicator Test

Saccharomyces cerevisiae as the most simple eukaryotic organism is broadly accepted as a laboratory model organism because it is easily propagated and manipulated in the laboratory and shares a common life cycle and cellular architecture with higher eukaryotes. The high degree of homology of essential cellular organization and metabolism shared by *S. cerevisiae* and higher eukaryotes has enabled the study of aspects of cellular toxicity and phenomena of relevance to human biology at the molecular level (Rasio et al., 1997, Castrillo and Oliver, 2004, Resnick and Cox, 2000). Such research has offered many insights into the complex mechanisms underlying the sensing and response to toxicant stressors (Gardner et al., 2003, Gasch et al., 2000). The degree to which gene expression profiles are conserved upon toxic stress and the regulation of key pathway elements enabled the identification of human signal transduction homologues (Bergmann et al., 2003, Gasch et al., 2001). Although gene expression profiling is not (yet) suited for high-throughput screening, insights in terms of hierarchic clustering of genetic stress-related networks potentially provides the means to identify surrogate markers that can be used to construct detection systems and prediction models. Thus, for the detection of potential toxic effects of pure compounds and complex composed samples like wastewater a miniaturised short-term cyto- and genotoxicity (indicator test) screening assay was constructed and characterized (Schmitt et al., 2002, 2004, 2005a, b, Lichtenberg-Fraté et al., 2003). The assay is based on genetically engineered *S. cerevisiae* cells deleted in the prominent multidrug ABC transporters PDR5 (pleiotropic drug resistance), SNQ2 (disruption confers sensitivity to 4-nitroquinoline N-oxide) and YOR1 (yeast oligomycin resistance) that facilitate pleiotropic drug resistance. The yeast strain devoid of these proteins that mediate the efflux of structurally diverse hydrophobic compounds exhibited an increased sensitivity to a variety of compounds and substances (Schmitt et al., 2004, 2005a, b). The DNA damage biomarker is based on the *RAD54* repair gene and serves as an indicator of general DNA injury in this strain. RAD54 is a DNA-dependent ATPase of the Snf2 family and is involved in repair of UV or chemically induced DNA damage. It is no essential gene with two nuclear localisation signals and one Leucin-Zipper motif. The promoter of RAD54 contains a 'DNA-damage-response' element ($-77 - -99$, TCGTAGCAGCTTTAGTGAAGAAAGAGAAA) that is fused to a yeast enhanced version of the green fluorescent protein of *Aequorea victoria* (Cormack et al., 1996), as a biomarker for genetic targets for broad toxicant-inducible DNA-integrity damage. Thereby the induction of green fluorescence serves as the genotoxic endpoint read-out. Simultaneously, due to its broad relevance the parameter "growth inhibition" as functional response is included by chronic toxicity assessments that are performed by growth behaviour assays to determine EC50 values. Obtained results (Schmitt et al., 2002, 2004, 2005a, b) indicated that the induction of green fluorescence is a suitable biomarker for the detection of genotoxic potentials irrespective of the underlying chemical mechanisms.

2.5 Metabolic Biomarker

The rapid detection of toxicant-induced stress on basic metabolism utilized a *S. cerevisiae* biosensor based on the housekeeping plasma membrane ATPase (*PMA1*) promoter for detection of early toxic effects. *PMA1*, one of the most prominent housekeeping genes in *S. cerevisiae*, encodes the major plasma membrane H^+-ATPase (Serrano et al., 1986) and is essential for viability. As a highly conserved member of the P-type

ATPases, the H^+-ATPase is a single 100-kDa polypeptide. The electrogenic proton pump is the major source of cytosolic proton extrusion and generation of the proton motive force across the cellular membrane. The proton motive force is responsible for secondary active transport mechanisms for a variety of nutrients and is also involved in pH homeostasis. A decrease in membrane potential has been suggested as primary cellular stress signal triggering the intracellular response (Moskvina et al., 1999). In being the major protein component of the plasma membrane (15 to 20% of total plasma membrane protein [Ambesi et al., 2000]), expression and activity of this proton pump are precisely regulated to match its numerous requirements (Capiaux et al., 1989, Fernandes and Sá-Correia, 2003, Kuo and Grayhack, 1994, Viegas et al., 1994). Rapid detection of toxicant-induced stress on basic metabolism was thus monitored by implementing the plasma membrane ATPase promoter (P_{PMA1})-reporter system. The P_{PMA1} is unique in its regulatory site composition and precise response to metabolic state. *PMA1* expression is adjusted according to the metabolic state of the cell, i.e., upregulated during exponential growth (Viegas et al., 1994) and on respiration (Fernandes and Sá-Correia, 2003). Under nonlimiting glucose conditions (fermentative metabolism), expression is potentially regulated by the TUF/RPG/RAP1 system (two consensus sites), known to regulate expression of essential genes involved in glycolysis and active transport (Capiaux et al., 1989, Quandt et al., 1995), and the MCM1 protein (one consensus site [Kuo and Grayhack, 1994, Quandt et al., 1995]). The P_{PMA1}-mediated transcriptional activation of the yeast optimized green fluorescent protein (yEGFP3; Cormack et al., 1996) results in the production of the green fluorescent protein (GFP). To monitor dynamic fluorescent changes however, the assay should indicate fluorescence emission decrease due to dose dependent intoxication. To that end, a destabilized version of GFP was constructed by coupling the PEST-rich C-terminal residues of the G1 cyclin CLN2 genetic element which has been reported to serve as a universal ubiquitin-targeting sequence and to reduce GFP half-life (t1/2) to approximately 30 min and thus also of the yeast-enhanced version of the yEGFP chimeric construct as a response to compound-mediated intoxication. Upon intoxication the shift of the steady-state turnover of P_{PMA1}-driven GFP transcription towards less transcription/translation and

Table 1 Data on three different toxicological endpoints detected in sensitized *Saccharomyces cerevisiae* strains

| Class | Compound | Short-term toxicity detection by metabolic biomarker | | Chronic toxicity | Genotoxicity | |
		EC_{20} in mg/L (4 h)	Confidence interval 95% in mg/L	EC_{20} EC_{50} in mg/L	Threshold Gentox 95% in mg/l	Growth inhibition at threshold in %
Metal Ions	Cd^{2+} as Cadmium chloride monohydrate CAS [35658-65-2]	2.5	1.7 – 3.6	0.18 0.80	0,7	60,2
	Co^{2+} as Cobalt(II) nitrate hexahydrate CAS [10026-22-9]	n. c.	n. c.	19.0 82.9	n. c.	–
	Cr(VI) as Potassium dichromate CAS [7778-50-9]	2.0	1.5–2.6	0.27 0.88	0,5	32,6
	Cu^{2+} as Cupric sulphate pentahydrate CAS [7758-98-8]	2.9	2.2–3.7	0.74 2.52	0,5	9,7
Pharmaceuticals	Diclofenac sodium salt CAS [15307-79-6]	140.7	122.9–156.7	115.7 246.6	75,0	12,3
	Hydroxyurea CAS [127-07-1]	n. c.	n. c.	1189 3311	312,5	3,7
Pesticide	Lindane CAS [58-89-9]	5.63	5.02–6.27	3.14 8.28	2,2	14,0
PCPs	3,5-Dichlorophenol CAS [591-35-5]	6.27	4.97–7.35	5.51 9.23	10,0	58,4
CNS stimulant	Caffeine CAS [58-08-2]	569.5	477.5–726.3	250.2 640.7	187,5	11,6
Genotoxic agent	MNNG CAS [70-25-7]	0.48	0.44–0.52	0.32 0.92	0,0625	3,4
Other	Sodium chloride CAS [7647-14-5]	7050	6360–7710	4870 11294	5000	19,1

proportionally increased PEST-mediated degradation and therefore decreased fluorescence signals was altered in a compound concentration-dependent manner. Thus, the modified construct with destabilized chromophore served to monitor metabolism directed toxic effects, comprising transcription and/or translation inhibition, by measuring the decrease in fluorescence (Schmitt et al., 2006). This approach, similar to the previous long-term cytotoxicity and genotoxicity tests deployed also a *S. cerevisiae* strain devoid of the pleiotropic drug transporters PDR5, SNQ2 and YOR1 (Hasenbrink et al., 2006) for increased sensitivity towards a broad spectrum of chemicals. Cells expressing the destabilized version of GFP (pP$_{PMA1}$-GFP/PEST) showed two thirds reduced fluorescence emission peak compared with cells expressing the pPPMA1-GFP construct. To evaluate the three different endpoint measurements, the growth-inhibition and genotoxicity assay results were compared with results assessed by determination of toxicant-induced stress (Table 1). Data on exemplary compounds and known toxicants show the *S. cerevisiae* system flexibility to monitor different effects and toxicological parameter.

3 New Technologies

The new and, in most cases high throuhgput technologies commonly referred to as ‚omics' offer significant potential in toxicological research. Genomics, transcriptomics, proteomics and metabolomics are key technologies with – on one hand – decreasing levels of throughput but – on the other hand – increasing level of physiological information. Such applications must be accompanied by profound statistical and bioinformatic resources. Where directed towards exposure and/or risk characterisation and assessment, geno- to phenotype association studies are indispensable. Particularly for this, model organisms like yeast are well suited. For instance, the determination of gene expression profiles in wild type and defined mutants following exposition towards single compounds or complex matrices enables the correlation of exposition (matrix, duration) with alterations in the transcriptom and thus insights into toxic mechanisms. Such changes in expression profiles can be understood as first signs of an intoxication, preceding manifestation in physiological responses or genotoxic endpoints like e.g. chromosomal losses or breaks. Examples for relevant families of genes that are associated with toxic expositions comprise but are not limited to detoxifying metabolic genes (e.g. P450), those involved in DNA damage and repair, oxidative stress, apoptotic events, cellular growth (proliferation) and cellular signalling. Combined with careful analysis of detectable physiological parameters like growth behaviour, morphology, oxygen consumption or ATP production, the correlation of all results enable integration into mechanistic based models and meaningful interpretation.

4 Summary

All three yeast indicator endpoint tests are sufficient to enable calculation of EC or threshold values for chemically unrelated compounds and complex matrices. Short term concentration-response measurements enable differentiation of quantity and quality of toxic effects of tested chemicals although obtained EC values are higher compared to long-term (chronic toxicity by growth inhibition) tests. For genotoxicity assessment the transactivation of green fluorescence driven by the DNA-damage inducible RAD54 promoter provides a reliable and sensitive system to monitor compounds with DNA-damage potential regardless of damaging mechanisms of the compounds. The advantage of *S. cerevisiae* lies in its simple handling and cultivation, its rapid response to toxicants and indication of bioavailability and furthermore, in the potential extrapolation of results for the evaluation of risks of chemicals to human health.

Toxicological risk assessment should predict potential environmental health risks posed by a wide range of products, including new or modified foods, dioxins, chemicals and waste materials. The results obtained with the *S. cerevisiae*-based metabolic biomarker, chronic toxicity and genotoxicity assays involving gene expression changes as an evolutionarily conserved response to injury indicate that tests based on this organism will help to define the toxic potential of a variety of compounds in a single assay with broad applications in toxicology, environmental monitoring, aquatic toxicology and the pre-screening of pharmaceutical products.

Acknowledgments This work was funded in part by EC QLK3-CT-2001-00401.

References

Afanassiev V, Sefton M, Anantachaiyong T, Barker G, Walmsley R, Wölfl S. (2000) Application of yeast cells transformed with GFP expression constructs containing the RAD54 or RNR2 promoter as a test for the genotoxic potential of chemical substances. Mutat Res 464:297–308

Ambesi A, Miranda M, Petrov VV, Slayman CW (2000) Biogenesis and function of the yeast plasma-membrane H(+)-ATPase. J Exp Biol 203:155–160

Ames B N, Lee F D, Durston W. E. (1973) An improved bacterial test system for the detection an classification of mutagens and carcinogens. Proc Nat. Acad Sci USA 70:782–786

Balzi E, Goffeau A. (1995) Yeast multidrug resistance: the PDR network. J Bioenerg Biomembr 27:71–76

Blackwell K J, Tobin J M, Avery S V, (1998) Manganese toxicity towards *Saccharomyces cerevisiae*: dependence on intracellular and extracellular magnesium concentrations. Appl Microbiol Biotech 49:751–757

Capieaux E, Vignais M L, Sentenac A, Goffeau A (1989) The yeast H+-ATPase gene is controlled by the promoter binding factor TUF. J Biol Chem 264:7437–46

Castrillo J I, Oliver S G (2004) Yeast as touchstone in postgenomic research: strategies for integrative analysis in functional genomics. Biochem Mol Biol 37:93–106

Cormack B P, Valdivia R H, Falkow S (1996) FACS-optimized mutants of the green fluorescent protein (GFP). Gene 173:33–38

Countryman P I, Heddle J A (1976) The production of micronuclei from chromosome aberrations in irradiated cultures of human lymphocytes. Mutat Res 41:321–332

Davies J (1994) Inactivation of antibiotics and the dissemination of resistance genes. Science 264:375–382

Eide D J (2000) Metal ion transport in eukaryotic microorganisms: insights from Saccharomyces cerevisiae. Adv Microb Physiol 43:1–38

Eide D J (2001) Functional genomics and metal metabolism. Genome Biol 2: 1028.1–1028.3

Fernandes A R, Peixoto F P, Sa-Correia I (1998) Activation of the H$^+$-ATPase in the plasma membrane of cells of *Saccharomyces cerevisiae* grown under mild copper stress. Arch Microbiol 171:6–12

Fish F, Lampert I, Halachmi A, Riesenfeld G, Herzberg M (1987) The SOS Chromotest kit: A rapid method for the detection of genotoxicity. Toxic Assess 2:135–147

Fortuniak A, Zadzinski R, Bilinski T, Bartosz G (1996) Glutathione depletion in the yeast *Saccharomyces cerevisiae*. Biochem Mol Biol Int 38:901–910

Gardner T S, di Bernardo D, Lorenz D, Collins J J (2003) Inferring genetic networks and identifying compound mode of action via expression profiling. Science 301:102–105

Gasch A P, Huang M, Metzner S, Botstein D, Elledge S J, Brown P O (2001) Genomic expression responses to DNA-damaging agents and the regulatory role of the yeast ATR homolog Mec1p. Mol Biol Cell 12: 2987–3003

Gasch A P, Spellman P T, Kao C M, Carmel-Harel O, Eisen M B, Storz G, Botstein D, Brown P O (2000) Genomic expression programs in the response of yeast cells to environmental changes. Mol Biol Cell 11: 4241–4257

Gee P, Maron D M, Ames B N (1994) Detection and classification of mutagens: a set of base-specific Salmonella tester strains. Proc Natl Acad Sci USA 91: 11606–11610

Hampsey M (1991) A tester system for detection each of the six base-pair substitutions in *Saccharomyces cerevisiae* by selecting for an essential cysteine in iso-1-cytochrome c. Genetics 128:59–67

Hasenbrink G, Sievernich A, Wildt L, Ludwig J, Lichtenberg-Fraté H (2006) Estrogenic effects of natural and synthetic compounds including tibolone assessed in *Saccharomyces cerevisiae* expressing the human estrogen α and ß receptors. FASEB J 20:1552–1554

Huggett R L, Kimerle R A, Mehrle P M Jr, Bergmann H L (1992) Bio-markers: Biochemical, Physiological, and Histological Markers of Anthropogenic Stress. Lewis, Boca Raton, USA

Kohn K W, Grimek-Ewig R A (1973) Alkaline elution analysis, a new approach to the study of DNA single-strand interruptions in cells. Cancer Res 33:1849–1853

Kolaczkowska A, Goffeau A (1999) Regulation of pleiotropic drug resistance in yeast. Drug Resist Updat 2:403–414

Kolaczkowski M, Goffeau A (1997) Active efflux by multidrug transporters as one of the strategies to evade chemotherapy and novel practical implications of yeast pleiotropic drug resistance. Pharmacol Ther 76:219–242

Kolaczkowski M, Kolaczowska A, Luczynski J, Witek S, Goffeau A (1998) In vivo characterization of the drug resistance profile of the major ABC transporters and other components of the yeast pleiotropic drug resistance network. Microb Drug Resist 4:143–158

Kuo M H, Grayhack E (1994) A library of yeast genomic MCM1 binding sites contains genes involved in cell cycle control, cell wall and membrane structure, and metabolism. Mol Cell Biol 14:348–359

Lichtenberg-Fraté H, Schmitt M, Gellert G, Ludwig J (2003). A yeast based method for the detection of cyto- and genotoxicity. Toxicol In Vitro 17:709–716

Mateus C, Avery S V (2000) Destabilized green fluorescent protein for monitoring dynamic changes in yeast gene expression with flow cytometry. Yeast 16: 1313–1323

McCarthy J F, Shugart L R (1990) Biomarkers of Environmental Contamination. Lewis, Boca Raton, USA

Moskvina E, Imre E M, Ruis H (1999) Stress factors acting at the level of the plasma membrane induce transcription via the stress response element (STRE) of the yeast Saccharomyces cerevisiae. Mol Microbiol 32:1263–1272

Müller D, Natarajan A T, Obe G, Röhrborn G (1982) Sister-Chromatid-Exchange-Test. Georg Thieme Verlag, Stuttgart & New York

Oda Y, Nakamura S I, Oki I, Kato T, Shinagawa H (1985) Evaluation of the new system (umu-test) for the detection of environmental mutagens and carcinogens. Mutat Res 147:219–229

Östling O, Johanson K J (1984) Microelectrophoretic study of radiation-induced DNA damages in individual mammalian cells. Biochem Biophys Res Commun 123:291–298

Otha T, Nakumara N, Moriya M, Shirasu T, Kada T (1984) The SOS-function-inducing activity of chemical mutagens in *Escherichia coli*. Mutat Res 131:101–109

Quandt K, Frech K, Karas H, Wingender E, Werner T (1995) MatInd and MatIn-spector: new fast and versatile tools for detection of consensus matches in nucleotide sequence data. Nucleic Acids Res 23:4878–4884

Quillardet P, Huisman O, D'Ari R, Hofnung M (1982) SOS chromotest, a direct assay of induction of an SOS function in *Escherichia coli* K-12 to measure genotoxicity. Proc Natl Acad Sci USA 79:5971–5975

Rasio D, Murakumo Y, Robbins D, Roth T, Silve A, Negrini M, Schmidt C, Burczak J, Fishel R, Croce C M (1997) Characterization of the human homologue of RAD54: a gene located on chromosome 1p32 at a region of high loss of heterozygosity in breast tumors. Cancer Res 57:2378–2383

REACH (EC 1907/2006) http://eur-lex.europa.eu/LexUriServ/LexUriServ.do?uri=CELEX:32006R1907:EN:NOT

Reifferscheid G, Heil J, Oda Y, Zahn R K (1991) A microplate version of the SOS/umu-test for rapid detection of genotoxins and genotoxic potentials of environmental samples. Mutat Res 253:215–222

Resnick M A, Cox B S (2000) Yeast as an honorary mammal. Mutat Res 451:1–11

Schmitt M, Gellert G, Kirberg B, Ludwig J, Lichtenberg-Fraté H (2002) Cy-Gene: Eine neue hefebasierte Methode zur Bestimmung des cytotoxischen und genotoxischen Potentials von Umweltgiften im Wasserbereich. Vom Wasser 99:111–118

Schmitt M, Gellert G, Lichtenberg-Fraté H (2005) The toxic potential of industrial effluents determined the *Saccharomyces cerevisiae* based assay. Water Res. 39:3211–3218

Schmitt M, Gellert G, Ludwig J, Lichtenberg-Fraté H (2005) Assessment of cyto- and genotoxic effects of a variety of chemicals using *Saccharomyces cerevisiae*. Acta Hydrochim Hydrobiol 33:56–63

Schmitt M, Gellert G, Ludwig J, Lichtenberg-Fraté H (2004) Phenotypic yeast growth analysis for chronic toxicity testing. Ecotoxicol Environ Safety 59:142–150

Schmitt M, Schwanewilm P, Ludwig J, Lichtenberg-Fraté H (2006) PMA1 as a housekeeping biomarker for the assessment of toxicant induced stress in *Saccharomyces cerevisiae*. Appl Environm Microbiol 72:1515–1522

Serrano R, Kielland-Brandt M, Fink G R (1986) Yeast plasma membrane H$^+$-ATPase is essential for growth and has homology with (Na^+-K^+)-, K^+-, and $Ca2^+$-ATPases. Nature 319:689–693

Spratt B G (1994) Resistance to antibiotics mediated by target alterations. Science 264:388–393

Stohs S J, Bagchi D (1995) Oxidative mechanisms in the toxicity of metal ions. Free Radic Biol Med 18:321–336

Van Bambeke F, Balzi E, Tulkens P M (2000) Antibiotic efflux pumps. Biochem Pharmacol 60:457–70

Viegas CA, Supply P, Capieaux E, Van Dyck L, Goffeau A, Sa-Correia I (1994) Regulation of the expression of the $H(^+)$-ATPase genes *PMA1* and *PMA2* during growth and effects of octanoic acid in *Saccharomyces cerevisiae*. Biochim Biophys Acta 1217:74–80

Walmsley R M, Billinton N, Heyer W D (1997) Green fluorescent protein as a reporter for the DNA damage-induced gene RAD54 in *Saccharomyces cerevisiae*. Yeast 13: 1535–1545

The Application of Cell Based Biosensor and Biochip for Environmental Monitoring

Junhong Min, Cheol-Heon Yea, Waleed Ahmed El-Said and Jeong-Woo Choi

Abstract The cell based biosensor has outstanding advantages in aspect of environmental monitoring, comparing with DNA or protein based biosensor. Various cell based biosensors to detect toxic chemicals have been researched in few decades and applied directly to environmental fields. In this study, some key technologies such as cell immobilization technology, on-chip cell cultivation technology, and cell viability detection technology required to develop cell based biosensors are categorized and briefly discussed. We suggests that cell based biochip can be applied to environmental monitoring by demonstrating that phenol can be successfully measured by two distinct cell based label-free biochips, SPR based cell chip and electric based cell chip, developed using key technologies.

Keywords Cell chip · Electrochemical detection · SPR · HepG2

1 Introduction

Environmental pollution affects human health has been rapidly considered by increment of environmental toxic substances such as endocrine disrupors, toxic chemicals, anti-biotic materials, heavy metals, and pathogenic microorganism in their environments, which is induced by dramatic development of various new industries such as new material synthesis, biotechnologies, and nanotechnologies. Various technologies have been researched and developed to measure environmental toxic substances and continuously monitor their increments. Enzymatic reactions in neural cells, Bio-affinity mechanism of antibodies, and specific structural properties of nucleic acids have been utilized to measure the presence of environmental toxic substances. Meanwhile, cell based sensors and chip system for environmentally usages have been emerged as new sensing tools in order to measure toxic substances because they can measure the effects of unknown analytes on physiological activities in human body as well as the fine detection capability with high specificity and sensitivity.

Various cell based biosensors to detect toxic chemicals have been researched in few decades and applied directly to environmental fields. Recently, cell chips for physiological research with animal cells including cancer cells also have fabricated. To develop cell based biosensors and biochip, the in vivo like environments is required to sustain their properties. Therefore, cell immobilization technologies and cell cultivation technologies as well as cell viability measurement technologies are also required in order to apply cell based biosensors and cell chip system to environmental monitoring fields.

In this paper, three technologies required in cell based biochip system, cell immobilization, cell cultivation in chip, and cell based biosensors are discussed and the new application of environmental cell chips are introduced.

J.-W. Choi (✉)
Departen of Chemical and Bomolecular Engineering, Sogang University; Interdisciplenary Program of Integrated Biotechnology, Sogang University, Seoul 121-742, Korea
e-mail: jwchoi@sogang.ac.kr

Y.J. Kim et al. (eds.), *Atmospheric and Biological Environmental Monitoring*,
DOI 10.1007/978-1-4020-9674-7_18, © Springer Science+Business Media B.V. 2009

2 On Chip Cell Cultivation Technologies

2.1 Cell Immobilization on Micro-and Nano-Structured Surface

The cell cultivation technology is one of the basic technologies in modern biology. Nevertheless, the controlling technologies of cell environments in chip including pH, nutrient, surface, air composition, and growth factor have not been sufficient to stabilize cell growing in vitro and to proliferate phenotype cell differentiation. Even though it is difficult to mimic tissue structure with mammalian cell in vitro (Bhatia et al. 1999), various technologies have been introduced to cell cultivation method. Nano-sized pattern technology using nano lithographical method has been introduced to single cell environmental optimization to stabilize cell immobilized on artificial surface and more ever to mimic in vivo environment. These nano-sized patterns on artificial surface have been achieved by conventional photolithographic technology, photo induced chemistry and several soft lithographic technology (nano imprint technology and micro fluidic patterning) (Folch and Toner 2000). With well-defined micro patterned surface, cell-ECM (Extracellular matrix) interaction can be controlled and results in well-defined whole cell matrix like tissue. Cell to cell adsorption and cell to surface adsorption processes are one of the key processes in most cell behaviors such as cell migration, cell differentiation, cell growth and cell apoptosis (Blau and Baltimore 1991, Ruoslahti and Obrink 1996). Focal adhesion (cell matrix adhesion) is sophisticated multi molecule complex existing in cell membrane. These are main part related to cell adhesion on ECM by integration of integrin and focal complex formed with several proteins due to integrin-ECM integration (Giancotti and Ruoslahti 1999). However, cell adhesion using ECM induces locally the accumulation of integrin, cytoplasmic protein. 3-dimensional structure fabricated using biodegradable and biocompatible materials patterned by lamination in nano scale, molding and photo induced polymerization can maximize the adhesion capability of cell on ECM (Balaban et al. 2001).

Cell immobilization on surface in vitro can be controlled by micro structured surface with different ways whereas nano structured surface can provide proper environments that cell adheres on surface to be cultivated on in vitro surface as mentioned above (Britland et al. 1992, Chen et al. 1997). For examples, we can control the cell spread by adjusting the hydrophilic properties of surface that cell adheres on and the grid pattern of substrates. This can be used to induce cell apoptosis when cell density is low whereas cell is dead when cell is placed in certain area with high density. If cell is located on specific region (intermediated connection between ECM and intergrin protein in cytoskeleton), adhered ECM can be fabricated to use other cytoskeleton tissue in order to form same cell morphology (Thery et al. 2005).

Adhesion force of cell on ECM can be measured by several ways. Particularly, measurement of distortion force of longitudinally aligned micro-sized arrays composed of elastic materials can be used to measure cell adhesion force. In case of soft muscle cell, the force on surface by its cell is about 100 nN and surface covered focal adhesion is also measured (Sentissi et al. 1986). Compared to conventional method to measure the force cell adhesion force, elastic material usage method has various advantages in aspects of its sensitivity and control ability. These kinds of force measuring method can be varied by changing post geometries without replacement of chemical properties (Tan et al. 2003).

2.2 Cell Immobilization Using Protein and Peptide

Various scientists in material engineering, surface chemistry, physics, biology, biochemistry, and pharmacology have had an effort to functionalize the specific interaction between surface and cell immobilization for past decades. In early stages, cell adhesion proteins such as fibronectin, collagen, laminin and so on were coated on surface to immobilize cell (Li et al. 1992, Miyata et al. 1991, Vohra et al. 1991, Thomson et al. 1991, Kaehler et al. 1989, Sentissi et al. 1986, Seeger and Klingman 1985). However, protein as surface property modification reagent has some problems in order to apply it to clinical fields. First, protein needs to be purified prior to use it for clinical purpose. Second, protein usually induces immuno-reaction in biological system if protein directly contact tissue or in body. Third, only few proteins can make cell oriented on surface. Therefore, it is impossible to use protein as adhesion reagent in long time usage because protein can be degraded as well as disadvantages mentioned above (Wei et al. 2001). Protein degradation is induced by hydrophobic interaction between surface and hydrophobic amino acid chain, which provides maximum adhesion force between them (Elbert and Hubbell 1996, Horbett and Lew 1994). These problems can be solved by introduction of peptide that can be immobilized on surface with nano-level. These have high thermal stability, durability under wide range of pH, and even long shelf life. And these can be modified and specialized with their structure changes (Ito et al. 1991, 23, Boxus et al. 1998, 24, Neff et al. 1998). Peptide can be integrated with high density because of its nano level size. Its high dense peptide surface can provide higher cell adhesion capabilities (Ruoslahti 1996). Peptide has only one cell recognition motif whereas ECM protein usually possesses various motifs. Therefore, peptide can react selectively with only one receptor in cell. This peptide containing characteristics mentioned above may be degraded slowly by enzymatic reaction as occurred in vivo. Cyclic peptide can be utilized instead of linear peptide in order to prevent its degradation as an inherent problem (Fields et al. 1998, Massia and Hubbell 1991, Aumailley et al. 1991, Gurrath et al. 1992, Ivanov et al. 1995). Cyclic peptide is perfectly stable against enzymatic reaction and sustained with original functionality for a long time. Most popular peptide contains RGD (arginine-glycine-aspartic acid) functional group known as cell adhesion inducing moiety in collagen, one of ECM protein. These characteristics can be applied directly to peptidomimetics considered nowadays (Sulyok et al. 2001, 32, Hersel et al. 2003). Therefore peptide in nano-sized level can play major role to develop biocompatible nano-platform by combination with nanotechnologies such as nanopatterning technology (Walter et al. 2006, Cavalcanti-Adam et al. 2006, Dalby et al. 2004)

Peptide can be modified as the characteristics of surface on which it is coated. Peptide can be self-assembled on Au surface to fabricate biocompatible layer by cystein is added on the end of peptides (Choi et al. 2005). Additionally, peptides can be modified and designed with certain cells and certain purpose peptide is used for (Hersel et al. 2003). Some peptides can be utilized as inducing materials to control cell motility, cell differentiation and so on.

3 Environmental Cell Based Biosensor

The essential technology to fabricate environmental biosensor is to quantitative measurement of the change of cell properties including cell death by nucleic acid or protein analysis. One of the popular tools to measure it is electrophoretic or chromatographic separation system composed of two major system, separator and detector. This system consists of microfluidic system, which has great advantages such as high sensitivity by the differences of chemical and physical properties. In case of chromatographic system, analytes start to be adsorbed and separated when sample is loaded in channel filled with adsorbent as stationary phase. Separated analyte is loaded into optical measuring system in which fluorescence and absorbance can be detected. This system can be also integrated with mass spectroscopy (Vilkner et al. 2004, Lion et al. 2003, Yin et al. 2005).

To simplify the analytical system to measure the effect of environment on cell properties by nucleic acid and protein, the microfluidic system has been applied to electrophoretic and chromatographic detection principles. For examples, specific capture system using antibody to analyze and measure specific analytes can make fluidic system simple (Verpoorte 2002). Antibody can be immobilized on the surface or solid structures inside channel of the integrated system (Sato et al. 2002). DNA can be also detected by DNA hybridization with similar system. For example, 1 μm of sample containing target DNA can be analyzed within 10 min by microfluidic system (Wei et al. 2005).

To measure such a DNA or protein for screening the properties of cell of interest, various optical dye such as fluorophor and light absorber are usually utilized. These labeling dye requires sophisticated process to conjugate dye to reporter probe. Label-free detection methods are one of the strong alternatives of fluorescence and luminescence based detection tools. For examples, Field effect transistor (FET) can be used to measure protein or DNA by the immobilization of target specific probe on its gate region. This FET measures the charge or capacity variation on antigen-antibody reaction or DNA hybridization (Gruner 2006, Zheng et al. 2005). Mechanical biosensing tools such as cantilever have been also introduced to measure their specific target in solution (Burg and Manalis 2003, Ziegler 2004).

Recently, the properties of cell have been directly measured to investigate the effect of chemicals of interest. Because these cell based biosensors use cells itself as target recognition modules, there are many advantages, comparing to DNA or protein detection. Cell contains complex biochemical pathways providing high sensitivity on various stimuli. Additionally, responses of cells on stimuli can provide functional analysis for a specific target. Because cell based biosensor detects physiological functions of cells, it can detect unknown target which never be measured by conventional biosensors utilizing DNA or protein. Such characteristics of cell based biosensors are outstanding advantages, totally different from biosensors using DNA hybridization, antigen-antibody reaction, and enzymatic reaction which depend on molecular events. These sensors can show high sensitivity and specificity on certain limited targets. Additionally, detection of target by molecular specific binding can not provide any information about the physiological effect of targets. For example, we can detect contaminant precisely with conventional biosensors if we know which contaminant is. However, when the sample is contaminated by unknown analytes, the effects of contamination on human bodies can be measured by only cell based biosensors. As mentioned above, high leveled information such as the physiological responses can be achieved by cell based biosensors.

Cell based biosensors can be categorized by secondary transduction method to monitor cell response. We explains various cell based biosensors in below.

3.1 Resistance Detection Based Environmental Biosensor

Resistance based biosensors utilize the difference of resistance between electrodes on which cell immobilizes by cell membrane adsorption. The difference of resistance between parallel electrodes can be used to detect cell adhesion, cell spread, and cell motility. Keese et al. mentioned about cell based biosensor to detect the degrees of cell motility and spread when it is exposed under certain material (Keese and Giaever 1994). Ehret used Interdigitated electrode structure to measure the difference of resistance by cell growth (Ehret et al. 1998). Borkholder fabricated new resistance based biosensor to measure

the difference of membrane resistance by target (Borkholder et al. 1997). This biosensor is based on resistance difference due to opening and closing of channel in cell membrane by a target. However, this biosensor can not distinguish the difference of membrane resistance by a target from resistance fluctuations due to cell movement and cell membrane flexibility.

New simple cell based biosensor is based on the resistance of whole cell layers immobilized on electrode surface. The resistance of immobilized cell layer can be varied by cell viability because cell response on target chemicals can affect their membrane structure. Using this simple biosensor, Cell viability and growth rate was analyzed without additionally equipment (Choi et al. 2007).

3.2 Metabolite Based Biosensor

Cell metabolites which can be measured induce the change of in vitro environment. Because acidification of cell medium is strongly related to cell metabolism, Measurement of pH of cell medium is typical metabolite based biosensor (Hafner 2000, Owicki and Parce 1992, Owicki et al. 1994). Other metabolite based biosensors are based on detection of oxygen consumption, generation of carbon dioxide and lactate, and thermal differentiation by microcalorimetry (Bousse 1996, Torisawa et al. 2003). The characteristic of this approach is that signal achieved from biosensor contains directly the metabolism effect of target. However, Metabolites biosensor detects are affected by other input such as temperature and nutrients as well as a target.

3.3 Optical Biosensor

Fluorescence detection is most popular method to measure cell behavior and response, combing their specific probe, which is utilized to measure membrane protein or specific region inside cell. Therefore the response on a target can be measured by fluorescent probe. Recent researches are focused on the development of cell based fluorescent method for high throughput screening (Torisawa et al. 2003, Biran and Walt 2002, Huang 2002, Prystay et al. 2001). Besides optical measurement by fluorescence probe, voltage sensitive dye can be measured to electrical action of cell (Chien and Pine 1991). Other optical cell based biosensors have been developed with recombinant bacteria which can illuminated by presence of toxic chemicals (Gil et al. 2000, Kim and Gu 2003). This fluorescence detection tools has great advantage that real responses within cells can be detected by fluorescence probe. However, potentially penetrable dye or reporter gene is required to develop fluorescence based optical biosensor (Daunert et al. 2000).

There is label free biosensor in optical biosensor category. Label free biosensor using SPR (Surface Rlasmon Resonance) and Ellipsometry can detect cell viability (Choi et al. 2005, Jyoung et al. 2006, Bae et al. 2006). This label free cell based biosensor using SPR is mentioned in the environmental application of cell based biochip.

3.4 Electrochemical Biosensor

Action potential is an indicator to detect cell function. The recording of transmembrane potential requires fine manipulations using micropipette type electrode and patch clamp. This is labor consuming and this penetration procedure is considered as the demerit to be overcome to develop robust biosensor (Connolly et al. 1990). Therefore, many researchers are focused on the measurement of action potential outside cell using microelectrodes. Action potential can generate an extracellular current which induces detectable local electric fields using electrode outside (Israel et al. 1984).

Tomas et al. published first electrical recording from electrically activated cells using micro electrode arrays (Thomas et al. 1972). They explained in situ extracellular action potential can be measured using multi electrodes from chick myocardial cell cluster. The important point of this research is non invasive micro electrode can be used to measure the behavior of cell cluster. These electrical responses were explained from neural tissue and dissociated neuron cell (Gross 1979, Gross et al. 1982, Mistry et al. 2002, Pine 1980). Other researchers have used micro electrode arrays to commercialize microbiosensors to measure a target (Gross et al. 1995, Pancrazio et al. 1998, DeBusschere and Kovacs 2001). Gross used neural network cultivated on micro electrode arrays to sense biochemical change

of spike trains (Gross et al. 1992). Borkholder suggested to use chick myocardial cell to analyze the action potential by chemical reaction (Borkholder et al. 1997).

This usage of microelectrodes to record extracellular response is now widely utilized in various fields. Commercial systems can be achieved from Plexon Inc. (Dallas, TX), Multichannel systems GmbH (Reutilingen, Germany), and Alpha MED Sciences, Co.(Tokyo, Japan).

We can realize the concept to sense toxic chemicals using extracellular potential monitoring by various researches mentioned above. There are, however, so many issues to be solved to develop such a cell based biosensor. For example, the reproducibility using several different cell groups to achieve same sensing result is too difficult (Banach et al. 2003). Offenhauser et al. suggested cell based biosensor consisting of cardiac myocytes cultivated on micro electrode array for pharmaceutical screening (Denyer et al. 1998). Sensitivity on single materials can be measured using single group of cells in this study. This study has also problem with reproducibility to achieve sufficient SNR (Signal to Noise Ratio) for morphological analysis. Therefore, versatility of living cell as cell density, cell kind and cell condition is required to be controlled for the development of cell based biosensor with high reproducibility.

4 The Environmental Application of cell Based Biosensor

We have mentioned about cell immobilization methods, cell cultivation methods, and various cell based biosensors. These technologies are required to achieve cell based biosensor to sense environmentally toxic chemicals. We explain two distinct cell based biosensors or chips which can detect chemicals without any reporter probes.

4.1 SPR Based Detection of Biological Toxicant on Cell Chip

Surface plasmon resonance (SPR) is an optical detection technique that measures molecular binding events at a metal surface by detecting changes in the local refractive index. SPR offers several advantages over conventional detection techniques such as fluorescence or ELISA (enzyme-linked immunosorbent assay) in biological analytical system. First, the detection of an analyte is label free and direct because the SPR measurements are based on refractive index changes. The analyte does not require any labels such as radioactive or fluorescent and can be detected directly without the need of sandwich structure. Second, the SRP measurements can be performed in real time. Last, SPR is a versatile technique to be able to detect analytes over a wide range of molecular weights and binding affinities. Because of its unique features, SPR has become a powerful tool for studying bioanalytical systems. For example, SPR has been used to investigate biological binding events such as protein–protein interactions, cellular ligation, protein–DNA interactions, and DNA hybridization, and to monitor in vitro the specific effects of environmental toxicants to biomolecules such as DNA, protein and cell, etc (Shliom et al. 2000, Quinn et al. 2000, Shumaker-Parry et al. 2004, Mischiati et al. 1999, Peterlinz and Georgiadis 1997, Peterson et al. 2002).

The understanding of modeling cell behavior based on only RNA or protein expression levels is much difficult, because cell is much more complicated system than sum of its components. Over the last several years, the analysis of living cells by itself has been increased for studying effects of drug and external stimuli on cell behavior. Because such analysis gave us invaluable information of cell with respect to toxicity of complex compounds and unexpected nonspecific effect, cell-based assay came to be an attractive method for those investigations above.

In this study, SPR based optical method for investigation of the effect of environmental toxicant such phenol on cells is introduced (Choi et al. 2005). An E. coli O157:H7 was immobilized on SPR surface using biocompatible chemical linker and its viability respect to phenol exposure concentration is investigated by the change of SPR angle. The experimental procedure of cell immobilization and biological toxicity detection is illustrated in Fig. 1.

SPR spectroscopy was carried out in the cited reference (Multiskop™, Optrel GbR, Germany) (Oh et al. 2003). A home-made flow cell with the Teflon™

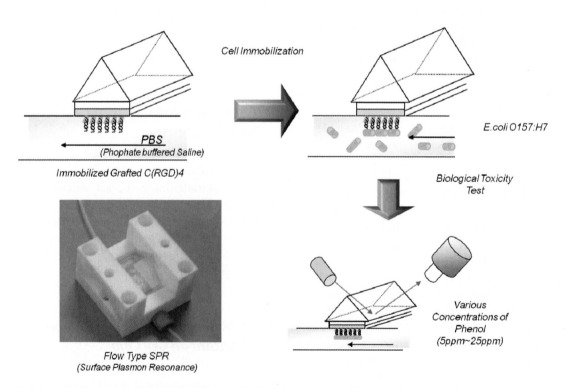

Fig. 1 Schematic diagram of cell immobilization for cell chip and biological toxicity detection of phenol

tubing and chamber was applied to biological toxicity detection. The flow-rate was controlled with peristaltic pump (MP-3N, EYELA, Japan). The SPR measurement system was composed of a He–Ne laser used as a light source to make a monochromatic light with a wavelength of 632.8 nm, a polarizer, and an analyzer, and photo-multiplier tube (PMT). Irrespective of detection type, a 90° glass prism same with the substrate was used as a configuration of Kretschmann ATR coupler (Kretschmann 1971). The plane face of the 90° glass prism was coupled to a glass slide via index matching fluid (Benzyl benzoate, Merck, Germany). The resolution of the angle reading of the goniometer was 0.001°.

The designed oligopeptide is utilized for cell immobilization and subsequent biological toxicity detection. It is a polypeptide network sequenced as C-R-G-D-R-G-D-R-G-D-R-G-D and has four branches (Grafted C(RGD)4). The oligopeptide immobilized on Au surface is anticipated to promote the binding of target cell. Because the chemical feature of oligopeptide may influence the binding capacity of cell immobilization. Gold (Au) substrate was prepared by DC magnetron sputtering on the cover glass (BK7, 18 mm × 18 mm, Superior, Germany) and designed glass (SF10, 18 mm × 18 mm × 1 mm, Korea Electro Optics, Korea) for flow type SPR, respectively. Before Au sputtering, chromium (Cr) was sputtered on the glasses to promote the adhesion of Au. The thickness of Au and Cr film was 43 and 2 nm, respectively. Before the fabrication of oligopeptide layer, the Au surface was cleaned using piranha solution as cited reference. A thin film of grafted C(RGD)4 on the Au surface was fabricated by submerging the substrate into the solutions for at least 24 h. And then, the prepared oligopeptide surfaces were washed with deionized distilled water and dried under N_2 gas.

The plasmon resonance characteristics of fabricated surfaces were investigated to monitor the biosurface fabrication and cell immobilization. Figure 2 shows the changes of SPR spectrum with respect to adsorbing synthetic peptides and subsequent binding of E. coli O157:H7 on Au substrate.

When the synthetic grafted peptide composed of (C(RGD)4) was absorbed on Au surface, the minimum position of plasmon angle was forwardly moved as shown in Fig. 2. The subsequent immobilization of E. coli O157:H7 made the plasmon angle significantly shifted. In principle, a surface plasmon is bound elec-

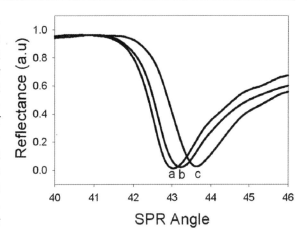

Fig. 2 Changes of SPR spectrum with respect to cell immobilization procedure. (**a**) bare gold, (**b**) grafted C(RGD)$_4$ network, (**c**) *E. coli* O157:H7

tromagnetic wave propagating at the metal-dielectric interface. The external laser field drives the free electron gas of metal in a distinct mode. The spatial charge distribution creates an electric field which is localized at the metal-dielectric interfaces. The Plasmon resonance is extremely sensitive to the interfacial architecture. An adsorption process leads to a shift in the Plasmon resonance and allows one to monitor the mass coverage at the surface with a high accuracy. The shift degree in the plamon resonance during adsorption process is also proportion to the mass coverage at the surface (Salamon et al. 1997). From the above results, it can be concluded that the prepared biosurface composed of synthetic oligopeptides was fabricated on Au surface and E. coli O157:H7 was immobilized on the fabricated oligopeptide platforms.

For the application to whole cell-based assay, the immobilized E. coli O157:H7 was applied to the exposure of phenol. Among the many criteria to distinguish between live and dead cell, one of the main features is the physical integrity of cellular plasma membrane (Huang et al. 2003). In case of the immobilized cell, live cell is expected to keep the physical integrity of the cellular membrane, and consequently, binding between cell and peptide-modified surface will be maintained. In contrast to that case above, the dead cell is expected to lose the physical integrity so that the amount of intracellular material will be decreased. When the immobilized cell is exposed to the toxic material such

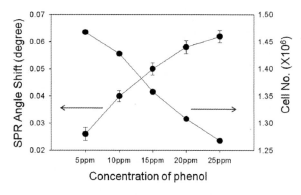

Fig. 3 Change of Plasmon angle shift and cell number of immobilized *E. coli* O157:H7 layer with respect to phenol exposure concentration for biological toxicity detection

as phenol, the change of mass and cellular density in accordance with the damage of membrane integrity can be occurred on the fabricated surface, which was investigated by SPR. E. coli O157:H7 immobilized on the grafted C(RGD)4 network was applied to biological toxicity detection. Figure 3 shows the change of plasmon angle shift with respect to the phenol concentration. As the concentration of phenol was increased, the plasmon angle shift was also increased. When the 5 ppm of phenol was exposed to the fabricated surface, the plasmon angle shift was slightly observed. Therefore, the detection limit of phenol was determined to be 5 ppm, at which the plasmon angle can be distinguished from the control. From these results, the biological toxicity detection can be successfully carried out based on immobilized cell and SPR. Also, the designed synthetic peptides could be applied to the fabrication of biocompatible surface for cell immobilization.

In conclusion, biosurface fabrication composed of synthetic oligopeptide was developed for the application to cell chip platform. Based on the principles of the interaction between peptide fragment (RGD) and cell surface, oligopeptide was designed to be self-assembled on Au surface, which was a cysteine-terminated oligopeptide. The designed oligopeptide was modified for the enhancement of immobilization degree. E. coli O157:H7 was immobilized on the designed oligopeptides and The E. coli O157:H7 was applied to biological toxicity detection such as phenol exposure. Because the damage following

exposure of lethal compound can be a significant impact on the total amount of intracellular biomaterials in the immobilized E. coli O157:H7, plasmon angle shift was successfully observed with respect to the exposure concentration of phenol. The detection limit was determined to be 5 ppm of phenol. The proposed cell immobilization method using self-assembly technique can be applied to construct the cell chip for the diagnosis, drug detection, and on-site monitoring.

4.2 Electrochemical Detection of Biological Toxicant on Cell Chip

Electrical detection methods have been developed to overcome the disadvantages or an optical biosensors, particularly SPR based biosensors because electrical detection requires only simple and small device whereas SPR needs desk top sized readers. Additionally, it is easily connected to other electronics.

In this study, electrical detection system for investigation of the effect of environmental toxicant such as phenol on animal cells is explained. The HepG2(Human hepatocellular liver carcinoma cell line) was used to measure phenol, as known as environmentally toxic chemicals and CV(Cyclic voltammetry, CHI 660A) was used for the electrical instrument for investigation of cells. Potential range of the CV is from 0.4 V to -0.2 V and the scan rate is 100 mV/s. The cells are incubated for 2 days in the DMEM solution. We introduced phenol containing sample solution to cell chip which consists of electrodes HepG2 cell immobilized on. And CV was utilized to record the response of cell on phenol as a voltage-current correlation. As a result, the unusual current peak was generated at certain potential when phenol is introduced, which current peak was proportional to cell density. In the Fig. 4, the optical microscopic picture was shown that the different morphology of cells as the addition of different concentration of phenol.

As the concentration of phenol is going up, the immobilized cells are going down. In the Fig. 5, electrochemical signal shows that the signal is linearly increased with increasing cell number. Thus, the electrical signal is correlation with concentration of cells. This result implies the cell viability affected by phenol

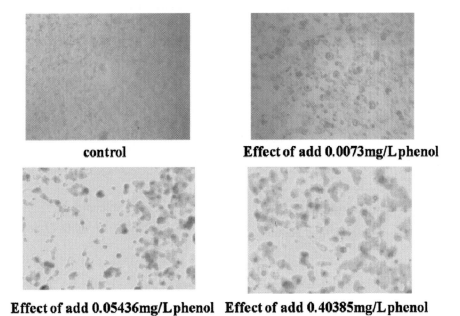

control   Effect of add 0.0073mg/L phenol

Effect of add 0.05436mg/L phenol   Effect of add 0.40385mg/L phenol

Fig. 4 The HepG2 cell morphology as the addition of phenol

concentration or cell density is measured by the magnitude of current peak generated at certain voltage. Therefore, we would get the viability of cells using electrical detection system. To apply this phenomenon, phenol was introduced to cell culture well in prepared concentration, 0.003 mg/L, 0.03 mg/L, 0.3 mg/L, 3 mg/L and 30 mg/L.

As a result, we can confirm the electrical signal in the Fig. 6, and the electrical peak current is going down, as the concentration of phenol is going up.

The electrical signal of cell is quantificated in the Fig. 7, then the correlation between cell number and concentration of toxicant, phenol, could be found. The proposed electrical detection system for biological

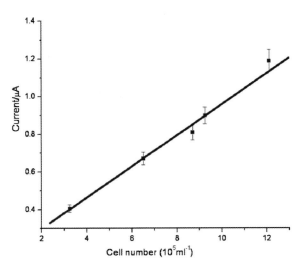

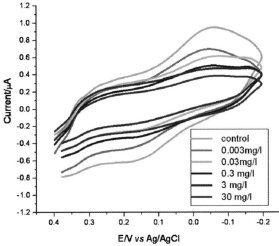

Fig. 5 Electrochemical analysis of HepG2 cell signal as the increasing number of cells

Fig. 6 Electrochemical analysis of environmental toxicant effect, phenol, of HepG2 cells

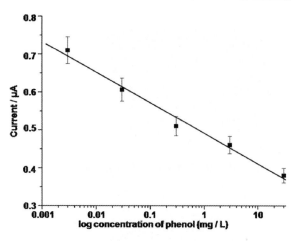

Fig. 7 Quantification of electrochemical HepG2 cell signaling as the environment l toxic effect

toxicant system can be applied to cell chip for detecting versatile type of toxicant effects and on-chip real time environmental monitoring system.

5 Prospects

Environmental cell based biochips have been researched in scientific and commercial fields because they have capabilities which can provide the physiological information by measuring cell behavior against stimuli such as toxic chemicals and electrical stress. However, such a cell based biosensors and biochips still have problems in aspects of reproducibility, signal interpretation, and durability for storage to realize them. To resolve these problems, the interdisciplinary researches among biology, mechanical engineering, chemical engineering, electrical engineering, chemistry, and so on are strongly required because the cell based biosensor or biochip can be developed with several key technologies as mentioned above. For examples, we require the information about cell membranes and surface, as well as the methods for nanopatterning, surface modification just for cell immobilization process.

However, because cell based biosensors and biochips have been continuously upgraded rapidly by focused researches and developments, we expect cell based biosensor can be utilized directly to monitor environmental toxicities at contaminated fields in near future.

Abbreviations

ECM:	ExtraCellular Matrix
DNA:	Deoxiribo Nucleic Acid
FET:	Field Effect Transistor
SNR:	Signal to Noise Ratio
SPR:	Surface Plasmon Resonance
ELISA:	Enzyme Linked ImmunoSorbent Assay
RNA:	Ribo Nucleic Acid
PMT:	Photo-Multiplier Tube
Au:	Gold
CV:	Cyclic Voltammetry
DMEM:	Dulbecco's Modified Essential Medium

References

Aumailley M, Gurrath M, Muller G, Calvete J, Timpl R, Kessler H et al. (1991) Arg-Gly Asp constrained within cyclic pentapeptides, strong and selective inhibitors of cell adhesion to vitronectin and laminin fragment P1. FEBS Lett 291:50–54

Bae YM, Park KW, Oh BK, Lee WH, Choi JW et al. (2006) Immunosensor for detection of Salmonella typhimurium based on imaging ellipsometry. Colloid Surf A 257–258: 19–23

Balaban NQ, Schwarz US, Rivelin D, Goichberg P, Tzur G, Sabanay L, Mahalu D, Safran S, Bershadsky A, Addadi L, Geiger B et al. (2001) Force and focal adhesion assembly: a close relationship studied using elastic micropatterned substrates. Nature Cell Biol 3:466–472

Banach K, Halbach MD, Hu P, Hescheler J, Egert U et al. (2003) Development of electrical activity in cardiac myocyte aggregates derived from mouse embryonic stem cells. Am J Physiol – Heart Circ Physiol 284:H2114–2123

Bhatia SN, Balis UJ, Yarmush ML, Toner M et al. (1999) Effect of cell–cell interactions in preservation of cellular phenotype: cocultivation of hepatocytes and nonparenchymal cells. FASEB J 13:1883–1900

Biran I, Walt DR (2002) Optical imaging fiber-based single live cell arrays: a high-density cell assay platform. Anal Chem 74:3046–3054

Blau HM, Baltimore D (1991) Differentiation requires continuous regulation. J Cell Biol 112:781–783

Borkholder DA, Bao J, Maluf NI, Perl ER, Kovas GTA et al. (1997) Microelectrode arrays for stimulation of neural slice preparations. J Neurosci Methods 77:61–66

Bousse L (1996) Whole cell biosensors. Sens Actu B 34: 270–275

Boxus T, Touillaux R, Dive G, Marchand-Brynaert J et al. (1998) Synthesis and evaluation of RGD peptidomimetics aimed at surface bioderivatization of polymer substrates. Bioorg Med Chem 6:1577–1595

Britland S, Clark P, Connolly P, Moores G et al. (1992) Micropatterned substratum adhesiveness – a model for morphogenetic cues controlling cell behavior. Exp Cell Res 198:124–129

Burg TP, Manalis SR (2003) Suspended microchannel resonators for biomolecular detection. Appl Phys Lett 83:2698–2700

Cavalcanti-Adam EA, Micoulet A, Blummel J, Auernheimer J, Kessler H, Spatz JP et al. (2006) Lateral spacing of integrin ligands influences cell spreading and focal adhesion assembly. Euro J Cell Biol 85:219–224

Chen CS, Mrksich M, Huang S, Whitesides GM, Ingber DE et al. (1997) Geometric control of cell life and death. Science 276:1425–1428

Chien CB, Pine J (1991) An apparatus for recording synaptic potentials from neuronal cultures using voltage-sensitive fluorescent dyes. J Neurosci Methods 38:93–105

Choi JW, Lee W, Lee DB, Park CH, Kim JS, Jang YH, Kim Y. et al. (2007) Electrochemical detection of pathogen infection using cell chip. Envion Monit Assess 129:37–42

Choi JW, Park KW, Lee DB, Lee W, Lee WH et al. (2005) Cell immobilization using self-assembled synthetic oligopeptide and its application to biological toxicity detection using surface plasmon resonance. Biosens Bioelectron 20:2300–2305

Connolly P, Clark P, Curtis ASG, Dow JAT, Wilkinson CDW et al. (1990) An extracellular microelectrode array for monitoring electrogenic cells in culture. Biosens Bioelectron 5:223–234

Dalby MJ, Pasqui D, Affrossman S et al. (2004) Cell response to nano-islands produced by polymer demixing: a brief review. IEE Proc -Nanobiotechnol 151:53–61.

Daunert S, Barrett G, Feliciano JS, Shetty RS, Shrestha S, Smith-Spencer W et al. (2000) Genetically engineered whole-cell sensing systems: coupling biological recognition with reporter genes. Chem Rev 100:2705–2738

DeBusschere BD, Kovacs GTA (2001) Portable cell-based biosensor system using integrated CMOS cell-cartridges. Biosens Bioelectron 16:543–556

Denyer MCT, Riehle M, Britland ST, Offenhauser A et al. (1998) Preliminary study on the suitability of a pharmacological bio-assay based on cardiac myocytes cultured over microfabricated microelectrode arrays. Med Biol Eng Comput 36:638–644

Ehret R, Baumann W, Brischwein M, Schwinde A, Wolf B et al. (1998) On-line control of cellular adhesion with impedance measurements using interdigitated electrode structures. Med Biol Eng Comput 36:365–370

Elbert DL, Hubbell JA (1996) Surface treatments of polymers for biocompatibility. Annu Rev Mater Sci 26:365–394

Fields GB, Lauer JL, Dori Y, Forns P, Yu Y-C, Tirrell M et al. (1998) Protein like molecular architecture: biomaterial applications for inducing cellular receptor binding and signal transduction. Biopolymers 47:143–151

Folch A, Toner M (2000) Microengineering of cellular interactions. Annu Rev Biomed Eng 2:227–256

Giancotti FG, Ruoslahti E (1999) Integrin Signaling. Science 285:1028–1033

Gil GC, Mitchell RJ, Chang ST, Gu MB et al. (2000) A biosensor for the detection of gas toxicity using a recombinant bioluminescent bacterium. Biosens Bioelectron 15:23–30

Gross GW (1979) Simultaneous single unit recording in vitro with a photoetched laser deinsulated gold multimicroelectrode surface. IEEE Trans Biomed Eng 26:273–279

Gross GW, Rhoades B, Jordan R et al. (1992) Neuronal networks for biochemical sensing. Sens Actu B 6:1–8

Gross GW, Rhoades BK, Azzazy HME, Wu M-C et al. (1995) The use of neuronal networks on multielectrode arrays as biosensors. Biosens Bioelectron 10:553–567

Gross GW, Williams AN, Lucas JH et al. (1982) Recording of spontaneous activity with photoetched microelectrode surfaces from mouse spinal neurons in culture. J Neurosci Methods 5:13–22

Gruner G (2006) Carbon nanotube transistors for biosensing applications. Anal Bioanal Chem 384:322–335

Gurrath M, Muller G, Kessler H, Aumailley M, Timpl R et al. (1992) Conformation/activity studies of rationally designed potent antiadhesive RGD peptides. Eur J Biochem 210:911–921

Hafner F (2000) Cytosensor Microphysiometer: technology and recent applications. Biosens Bioelectron 15:149–158

Hersel U. Dahmen C, Kessler H et al. (2003) RGD modified polymers: biomaterials for stimulated cell adhesion and beyond. Biomaterials 24:4385–4415

Horbett TA, Lew KR (1994) Residence time effects on monoclonal antibody binding to adsorbed fibrinogen. J Biomater Sci Polym Ed 6:15–33

Huang SG (2002) Development of a high throughput screening assay for mitochondrial membrane potential in living cells. J Biomol Screen 7:383–389

Huang Y, Sekhon NS, Borninski J, Chen N, Rubinsky B et al. (2003) Instanteous, quantitative single-cell viability assessment by electrical evaluation of cell membrane integrity with microfabricated devices, Sens Actu A 105: 31–39

Israel DA, Barry WH, Edell DJ, Mark RG et al. (1984) An array of microelectrodes to stimulate and record from cardiac cells in culture. Am J Physiol 247:H669–H674

Ito Y, Kajihara M, Imanishi Y et al. (1991) Materials for enhancing cell adhesion by immobilization of cell-adhesive peptide. J Biomed Mater Res 25:1325–1337

Ivanov B, Grzesik W, Robey FA et al. (1995) Synthesis and use of a new bromoacetyl-derivatized heterotrifunctional amino acid for conjugation of cyclic RGD-containing peptides derived from human bone sialoprotein. Bioconjug Chem 6:269–277

Jyoung JY, Hong S, Lee W, Choi JW et al. (2006) Immunosensor for the detection of Vibrio cholerae O1 using surface plasmon resonance. Biosens Bioelectron 21:2315–2319

Kaehler J, Zilla P, Fasol R, Deutsch M, Kadletz M et al. (1989) Precoating substrate and surface configuration determine adherence and spreading of seeded endothelial cells on polytetrafluoroethylene grafts. J Vasc Surg 9:535–541

Keese CR, Giaever IA (1994) Biosensor that monitors cell morphology with electrical fields. IEEE Eng Med Biol Magazine 13:402–408

Kim BC, Gu MB (2003) A bioluminescent sensor for high throughput toxicity classification. Biosen Bioelectron 18:1015–1021

Kretschmann E (1971) Die bestimmung optischer konstanten von metallen durch anregung von oberfachenplasmaschwingungen, Zeitschrift fur Physik 241: 313–324

Li JM, Menconi MJ, Wheeler HB, Rohrer MJ, Klassen VA, Ansell JE, Appel MC et al. (1992) Precoating expanded polytetrafluoroethylene grafts alters production of endothelial cell-derived thrombomodulators. J Vasc Surg 15:1010–1017

Lion N, Rohner TC, Dayon L, Arnaud IL, Damoc E, Youhnovski N, Wu Z-Y, Roussel C, Josserand J, Jensen H, Rossier JS, Przybylski M, Girault HH et al. (2003) Microfluidic systems in proteomics. Electrophoresis 24:3533–3562

Massia SP, Hubbell JA (1991) An RGD spacing of 440 nm is sufficient for integrin – 3-mediated fibroblast spreading and 140 nm for focal contact fiber formation. J Cell Biol 114:1089–1100

Mischiati C, Borgatti M, Bianchi N, Rutigliano C, Tomassetti M, Feriotto G, Gambari R (1999) Interaction of the human NF-kappa B p52 transcription factor with DNA-PNA hybrids mimicking the NF-kappa B binding sites of the human immunodeficiency virus type 1 promoter, J Biol Chem 274:33114–33122

Mistry SK, Keefer EW, Cunningham BA, Edelman GM, Crossin KL et al. (2002) Cultured rat hippocampal neural progenitors generate spontaneously active neural networks. Proc Natl Acad Sci 99:1621–1626

Miyata T, Conte MS, Trudell LA, Mason D, Whittemore AD, Birinyi LK et al. (1991) Delayed exposure to pulsatile shear stress improves retention of human saphenous vein endothelial cells on seeded ePTFE grafts. J Surg Res 50:485–493

Neff JA, Caldwell KD, Tresco PA et al. (1998) A novel method for surface modification to promote cell attachment to hydrophobic substrates. J Biomed Mater Res 40:511–519

Oh BK, Lee W, Lee WH, Choi JW et al. (2003) Nano-scale probe fabrication using self-assembly technique and application to detection of Escherichia coli O157:H7. Biotechnol Bioprocess Eng 8:227–232

Owicki JC, Bousee LJ, Hafeman DG, Kirk GL, Olson JD, Wada G, Parce JW et al. (1994) The light-addressable potentiometric sensor: principles and biological applications. Annu Rev Biophys Biomol Struct 23:87–113

Owicki JC, Parce JW (1992) Biosensors based on the energy metabolism of living cells: the physical chemistry and cell biology of extracellular acidification. Biosen Bioelectron 7:255–272

Pancrazio JJ, Bey PP Jr, Cuttino DS, Kusel JK, Borkholder DA, Shaffer KM, Kovacs GTA, Stenger DA et al. (1998) Portable cell-based biosensor system for toxin detection. Sens Actu B 53:179–185

Peterlinz KA, Georgiadis RM (1997) Observation of hybridization and dehybridization of thiol-tethered DNA using two-color surface plasmon resonance spectroscopy. J Am Chem Soc 119:3401–3402

Peterson AW, Wolf LK, Georgiadis RM et al. (2002) Hybridization of mismatched or partially matched DNA at surfaces. J Am Chem Soc 124:14601–14607

Pine J (1980) Recording action potentials from cultured neurons with extracellular microcircuit electrodes. J Neurosci Methods 2:19–31

Prystay L, Gagne A, Kasila P, Yeh LA, Banks P et al. (2001) Homogeneous cell-based fluorescence polarization assay for the direct detection of cAMP. J Biomol Screen 6:75–82

Quinn JG, O'Neill S, Doyle A, McAtamney C, Diamond D, MacCraith BD, O'Kennedy R et al. (2000) Development and application of surface plasmon resonance-based biosensors for the detection of cell-ligand interactions. Anal Biochem 281:135–143

Ruoslahti E (1996) RGD and other recognition sequences for integrins. Annu Rev Cell Dev Biol 12:697–75

Salamon Z, Macleod HA, Tollin G et al. (1997) Surface Plasmon resonance spectroscopy as a tool for investigating the biochemical and biophysical properties of membrane protein systems. II: Application to biological systems. Biochimica et Biophysica Acta 1331:131–152

Sato K, Yamanaka M, Takahashi H, Tokeshi M, Kimura H, Kitamori T et al. (2002) Microchip-based immunoassay system with branching multichannels for simultaneous determination of interferon-γ. Electrophoresis 23:734–739

Seeger JM, Klingman N (1985) Improved endothelial cell seeding with cultured cells and fibronectin-coated grafts. J Surg Res 38:641–647

Sentissi JM, Ramberg K, O-Donnell Jr TF, Connolly RJ, Callow AD et al. (1986) The effect of flow on vascular endothelial cells grown in tissue culture on polytetrafluoroethylene grafts. Surgery 99:337–343

Shliom O, Huang M, Sachais B, Kuo A, Weisel JW, Nagaswami C, Nassar T, Bdeir K, Hiss E, Gawlak S et al. (2000) Novel interactions between urokinase and its receptor. J Biol Chem 275:24304–24312

Shumaker-Parry JS, Aebersold R, Campbell CT et al. (2004) Parallel, quantitative measurement of protein binding to a 120-element double-stranded DNA array in real time using surface plasmon resonance microscopy. Anal Chem 76:2071–2082

Sulyok GAG, Gibson C, Goodman SL, Holzemann G, Wiesner M, Kessler H et al. (2001) Solid-phase synthesis of a nonpeptide RGD mimeticlibrary: New selective – 3 integrin antagonists. J Med Chem 44:1938–1950

Tan JL, Tien J, Pirone DM, Gray DS, Bhadriraju K, Chen CS et al. (2003) Cells lying on a bed of microneedles: an approach to isolate mechanical force. Proc Natl Acad Sci USA 100:1484–1489

Thery M, Racine V, Pepin A, Piel M, Chen Y, Sibarita J-B, Bornens M et al. (2005) The extracellular matrix guides the orientation of the cell division axis. Nature Cell Biol 7:947–953

Thomas CA Jr, Springer PA, Loeb GE, Berwald-Netter Y, Okun LMA et al. (1972) Miniature microelectrode array to monitor the bioelectric activity of cultured cells. Exp Cell Res 74:61–66

Thomson GJ, Vohra RK, Carr MH, Walker MG et al. (1991) Adult human endothelial cell seeding using expanded polytetrafluoroethylene vascular grafts: a comparison of four substrates. Surgery 109:20–27

Torisawa YS, Kaya T, Takii Y, Oyamatsu D, Nishizawa M, Matsue T et al. (2003) Scanning electrochemical microscopy-based drug sensitivity test for a cell culture integrated in silicon microstructures. Anal Chem 75:2154–2158

Verpoorte E (2002) Microfluidic chips for clinical and forensic analysis. Electrophoresis 23:677–712

Vilkner T, Janasek D, Manz A et al. (2004) Micro total analysis systems. Recent developments. Anal Chem 76:3373–3385

Vohra R, Thomson GJ, Carr HM, Sharma H, Walker MG et al. (1991) Comparison of different vascular prostheses and matrices in relation to endothelial seeding. Br J Surg 78:417–420

Walter N, Selhuber C, Kessler H, Spatz JP et al. (2006) Cellular unbinding forces of initial adhesion processes on nanopatterned surfaces probed with magnetic tweezers. Nanolett 6:398–402

Wei CW, Cheng JY, Huang CT, Yen MH, Young TH et al. (2005) Using a microfluidic device for 1 μl DNA microarray hybridization in 500 s. Nucleic Acids Res 33:e78

Wei N, Klee D, Hocker H et al. (2001) Konzept zur bioaktiven Ausr.ustung von Metallimplantatoberf.achen. Biomater 2:81–86

Yin NF et al. (2005) Microfluidic chip for peptide analysis with an integrated HPLC column, sample enrichment column, and nanoelectrospray tip. Anal Chem 77:527–533

Zheng GF, Patolsky F, Cui Y, Wang WU, Lieber CM et al. (2005) Multiplexed electrical detection of cancer markers with nanowire sensor arrays. Nature Biotechnol 23:1294–1301

Ziegler C (2004) Cantilever-based biosensors. Anal Bioanal Chem 379:946–959

Fabrication of Electrophoretic PDMS/PDMS Lab-on-a-chip Integrated with Au Thin-Film Based Amperometric Detection for Phenolic Chemicals

Hidenori Nagai, Masayuki Matsubara, Kenji Chayama, Joji Urakawa, Yasuhiko Shibutani, Yoshihide Tanaka, Sahori Takeda and Shinichi Wakida

Abstract A new microfluidic device with a fully integrated electrochemical detection (ECD) system was fabricated using soft lithography. The electrophoretic device made of polydimethylsiloxane (PDMS) comprises an off-channel ECD with a narrow channel decoupler using a duplex replication process. We fabricated a PDMS device which was composed of a PDMS microchannel and a PDMS-coated glass integrated with Au thin-film based electrodes, using an original fabrication process. As an application of the fabricated device based on microchip capillary electrophoresis with ECD, we investigated the simultaneous detection of phenolic chemicals, known to be endocrine disruptors. We achieved simultaneous detection of 4-t-butylphenol and bisphenol A using the micellar electrokinetic chromatographic (MEKC) mode.

Keywords Microchip · Micellar electrokinetic chromatography · Electrochemical detection

1 Introduction

Several chemicals have been suspected of having endocrine disrupting effects (Corborn et al. 1996). Among the 67 candidates as endocrine disruptors (EDs) postulated in the strategic program in Japan, only 3 types of phenolic chemical such as 4-nonylphenol (4-NP), 4-(1,1,3,3-tetramethylbutyl) phenol (4-tOP) and bisphenol A (BPA) were postulated to have strong endocrine disrupting effects on fish. The other phenolic chemical in the candidate EDs, 4-tert-butylphenol (4-tBP), is under detailed investigation for possible study of endocrine disrupting effects in Japan. The development of on-site monitoring for these phenolic chemicals has become more important recently.

For on-site monitoring of the phenolic chemicals in water, microfluidic devices based on microchip capillary electrophoresis (MCE), coupled with electrochemical detection (ECD) (Chen and Wang 2006), have attracted much attention because of the small instrumentation requirements with the advantages of not requiring reagents, a high-throughput screening and μl levels of waste volumes. MCE is an advanced method in a conventional capillary electrophoresis (CE). In the case of CE, charged ions are separated in a hollow fused silica capillary filled with buffer solution. Since the narrow capillary is able to prevent thermal diffusion and convection in itself, the separation efficiency is extremely high. The separation by electrophoresis is based on differences in electrophoretic mobility μ_{ep} in an electric field. In addition, a fundamental constituent of CE separation is electroosmotic flow (EOF). EOF is the bulk flow of liquid in the capillary and is a consequence of surface charge on the interior capillary wall. The EOF results from the effect of the applied electric field on the solution double-layer at the wall. The magnitude of the EOF can be expressed in term of EOF mobility μ_{EOF} by

$$\mu_{EOF} = \varepsilon\xi/\eta \tag{1}$$

where ξ is zeta potential, ε and η are dielectric constant and viscosity of the solution, respectively. The migra-

S. Wakida (✉)
Human Stress Signal Research Center (HSS), National Institute of Advanced Industrial Science and Technology (AIST), Midorigaoka 1-8-31, Ikeda, Osaka 563-8577, Japan
e-mail: s.wakida@aist.go.jp

Y.J. Kim et al. (eds.), *Atmospheric and Biological Environmental Monitoring*, DOI 10.1007/978-1-4020-9674-7_19, © Springer Science+Business Media B.V. 2009

tion time t and other experimental parameters can be used to calculate the apparent solute mobility μ_a using

$$\mu_a = \mu_{ep} + \mu_{EOF} = l/tE \qquad (2)$$

where l, t, and E are capillary effective length, migration time, and field strength, respectively.

When two solutes are completely separated in CE, the difference in the electrophoretic mobility have to be enough large to divide the two peaks. For a Gaussian peak, peak width at inflection point is depends on each standard deviations by diffusion and injection width and the interaction between solute and the interior wall of capillary. The peak width w_{tot} is described as

$$w_{tot} = 2(\sigma_{dif}^2 + \sigma_{inj}^2 + \sigma_{int}^2)^{1/2}$$
$$= 2(2Dt + w_{inj}^2/12 + \sigma_{int}^2)^{1/2} \qquad (3)$$

where D is diffusion coefficient of the solute and σ is standard deviation of the peak. Subscripts dif, inj, and int indicates diffusion, injection, and interaction, respectively. When the interaction with the interior wall is negligible degree, the number of theoretical plates, N, can be obtained by

$$N = (l/\sigma)^2 = \frac{l^2}{2Dt + w_{inj}^2/12} \qquad (4)$$

Thus, the injected sample width is important for the performance of separation. In the case of MCE, sample plug is miniaturized within several hundred μm. Therefore, the number of theoretical plates is extremely enhanced in MCE system. The complete separation would be also carried out even though the channel length is short.

We have already investigated several types of on-chip ECD for microfluidic devices with potentiometric detection (Masadome et al. 2005). We have also investigated a preliminary separation of the phenolic chemicals, BPA and three alkylphenols employing MCE/UV detection using micellar electrokinetic chromatographic (MEKC) mode (Wakida et al. 2006).

MEKC is ultimately a useful method for the separation of relatively hydrophilic neutral-analytes as a mode of capillary electrophoresis (Kim and Terabe 2003, Pappas et al. 2005). Surfactants, i.e., sodium dodecyl sulfate (SDS) (Monton et al. 2003), cethyltriammonium bromide (CTAB) (Muijselaar et al. 1997), and cethyltriammonium chloride (CTAC)

(Kim et al. 2001), were used as pseudo-stationary phases by mixing above the critical micelle concentration into a running buffer. The chromatographic principle of MEKC is that the formed micelle captures analytes, depending upon the capacity factor between the micelle and the analytes (Wakida et al. 2006).

The microfluidic devices for electrophoresis coupled with ECD have been fabricated in glasses and polymers using a variety of micromaching technologies. Most of the electrophoretic devices coupled with ECD encorporate hybrid type devices composed of a PDMS top plate and a glass base plate (Vandaveer et al. 2004, Wu et al. 2003). Although the PDMS/glass hybrid devices are easily fabricated, the EOF is non-uniform and results in incorrect separation, essentially due to the difference in zeta potential between the PDMS and glass surfaces as expressed in Eq.1. Therefore, the application is limited to within gel electrophoresis in suppressed EOF for the separation of macromolecules. To apply the electrophoretic devices to the separation of small molecules using EOF, the microchannel should be constructed with a single material.

In the fabrication of the glass-based devices it is well-known to be difficult to cover the microchannel avoiding roughness at the boundary surface, in particular due to the thickness of electrode layer. To fabricate glass-based microfluidic devices coupled with ECD, several micromachining technologies were developed (Schoning et al. 2005, Zeng et al. 2002). However, complications in the construction and fabrication processes are unavoidable.

Polymer-based microfluidic devices coupled with ECD have potential advantage of relatively simple fabrication, with low cost with respect to glass-based devices. While PDMS is especially suitable to fabricate the microchannel structure due to the elastomeric property, as well as the fact that the PDMS/glass hybrid devices are bonded using elastomeric adhesion, there are only a few reports of PDMS-based electrophoretic devices coupled with ECD. Thin-film electrodes can easily be deposited onto PDMS. However, PDMS is such a soft material that the thin-film electrodes will sustain a lot of cracking during the handling of the device. Consequently, the cracks in the electrode lead to electrical disconnection.

Moreover, such a metal layer only weekly adheres to PDMS surface (Schmid et al. 2003, Loo et al. 2002), and is easily peeled off during the fabrication processes. Nuzzo et al. overcame the drawback of metal thin layer onto PDMS by embedding the electrodes into PDMS matrix, and demonstrated MEKC in PDMS-based electrophoretic devices (Lee et al. 2005). Moreover, high electric field and electrophoretic current can give rise to drastic damage to the thin-film electrodes, such as fragmentation and delamination from the PDMS.

In addition, such high voltages interfere strongly with the detection of ECD, resulting in a raised background current and shifted redox potential of analytes. When the electrophoretic current is large, it is necessary to minimize the interference and to prevent the detector from damage due to surge spikes. To resolve the serious interference, end-channel detection has been demonstrated (Zeng et al. 2002, Lee et al. 2005), and applied to most separations of phenolic chemicals (Scampicchio et al. 2004, Wang et al. 2003, 2002, Schwarz et al. 2001). The working electrode (WE) for ECD is located beyond the end of microchannel. Using end-channel detection, the current density in the electrophoretic operation obviously decreases around the WE with microchannel, because the resistance is inversely proportional to cross-sectional area of running buffer reservoir. As the analyte is diffused and diluted into the wide outlet beyond the microchannel, the sensitivity and resolution of separation obviously decrease with increasing distance between the WE and the microchannel end (Vandaveer et al. 2004). Therefore, the conformation of end-channel detection is not ideal for electrophoretic separation. It is considered that the ideal construction of ECD involves off-channel conformation, which implies that the WE is positioned inside the microchannel. As an interface to isolate the high electrophoretic current and ECD inside microchannel, several types of decoupler have already been developed for microfluidic devices coupled with ECD. The excellent decoupler can be fabricated with a Pd electrode (Schoning et al. 2005, Chen et al. 2001). However, with the Pd-based decoupler, it may be inconvenient and difficult to fabricate the electrodes incorporated inside the microchannel. Instead of a Pd-based decoupler, small holes were adopted to isolate ECD from high voltage (Rossier et al. 2000). In this case, the holes were perforated vertically along the separation channel. Therefore, the preparation of holes requires an advanced technique, such as UV ablation.

In this paper, we demonstrate a new decoupler employing a narrow channel, fabricated using a new soft-lithography. As this soft-lithography of PDMS will be suitable for fabrication of narrow channels, the use of PDMS might have considerable impact, should we develop the fabrication of thin-film electrodes with a narrow channel decoupler. We also describe a new PDMS/PDMS device comprising a durable PDMFS base plate, integrated with Au thin-film electrodes using a simple modification of the conventional fabrication procedure. The Au thin-film was deposited directly onto soft PDMS surface, and then, patterned to the electrode shape without any mechanical stress to cause cracking. The fabricated PDMS-based electrophoretic device was also demonstrated for use in the simultaneous detection of phenolic chemicals.

2 Experimental

2.1 Materials

Commercial PDMS (Sylgard 184) was purchased from Dow Corning (Midland, MI, USA). AZ 4620 photoresist and AZ 300 MIF were obtained from Clariant (Somerville, NJ, USA). SU-8 50 photoresist and SU-8 developer were purchased from Microchem (Newton, MA, USA). Silicon wafers and quartz glasses wafer with a diameter of 4 in were obtained from Shin-etsu chemical (Tokyo, Japan). As photomasks for microchannel and electrode, two transparency films printed with the desired pattern using a resolution 2,400 dpi were obtained from Kansai Graphics (Osaka, Japan) and attached to 5 in square and 0.09 in thick quartz plates (from Shin-etsu Chemicals). Gold wire (99.95 %) was purchased from Nilaco (Tokyo, Japan). Chromium powder (99.9 %) was purchased from Mitsuwa Chemicals (Osaka, Japan). The other chemicals were purchased from Kanto chemicals (Tokyo, Japan). 25 mg/ml iodine and 100 mg/ml potassium iodate solution was prepared for Au etchant. 170 mg/ml diammonium cerium(IV) nitrate in 5 %v/v perchloric acid solution was prepared for Cr etchant.

2.2 Instruments

A potentiostat HSV-100MCE (Hokuto Denko, Japan) was used for electrochemical characterization of the Au thin-film electrodes andECD. A programmable

8-channel High Voltage Sequencer, HVS448 (Lab-Smith, USA) was used for electrophoretic separation. A vacuum evaporator (VC-21, Yamato, Japan) was used for the coating of Au or Cr onto the substrate. A laser beam direct drawing (LBDD) system (Hakuto, Japan) equipped with a He-Cd laser, which wavelength was 442 nm, was used for the fabrication of photomask for narrow channel.

2.3 Fabrication of Photomask for the Narrow Channel

As a photomask for narrow channel, one side of a 5 in square and 0.09 in thick quartz plate was coated with Cr and Au subsequently using vacuum evaporator and patterned with almost the same condition described above. The thicknesses of the Cr and Au layers were 50 nm and 200 nm, respectively. The substrate was spin-coated with AZ 4620 photoresist. A hairline was drawn onto the photoresist layer using LBDD system; the scan speed was 0.5 mm/sec and the laser power was 5 mW. After development and wet etching of Au, the Cr layer along the hairline pattern was removed for 7 min with Cr etchant. The quartz substrate with negative hairline pattern was used for the patterning of narrow channel for decoupler.

2.4 Fabrication of an Amperometric Sensor Chip on the PDMS Surface

Figure 1 shows the scheme of the electrophoretic device. The sensor electrode was fabricated on quartz glass substrates using improved silicon microfabrication processes. A quartz glass wafer was cut to 50 mm × 60 mm, and washed. PDMS prepolymer and curing agent were mixed with the weight ratio of 10 : 1, and then spin-coated onto the substrate using a final spin speed of 5,000 rpm for 30 s, after an initial spin for 10 s at 1,000 rpm. The PDMS layer was cured at 70°C for 60 min followed by 120°C for 30 min in an oven (MOV-112U, Sanyo, Japan). An Au layer was deposited onto the PDMS-coated substrate by vacuum evaporation. The thickness of the Au thin-film was 200 nm. The coating rate was 25.6 nm/min. The substrate was covered with AZ 4620 photoresist by spincoating at 4000 rpm for 30 s; the resist was hardened at 85°C for 20 min in the oven. After cool down at room temperature, The substrate was exposed to UV light for 40 s through a photomask for the electrode pattern, using a mask aligner (MA-10, Mikasa, Japan). The developing was performed by immersing into AZ 300 MIF developer for 1 min. The patterned electrodes were cleaned with water and dried with nitrogen. The resist layer was finally stripped in an O_2 plasma cleaner (PDC210, Yamato, Japan); the RF power was 200 W, O_2 gas flow rate was 40 ml/min, Ar gas flow rate was 8 ml/min, and operation time was 20 min.

2.5 Fabrication of PDMS Microchannel

The PDMS microchannel was fabricated using a conventional molding process. At first, a template for molding was fabricated using a double mask process. The SU-8 50 negative photoresist was spin-coated on the silicon wafer at 6,000 rpm for 30 s. After a preexposure 'bake' for 5 min at 65°C, followed by 10 min at 95°C, the photomask for electrophoretic channel, was placed on the coated silicon wafer and exposed to UV light for 10 s. SU-8 50 photoresist was spincoated for a second time at 4,000 rpm for 30 s and baked for 15 min at 95°C, prior to second UV exposure through photomask for narrow channel for 40 s.

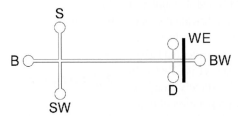

Fig. 1 Scheme of electrophoretic device coupled with narrow channel as a decoupler and electrode applied to ECD. The distance between the narrow channel and the working electrode (WE) was fixed at 1 mm. S: sample reservoir, SW: sample waste, B: buffer reservoir, BW: waste, D: reservoir for discharging the current to the ground as the decoupler

A post-exposure 'bake' was carried out at 65°C for 2 min, followed by at 95°C for 4 min. Subsequently, the wafer was developed using SU-8 developer until the unexposed SU-8 50 photoresist was completely removed, cleaned with the fresh developer, and dried with nitrogen.

Using the fabricated template, PDMS was molded into a cross-shaped microchannel crossing with a narrow channel. PDMS prepolymer solution and curing agent were mixed as mentioned condition above, poured onto the template and cured for 60 min at 70°C. The PDMS layer with the microchannel pattern was removed then from the template. To provide a reservoir to connect the electrophoretic solution, six holes with a diameter of 1.5 mm were punctured. The PDMS microchannel plate and the amperometric sensor chip were bonded using the O_2 plasma cleaner; the RF power was 75 W, O_2 gas flow rate was 150 ml/min, and the operation time was 10 s. The distance between the narrow channel and the WE was fixed at less than 1 mm. The MCE-ECD device was constructed as shown in Fig. 1.

2.6 Electrochemical Characterization and Detection

Linear potential sweep voltammetry of 4-NP, 4-tOP, BPA, and 4-tBP was individually performed using the potentiostat and acquired data was stored on a PC. 5 mM $K_3Fe(CN)_6/K_4Fe(CN)_6$ in 0.2 M Na_2SO_4 was prepared as a voltammetric test solution. An Au thin-film electrode, an Ag/AgCl electrode, and a Pt wire counter electrode (CE) were used as a working electrode, a reference electrode (RE), and an auxiliary electrode, respectively.

Electrophoretic separation was performed in the uncoated channels on the MCE device. As a standard sample, a mixture of 4-NP, 4-tOP, BPA, and 4-tBP was prepared. Electrophoresis was performed with a pH 8.0 buffer containing 20 mM $Na_2B_4O_7$, 20 mM NaH_2PO_4, and methanol, mixed in various concentrations. The sample was electrokineticaly introduced using pinched injection, by applying a voltage to the sample, buffer, and decoupler reservoirs, with the sample waste reservoir grounded. Once injection was completed, a separation potential was applied to the buffer reservoir, while the detection end remained grounded.

Pt wire was inserted in the waste reservoir and connected to both RE and CE lines of the potentiostat. During the application of the separation potential, the ECD for the phenolic chemicals were carried out using a two-electrode setup with Au thin-film electrode.

3 Results and Discussion

3.1 Fabrication of the MCE-ECD Device

From several configurations of aligning the working electrode in amperometric detection under an isolated separation voltage, we fabricated an off-channel detection type MCE-ECD with a narrow channel decoupler as shown in Fig. 2.

The Au thin-film electrodes were incorporated between both the PDMS surfaces. On base side of PDMS layer, the thickness of spin-coated PDMS on substrate was 40 μm. To prevent the generation of cracks in the Au thin-film, conventional Au patterning procedure was reconsidered. First, pretreatment of Au surface with 1,1,1,3,3,3-hexamethyldisilazane (HMDS) was excluded. When HMDS was dropped onto the substrate, exfoliation of the Au thin-film was immediately observed. As adhesion between the Au film and PDMS was weaker than that of PDMS and hydrophobic solution, such as HMDS, it seems that

Fig. 2 Photograph of the MCE-ECD device constructed from the PDMS-based microchannel plate and Au thin-film electrodes. The electrodes were deposited onto the PDMS layer supported with a glass backing substrate

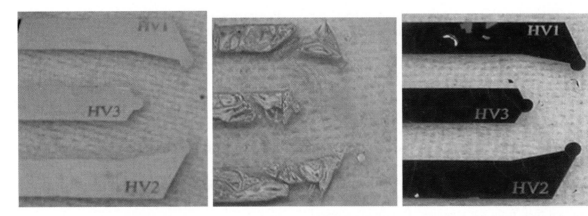

Fig. 3 The configuration of Au thin-film electrodes produced using simple modification of the conventional processes; (**a**) without post-exposure "baking" and exposure to HMDS and stripping solution; (**b**) using stripping solution; (**c**) using post-exposure "bake"

HMDS solution produced an internal stress and permeated into the interface, so that the Au thin-film could be peeled off the PDMS surface. In the same manner as HMDS, the wet stripping solution for photoresist may be unsuited for Au thin-films on PDMS. Figure 3 shows the comparison of the surfaces of Au thin-films processed by both the present and the conventional procedures. While the thin-film was successfully protected by the covering resist layer during both development and wet-etching processes, the thin-film was directly exposed into the stripping solution in the following resist removal and led to the delamination (Fig. 3b). As a result, we finally adopted resist removal using a dry process with O_2 plasma ashing instead of the wet process with the stripping solution (Fig. 3a).

Secondly, the process of post-exposure baking, the heat treatment of resist layer was also deleted from the conventional procedure. As there is a difference in coefficients of thermal expansion between Au and PDMS, the internal stress would affect the thin-film during heating and cooling. If the stress exceeds the tensile strength of the thin-film, it is anticipated that the thin-film undergoes fatal fragmentation as shown in Fig. 3c. Unlike the post-exposure baking, no cracking was observed in the case of pre-exposure baking. The effective temperatures of pre- and post-exposure baking were 85 and 120°C, respectively in the results of the cracking observation of the Au thin-film on PDMS.

Hence, several processes in the conventional procedure were considered to result in critical stress in the Au thin-film and consequent pattern damage. Using an original procedure and a backing glass, which was essential to avoid the stress in the thin-film during the

Fig. 4 Expanded view of the narrow channel decoupler. The vertical channel was separation channel 110 μm in width and 32 μm in depth and the horizontal channel was narrow channel 7 μm in width and 22 μm in depth

handling (Lee et al. 2005), we were able to obtain durable MCE-ECD devices.

As shown in Fig. 4, The fabricated microchannels had two different dimensions for electrophoresis and decoupling. The electrophoretic channel, which was the horizontal channel in Fig. 4, was 110 μm in width and 32 μm in depth, and the narrow channel for decoupling, which was the right vertical channel in Fig. 4, was 7 μm in width and 22 μm in depth.

3.2 Electrokinetic Control of Microfluids

For an electrophoretic analysis on the microfabricated device, sample and buffer were electrokinetically controlled using EOF. At first, the sample flow out of the sample reservoir S was spatially confined by incoming

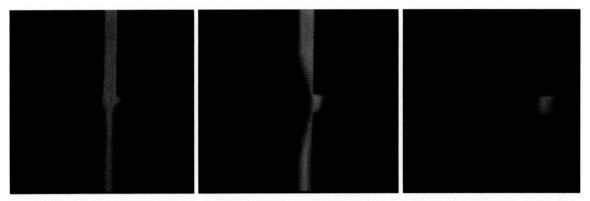

Fig. 5 Fluorescence images during the pinched injection of sample solution. The fluorescence sample was injected from the upper channel and electrokinetically confined by the buffer streams from both horizontal channels (**a**). The voltage profile was changed to inject a portion of sample in the cross intersection to the separation channel (**b** and **c**)

buffer streams from the horizontal channels between B and BW at the cross intersection in Fig. 1 before dispensing it into the separation channel. The sample manipulation method is called 'pinched injection', that is able to generate well-defined short axial extent sample plugs suitable for high performance separation based on Eq. 4. When high voltage was applied to the MCE-ECD device, the electrokinetic flow of sample solution was observed using a microscope (FZIII, Leica, Germany) in Fig. 5. The sample, which comprised 10 μg/ml fluorescein solution, was loaded into the cross intersection by applying a positive potential of 230 V (corresponding to 135 V/cm) to the sample reservoir and keeping the sample waste reservoir at ground. In addition, the sample stream was electrokinetically confined by applying an electric field (~100 V/cm) to both buffer reservoir B and the separation channel through the narrow channel as shown in Fig. 5a. Although the voltage was applied to the reservoir D connected to a narrow channel, buffer waste reservoir BW was electrically floating. In all experiments, an injection voltages were applied for 30 sec and then changed the each voltage to electrophoretic separation (Fig. 5b and c). The small sample plug was generated by applying a positive voltage 500 V (corresponding to 188 V/cm) to the separation channel and leakage of excess sample from the cross intersection into the separation channel was prevented with a small electric pull back voltage (~60 V/cm) applied to both S and SW reservoirs. The sample volume in the cross-intersection corresponded to approximately 100 pL.

On the other hands, the sample transportation around the narrow channel decoupler was also observed (Fig. 6). Figure 6 obviously shows injected fluorescent sample transported to the detection end through the separation channel across the junction with narrow channel. It was considered that the flow to the detection end was normally caused by diffusion as well as a conventional floating injection technique. In addition, the miniaturization of the cross section resulted the increase in flow resistance resulting from vena contracta. The flow resistance in the narrow channel was higher than that on the detection side, leading to transportation of the sample to the detection end. As the leakage of sample into the narrow channel was minimal as shown in Fig. 6, it was expected that the narrow channel was useful for use as a decoupler.

3.3 Application of Detection for Phenolic Chemicals

The fabricated Au thin-film electrode was first tested for electrode stability for the measurement of linear potential sweep voltammetry s of each analyte. For detection using the MCE-ECD device, the electrode should selectively detect the phenolic chemicals without interference from the running buffer solution for electrophoresis. The linear potential sweep voltammetry s of each analyte for the Au thin-film electrode system are shown in Fig. 7. The measured linear potential sweep voltammetry s were reproducible and fully correspond to those expected for these phenolic chemicals.

Control tests conclusively demonstrated that the Au thin-film electrode was quite effective for linear

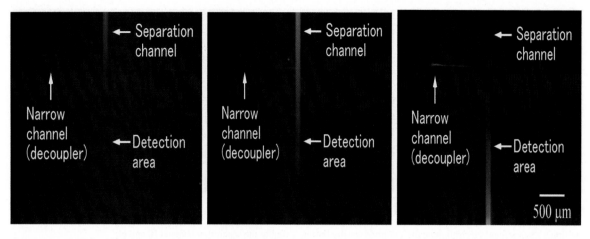

Fig. 6 Expanded views of electrophoretic channel at the junction with the narrow channel (**a**). A fluorescent sample was electrokinetically injected to the separation channel (**b**), and transported through the junction (**c**), to the detection side (**d**). The leakage into the narrow channel was almost completely prevented due to the flow resistance

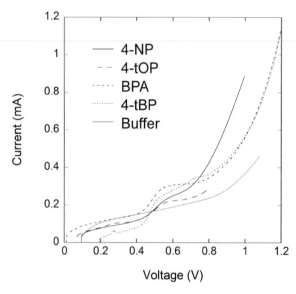

Fig. 7 Linear potential sweep voltammograms of four types of phenolic chemical in the buffer solution containing 20 mM $Na_2B_4O_7$, 20 mM NaH_2PO_4, and 5 %v/v methanol, pH 8.0. The concentration of each phenolic chemical was 10 mM. The working, the reference, and the counter-electrodes were a Au thin-film electrode, a Ag/AgCl, and a Pt wire, respectively. The scanning rate was 100 mV s^{-1}

potential sweep voltammetry and it was found that the most suitable detection potentials for 4-NP, 4-tBP, BPA, and 4-tOP were over 0.58, 0.73, 0.54, and 0.56 V, respectively.

We investigated electrophoretic separation by ECD of phenolic chemicals using the MEKC mode. A Pt

wire was inserted into the waste reservoir and used as a pseudo-reference. The detection potential was fixed at 1.0 V (vs Pt). Using 10 mM SDS in 20 mM borate-phosphate (pH 8.0) and adding 10% (v/v) methanol

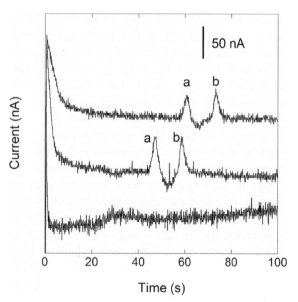

Fig. 8 Electropherograms of phenolic chemicals using the MCE-ECD device: (**a**) 4-tBP and BPA, and (**b**) 4-tOP and 4-NP. The concentration of each standard sample was 25 mM. As a running buffer, 20 mM $Na_2B_4O_7$ and 20 mM NaH_2PO_4 was mixed with methanol (MeOH) to examine the improvement of separation. Device: cross channel PDMS chip with 2.5 cm separation channel; separation distance = 2.0 cm and applied electric field = 333 V cm^{-1}; detection potential = 1.0 V. The polarity of the separation was from the anode (sample injector) to the cathode (separation waste)

as running buffer, we obtained two overlapping peaks, one overlapping peak was assigned to 4-tBP and BPA and the other was assigned to 4-tOP and 4-NP within 1.5 min as shown in Fig. 8.

We obtained an improved peak separation of 4-tBP and BPA by adding 5 mM β-cyclodextrin (CD) to the running buffer (Fig. 9).

The limits of detection (LOD) at a signal-to-noise ratio of 3 (S/N=3) obtained for 4-tBP and BPA were 18 and 20 mM, respectively. During the detection using the MCE-ECD device, the Au thin-film electrode withstood the high voltages applied to drive the electrophoresis. For PDMS-based MCE-ECD device, there had previously never been a simple modification of the fabrication procedures which entirely prevented the failures, such as cracking, fragmentation, and delamination.[18] We have successfully developed a durable MCE-ECD device using a very simple procedure.

In the present MCE-ECD device, the distance between the narrow channel and the WE has not been optimized. This distance influences noise level and peak separation.[24] Furthermore, the used apparatus used was not apprproiate to minimize the noise. Such high noise levels encountered required a high concentration of samples, so that the peak separation was degraded due to diffusion. The improvement in noise level and perfect separation in the MCE-ECD device is now being investigated and will be reported in future publications.

4 Conclusion

We have demonstrated new fabrication processes for a PDMS/PDMS device, coupled with a durable Au thin-film electrode for ECD. These processes contributed to the minimization of the internal stress in the thin-film on PDMS and realized a durable MCE-ECD device for high voltages in electrophoresis as a first demonstration. To isolate the ECD from the electrophoretic high voltage, a narrow channel was applied as a decoupler and this successfully operated to protect the apparatus from surge, but the effect of decoupling was not enough to decrease the noise level. The fabricated PDMS-only device is suitable for maintaining an homogeneous velocity in EOF so as to separate relatively hydrophilic neutral analytes using the MEKC mode. As a model assay using the MCE-ECD device, phenolic chemicals were examined and 4-tBP and BPA could be perfectly separated using β-CD. The durable MCE-ECD device will be suitable for commercial use and realizes a miniaturized total system for assay.

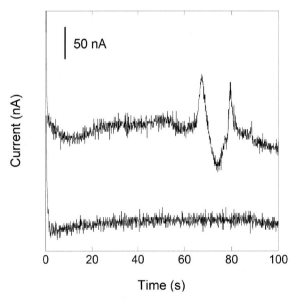

Fig. 9 Effect of β-cyclodextrin (CD) on electrophoretic separation of phenolic chemicals: (**a**) 4-tBP and (**b**) BPA. The concentration of each standard sample was 25 mM. As a running buffer, 20 mM $Na_2B_4O_7$, 20 mM NaH_2PO_4, 10 %v/v methanol, and 5 mM β-CD, pH 8.0 was used. Device: cross channel PDMS chip with 2.5 cm separation channel; separation distance = 2.0 cm and applied electric field = 333 V cm^{-1}; detection potential = 1.0 V. The polarity of the separation was from the anode (sample injector) to the cathode (separation waste)

References

Chen, D., Hsu, F.-L., Zhan, D.-Z., Chen, C. (2001). Palladium film decoupler for amperometric detection in electrophoresis chips. *Analytical Chemistry*, 73, 758–762.

Chen, G., Wang, Y.H. (2006). Microchip capillary electrophoresis with electrochemical detection for monitoring environmental pollutants. *Current Analytical Chemistry*, 2, 43–50.

Corborn, T., Dumanoski, D., Myers, J.P. (1996). *Our Stolen Future*. New York: Penguin.

Kim J.-B., Terabe, S. (2003). On-line sample preconcentration techniques in micellar electrokinetic chromatography. *Journal of Pharmaceutical and Biomedical Analysis*, 30, 1625–1643.

Kim, J.-B., Otsuka, K., Terabe, S. (2001). Anion selective exhaustive injection-sweep–micellar electrokinetic chromatography. *Journal of Chromatography A*, 932, 129–137.

Lee, K.J., Fosser, K.A., Nuzzo, R.G. (2005). Fabrication of stable metallic patterns embedded in poly(dimethylsiloxane) and model applications in non-planar electronic and lab-on-a-chip device patterning. *Advanced Functional Materials*, 15, 557–566.

Loom Y.-L., Willett, R.L., Baldwin, K.W., Rogers, J.A. (2002). Additive, nanoscale patterning of metal films with a stamp and a surface chemistry mediated transfer process: Applications in plastic electronics. *Applied Physics Letter*, 81, 562–564.

Masadome, T., Kugoh, S., Ishikawa, M., Kawano, E., Wakida, S. (2005). Polymer chip incorporated with anionic surfactant-ISFET for microflow analysis of anionic surfactants. *Sensors and Actuators B*, 108, 888–892.

Monton, M.R.N., Otsuka, K., Terabe, S. (2003), On-line sample preconcentration in micellar electrokinetic chromatography by sweeping with anionic–zwitterionic mixed micelles. *Journal of Chromatography A*, 985, 435–445.

Muijselaar, P.G., Otsuka, K., Terabe, S. (1997), Micelles as pseudo-stationary phases in micellar electrokinetic chromatography. *Journal of Chromatography A*, 780, 41–61.

Pappas, T.J., Gayton-Ely, M., Holland, L.A. (2005). Recent advances in micellar electrokinetic chromatography. *Electrophoresis*, 26, 719–734.

Rossier, J.S., Ferrigno, R., Girault, H.H. (2000). Electrophoresis with electrochemical detection in a polymer microdevice. *Journal of Electroanalytical Chemistry*, 492, 15–22.

Scampicchio, M., Wang, J., Mannino, S., Chatrathi, M.P. (2004). Microchip capillary electrophoresis with amperometric detection for rapid separation and detection of phenolic acids. *Journal of Chromatography A*, 1049, 189–194.

Schmid, H., Wolf, H., Allenspach, R., Riel, H., Karg, S., Michel, B., Delamarche, E. (2003). Preparation of metallic films on elastomeric stamps and their application for contact processing and contact printing. *Advanced Functional Materials*, 13, 145–153.

Schoning, M.J., Jacobs, M., Muck, A., Knobbe, D.-T., Wang, J., Chatrathi, M., Spillmann, S. (2005). Amperometric PDMS /glass capillary electrophoresis-based biosensor microchip for catechol and dopamine detection. *Sensors and Actuators B*, 108, 688–694.

Schwarz, M.A., Galliker, B., Fluri, K., Kappes, T., Hauser, P.C. (2001). A two-electrode configuration for simplified amperometric detection in a microfabricated electrophoretic separation device. *Analyst*, 126, 147–151.

Vandaveer, W.R., Pasas-Farmer, S.A., Fischer, D.J., Frankenfeld, C.N., Lunte, S.M. (2004). Recent developments in electrochemical detection for microchip capillary electrophoresis. *Electrophoresis*, 25, 3528–3549.

Wakida, S., Fujimoto, K., Nagai, H., Miyado, T., Shibutani, Y., Takeda, S. (2006). On-chip micellar electrokinetic chromatographic separation of phenolic chemicals in waters. *Journal of Chromatography A*, 1109, 179–182.

Wang, J., Chen, G., Chatrathi, M.P., Musameh, M. (2003). Capillary electrophoresis microchip with a carbon nanotube-modified electrochemical detector. *Analytical Chemistry*, 76, 298–302.

Wang, J., Pumera, M., Chatrathi, M.P., Escarpa, A., Konrad, R., Griebel, A., Dörner, W., Löwe, H. (2002). Towards disposable lab-on-a-chip: Poly(methylmethacrylate) microchip electrophoresis device with electrochemical detection. *Electrophoresis*, 23, 596–601.

Wu, C.-C., Wu, R.-G., Huang, J.-G., Lin, Y.-C., Chang, H.-C. (2003). Three-electrode electrochemical detector and platinum film decoupler integrated with a capillary electrophoresis microchip for amperometric detection. *Analytical Chemistry*, 75, 947–952.

Zeng, Y., Chen, H., Pang, D.-W., Wang, Z.-L., Cheng, J.-K. (2002). Microchip capillary electrophoresis with electrochemical detection. *Analytical Chemistry*, 74, 2441–2445.

Swimming Behavioral Toxicity in Japanese Medaka (*Oryzias latipes*) Exposed to Various Chemicals for Biological Monitoring of Water Quality

Ik Joon Kang, Junya Moroishi, Mitoshi Yamasuga, Sang Gyoon Kim and Yuji Oshima

Abstract We conducted a short-term behavioral toxicity test in medaka (*Oryzias latipes*). Fish were exposed to toxicants (potassium cyanide [1 or 5 mg/L], phenol [12.5 or 25 mg/L], fenitrothion [10 or 20 mg/L], or benthiocarb [10 or 20 mg/L]), and swimming behavior was recorded and evaluated for 1 h. The medaka were placed in an exposure chamber with a continuous flow-through system. Two cameras tracked the fish at positions to the front and side of the test chamber, and images from the cameras were used to calculate the positions of the medaka in three dimensions (3D); the 3D data were processed by computer and analyzed as swimming activity (swimming speed and surfacing behavior). The swimming behavior of medaka was affected by exposure to toxic chemicals. The frequency of fast swimming was remarkably increased in medaka treated with potassium cyanide (5 mg/L), phenol (25 mg/L), or fenitrothion (10 or 20 mg/L). An increase in the time spent close to the water surface was observed in fish exposed to potassium cyanide (1 or 5 mg/L), fenitrothion (10 or 20 mg/L), or benthiocarb (20 mg/L). We concluded that pollution of water with these toxic chemicals at high concentrations can be detected by monitoring the swimming behavior of medaka for 1 h.

Keywords Medaka · Swimming behavior · Water quality · Biomonitoring

1 Introduction

Today, increasing numbers of chemicals are being used in a wide range of fields, and the threat posed to aquatic systems by chemical spills is of concern. In fact, a number of accidental spills of toxic chemicals into rivers or streams have been reported around the world. In 1986, 30 tonnes of pesticides was discharged into the Rhine River by a fire accident at an industrial complex (Capel et al. 1988). In 1991, a huge amount of phenol was run off into the Naktong River in South Korea, and aquatic organisms were killed (Yoshida 2003). An accidental spillage of cyanides occurred in northeast Romania in 2000 (Soldán et al. 2001). These water incidents advocate the need for online monitoring systems for evaluation of water quality in the early stages of toxicant spills.

In general, modern analytical methods are used to monitor water quality in aquatic environments, including in water sources. However, the use of instrument analysis as part of online monitoring can be difficult for detecting unexpected toxic substances and complexes of chemicals: instrument analysis is still unable to efficiently detect such toxicants in real time. Therefore, there is a need to develop online monitoring systems that use aquatic organisms. In fact, several online biological monitoring systems that use aquatic organisms have been developed and are being used to monitor water sources and supplies. Algae, daphnia, bivalves, and fish have been chosen as test organisms (Gerhardt et al. 2006).

We focused here on the use of fish behavior to detect contamination of water with toxicants. Fish are ideal test organisms for investigating the behavioral toxicity of chemicals in water (Little et al. 1990). Behavior

I.J. Kang (✉)
Aquatic Biomonitoring and Environmental Laboratory, Division of Bioresource and Bioenvironmental Sciences. Kyushu University Graduate School, Hakozaki 6-10-1, Higashi-ku, Fukuoka 812-8581, Japan

Y.J. Kim et al. (eds.), *Atmospheric and Biological Environmental Monitoring*,
DOI 10.1007/978-1-4020-9674-7_20, © Springer Science+Business Media B.V. 2009

change in fish is a valuable endpoint for immediate evaluation of the presence of toxicants in water. When fish are exposed to high levels of toxicants they show altered behavior until death. Thus, the analysis of such abnormal behavior in fish is valuable for monitoring water quality in the short-term.

Previous studies have reported alterations in behavioral parameters in fish; two-dimensional data have mostly been used for analysis of swimming behavior (Kane et al. 2004, Park et al. 2005, Suzuki et al. 2003). Fish have spatial reasoning capacity, and their swimming performance is stereoscopic in the test chamber. We therefore considered that it would be useful to collect and analyze three-dimensional (3D) data on swimming behavior.

Japanese medaka was chosen as the test organism and has been recommended as a reference fish (OECD 1999) for several ecotoxicological tests. It is easy to maintain, and the adults are small (average length ca. 3 cm). We conducted short-term behavioral toxicity testing in medaka and examined the changes in their behavior as a way of monitoring water quality.

2 Material and Methods

2.1 Test Chemicals

The test substances potassium cyanide (KCN, >98% purity) and phenol (>99% purity) were purchased from Katayama Chemical Industries Co. Ltd. (Osaka, Japan) and Kanto Chemical Co. Inc. (Tokyo, Japan), respectively. Fenitrothion (MEP, > 98% purity) and benthiocarb (>99% purity) were purchased from Wako Pure Chemical Industries (Osaka, Japan). KCN and phenol stock solution (100 mg/L) were prepared by dissolution in dechlorinated tap water. Stock solutions of MEP and benthiocarb were prepared by dissolution in dimethyl sulfoxide (Wako Pure Chemical Industries, Osaka, Japan.), followed by dilution with dechlorinated tap water.

2.2 Test Fish and Exposure Test

About 300 adult medaka (4–6 months post-hatch; mean [±SD] body weight, 270.0 ± 40.0 mg; total length, 31.2 ± 1.9 mm) were selected from broodstock that had been maintained for 6 years in the laboratory. The fish were kept under a photoperiod (14:10-h light:dark) and fed with *Artemia* nauplii (<24 h after hatching) twice a day. The water temperature was maintained at $22 \pm 1°C$, and the dissolved oxygen concentration (mean $\pm$ SD) was 7.0 ± 0.2 mg/L. We exposed the fish to the test chemicals KCN (1 or 5 mg/L; four fish/treatment), phenol (12.5 or 25 mg/L; four fish/treatment), MEP (10 or 20 mg/L; four fish/treatment), and benthiocarb (10 or 20 mg/L; four fish/treatment) for 1 h. Figure 1 is a schematic diagram of the exposure test. The fish were placed into test chamber (length, 10 cm; width, 10 cm; height, 15 cm), which contained about 1.8 L of the test solution supplied by a flow-through system (flow rate, 400 ml/min). Test solution was delivered to the test chamber by a roller pump (Eyela RP-1000, Tokyo Rikakikai Co. Ltd., Tokyo, Japan). The flow rates of dechlorinated tap water and test solutions were checked by using a graduated cylinder. The test solution in the test chamber was renewed about 12 times/h.

Before the initiation of exposure, the test fish was placed in the test chamber and only dechlorinated tap water was run through the chamber for 30 min, for acclimatization. After the 30-min acclimatization, dechlorinated water was run through the test chamber for another 30 min while we recorded the pre-exposure behavior of the medaka. The test solution

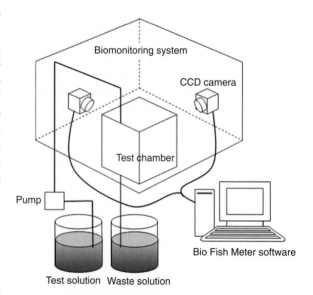

Fig. 1 Schematic diagram of the exposure test

was then introduced into the test chamber for 60 min. The swimming behavior of the medaka was recorded for 90 min, including the pre-exposure period. The behavioral parameters of the test fish were recorded and quantified with Bio Fish Meter Lab equipment (Seiko Electric Co., Ltd., Fukuoka, Japan). Two cameras tracked the fish at positions to the front and side of the test chamber. The swimming behavior of the fish was recorded by two cameras and the images were sent to a computer. The movement was recorded in 3D (*x, y,* and *z* coordinates). From these data, we evaluated the behavior of the medaka from their 3D swimming speed and 3D surfacing behavior. After the exposure test, all fish were anesthetized in FA-100 solution (dilution, 2000-fold; Tanabe Seiyaku Co. Ltd., Osaka, Japan), drained on filter paper, and measured for body weight and length.

2.3 Data Calculation and Statistical Analysis

Previous studies have demonstrated that the maximum swimming speed of medaka is approximately 300 mm/s under normal conditions. Iwamatsu (2006) reported that the maximum speed of fish is 10 times their total length per second. Thus, data on 3D swimming speed (S) were divided into 6 speed groups ($0 \leq S \leq 50$, $50 < S \leq 100$, $100 < S \leq 150$, $150 < S \leq 200$, $200 < S \leq 250$, and $250 < S \leq 300$ mm/s), and the frequencies of each swimming speed were evaluated. In addition, it is known that fish stay near the water surface when they are exposed to toxicants at high concentrations (surfacing behavior). We therefore evaluated the duration of surfacing behavior (i.e. the time spent at no more than 20 mm below the surface).

Differences in the frequencies of each swimming speed and in the durations of surfacing behavior were statistically compared between unexposed and exposed conditions. All data were checked for assumptions of homogeneity of variance across treatments by using Levene's test. Data were analyzed by one-way analysis of variance (ANOVA) and were then tested by Dunnett's test. When no homogeneity was observed, nonparametric statistical comparisons were used to detect differences among treatments (Kruskal-Wallis test). Differences between unexposed and exposed conditions in treatment groups were identified by using individual Mann-Whitney *U*-tests. A *P* value of < 0.05 was considered to indicate significance; Bonferroni's *P* was used in nonparametric tests. All statistical analyses were performed with SPSS Base 11.0J (SPSS Inc., Chicago, IL, USA).

3 Results

3.1 Frequencies of 3D Swimming Speeds

Throughout the exposure period, no significant changes in frequencies of swimming speeds were observed in fish exposed to KCN 1 mg/L (Fig. 2A). In contrast, a significant increase in the frequency of swimming at $150 < S \leq 250$ mm/s occurred at between 30 and 60 min of exposure to KCN (5 mg/L), compared with unexposed conditions (Fig. 2B, $P < 0.05$, Dunnett's test).

Phenol at 12.5 mg/L induced an increase in the frequency of fast swimming ($150 < S \leq 300$ mm/s) in medaka (Fig. 3A), but the differences in speed frequency between unexposed and exposed conditions were not significant. Exposure to phenol at 25 mg/L significantly increased the frequencies of swimming speeds of $100 < S \leq 250$ mm/s during 0 to 30 min of exposure, compared with unexposed conditions ($P < 0.05$, Fig. 3B, Dunnett's test).

Additionally, at the same level of exposure there was a statistically significant increase in the frequency of swimming speeds of $250 < S \leq 300$ mm/s compared with unexposed conditions ($P < 0.05$, Kruskal-Wallis and Mann-Whitney tests).

Exposure to MEP at 10 mg/L increased the frequency of swimming speeds of $100 < S \leq 150$ mm/s at between 0 and 30 min of exposure (Fig. 4A, $P < 0.05$, Kruskal-Wallis and Mann-Whitney tests). A significant increase in the frequency of fast swimming ($150 < S \leq 200$ mg/L) was observed at between 30 and 60 min of exposure to MEP 20 mg/L compared with under control conditions (Fig. 4B, $P < 0.05$, Dunnett's test).

Similarly, we found an increase in the frequency of fast swimming ($200 < S \leq 250$ mg/L) at between 30 and 60 min of MEP exposure ($P < 0.05$, Kruskal-Wallis and Mann-Whitney tests).

Fig. 2 Frequencies of
swimming speeds of medaka:
(**A**) fish exposed to KCN
1 mg/L and (**B**) fish exposed
to KCN 5 mg/L during the
1.5-h behavior test (*clear
bars*: unexposed [control]
30 min; *gray bars*: exposed 0
to 30 min; *black bars*: exposed
30 to 60 min). * P < 0.05
compared with control. Data
are shown as means ± SE

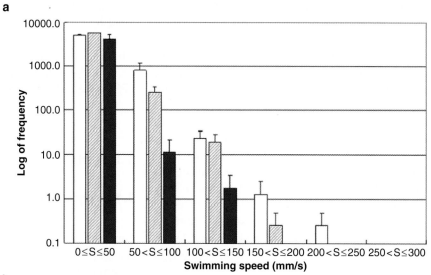

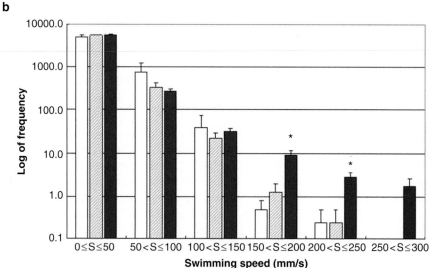

We examined the frequencies of the swimming speed in medaka during the benthiocarb exposure test (Fig. 5A). Throughout the benthiocarb 20 mg/L exposure, there was a trend toward increased frequencies of fast swimming (150<S≤300 mm/s), but no significant differences were found between control and exposed conditions (Fig. 5B).

3.2 Duration of 3D Surfacing Behavior

We found changes in the duration of surfacing behavior in medaka exposed to 3 of these test chemicals (Fig. 6).

KCN (1 mg/L) exposure increased the duration of surfacing behavior, and there was a significant differences between unexposed and exposed conditions (30 to 60 min exposure, P < 0.05, Dunnett's test). The duration of surfacing behavior was also significantly increased after 0 to 60 min exposure to KCN at 5 mg/L (P < 0.05, Kruskal-Wallis and Mann-Whitney tests). The duration of surfacing behavior in medaka treated with MEP increased in the 10-mg/L group at 30 to 60 min exposure and in the 20-mg/L group at 0 to 60 min exposure, respectively (P < 0.05, Dunnett's test). With benthiocarb exposure, an increase in the duration of surfacing behavior was found in the 20-mg/L group at 30 to 60 min exposure (P < 0.05,

Fig. 3 Frequencies of swimming speeds of medaka: (**A**) fish exposed to phenol 12.5 mg/L and (**B**) fish exposed to phenol 25 mg/L during the 1.5-h behavior test (*clear bars*: unexposed [control] 30 min; *gray bars*: exposed 0 to 30 min; *black bars*: exposed 30 to 60 min). * P < 0.05 compared with control. Data are shown as means ± SE

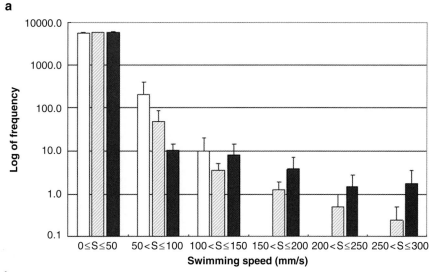

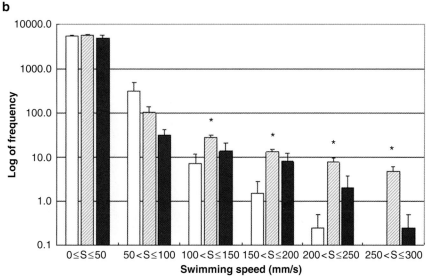

Dunnett's test), but no significant change was observed in the 10-mg/L group. There was no increase or decrease in the duration of surfacing behavior in fish exposed to phenol.

4 Discussion

Exposure to test chemicals at high concentrations induced significant changes in the 3D swimming behavior of medaka during the 1-h exposure period. We focused on two kinds of behavioral parameter—swimming speed and surfacing behavior—and these two parameters were affected to varying degrees by KCN, phenol, MEP, and benthiocarb (Table 1).

Previous studies have reported that fish show altered swimming performance soon after exposure to toxic chemicals such as phenol and MEP (Bull and McInerney 1974, Smith and Bailey 1988). Medaka treated with KCN or MEP showed significant increases in both the frequency of fast swimming and the duration of surfacing behavior, compared with under unexposed conditions. In contrast, phenol exposure induced a significant increase in the frequency of fast swimming, but there were no changes in the duration of surfacing behavior. In the benthiocarb group, swimming speed was not affected but duration of surfacing behavior

Fig. 4 Frequencies of
swimming speeds of medaka:
(A) fish exposed to MEP
10 mg/L and (B) fish exposed
to MEP 20 mg/L during the
1.5-h behavior test (*clear
bars*: unexposed [control]
30 min; *gray bars*: exposed 0
to 30 min; *black bars*: exposed
30 to 60 min). * P < 0.05
compared with control. Data
are shown as means ± SE

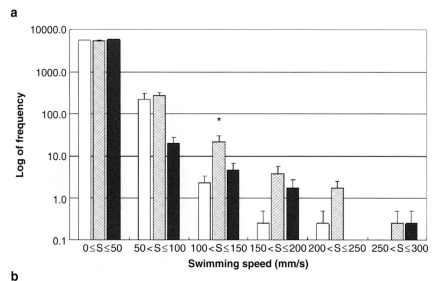

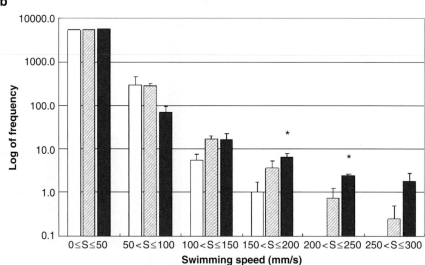

was increased. Thus, the patterns of effect on behavioral parameters differed among the toxic chemicals and their concentrations.

We found a significant increasing in the frequency of 3D swimming speeds greater than 100 mm/s upon toxic chemical exposure except benthiocarb; the swimming speeds of medaka exposed to test chemicals at high concentrations were frequently high (150 < S ≤ 300 mm/s). In a preliminary test we evaluated the 3D swimming behavior of 25 medaka in control water for 90 min. The frequency distribution of swimming speeds in these controls was 89.5% 0 ≤ S ≤ 50 mm/s, 10.05% 50 < S ≤ 100 mm/s, 0.41% 100 < S ≤ 150 mm/s, 0.04% 150 < S ≤ 200 mm/s, 0.01%

200 < S ≤ 250 mm/s, and 0% 250 < S ≤ 300 mm/s. Thus, swimming speeds of 150 < S ≤ 300 mm/s do not often occur in unexposed medaka; we concluded that fast swimming by medaka (150 < S ≤ 300 mm/s) is an efficient endpoint for the observation of abnormal swimming activity in response to toxins.

We compared these behavioral toxicity data with LC_{50} values from previous studies. Abnormalities in swimming behavior were detected faster than it took to detect fish mortality. The 96-h LC_{50} for phenol in medaka is 29.3 mg/L (Shigeoka et al. 1988), but we found an alteration in swimming speed upon exposure to phenol at 25 mg/L at 0 to 30 min exposure. For MEP, the 48-h LC_{50} value is 3.5 mg/L (Tsuda et al. 1997);

Fig. 5 Frequencies of swimming speeds of medaka: (**A**) fish exposed to benthiocarb 10 mg/L and (**B**) fish exposed to benthiocarb 20 mg/L during the 1.5-h behavior test (*clear bars*: unexposed [control] 30 min; *gray bars*: exposed 0 to 30 min; *black bars*: exposed 30 to 60 min). Data are shown as means ± SE

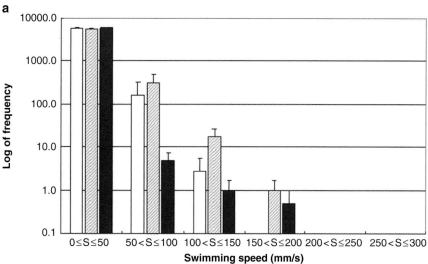

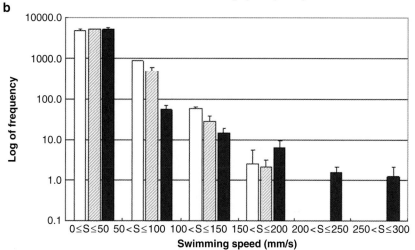

Fig. 6 Time spent in surface water by medaka exposed to test chemicals during the behavior test (*clear bars*: unexposed [control] 30 min; *gray bars*: exposed 0 to 30 min; *black bars*: exposed 30 to 60 min). * $P < 0.05$ compared with control. Data are shown as means ± SE

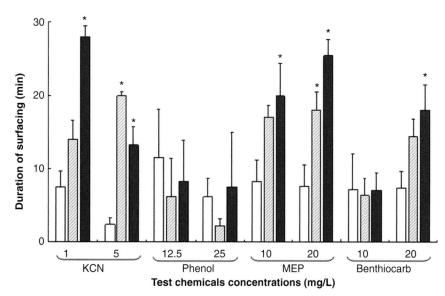

Table 1 Summary of effects observed on 3D swimming speed and surfacing behavior of medaka exposed to test chemicals for 1 h

Test chemical	Nominal concentration (mg/L)	Behavioral parameters						Duration of surfacing behavior (min)
		Frequency of swimming speed (mm/s)						
		$0 \leq S \leq 50$	$50 < S \leq 100$	$100 < S \leq 150$	$150 < S \leq 200$	$200 < S \leq 250$	$250 < S \leq 300$	
KCN	1	–	–	–	–	–	–	+
	5	–	–	–	+	+	–	+
Phenol	12.5	–	–	–	–	–	–	–
	25	–	–	+	+	+	+	–
MEP	10	–	–	+	–	–	–	+
	20	–	–	–	+	+	–	+
Benthiocarb	10	–	–	–	–	–	–	–
	20	–	–	–	–	–	–	+

+ significant effects were observed during the exposure.
– no effects were observed during the exposure.

even though the 48-h LC_{50} data for MEP were three times lower than the value at which alteration in swimming activity occurred (10 mg/L), we found swimming behavioral abnormalities within 1 h of exposure. Thus, we concluded that contamination with toxicants in a similar range of LC_{50} values could be detected by monitoring the swimming behavior of the medaka. Our results also suggested that observation of these behavioral parameters (swimming speed and surfacing behavior) can help detect the contamination of water by toxic chemicals at an early stage.

Acknowledgments We thank Yoko Yamasaki and Ayumi Nakamura, Bio monitoring Group, SEIKO Electric Co., Ltd., for their technical help in conducting this study.

References

Bull CJ, McInerney JE (1974) Behavior of juvenile coho salmon (*Oncorhynchus kisutch*) exposed to Sumithion (fenitrothion), an organophosphate insecticide. J Fish Res Board Can 31:1867–1872.

Capel PD, Giger W, Reichert P, Wanner O (1988) Accidental input of pesticides into the Rhine River. Environ Sci Technol 22:992–997.

Gerhardt A, Ingram MK, Kang IJ, Ulitzur S (2006) In situ online toxicity biomonitoring in water: recent developments. Environ Toxicol Chem 25:2263–2271.

Iwamatsu T (2006) The integrated book for the biology of the medaka. Daigaku-kyoiku-shuppan. Okayama, Japan (*in Japanese*).

Kane AS, Salierno JD, Gipson GT, Molteno TCA, Hunter C (2004) A video-based movement analysis system to quantify behavioral stress responses of fish. Water Res 38:3993–4001.

Little EE, Archeski RD, Flerov BA, Kozlovskaya VL (1990) Behavioral indicators of sublethal toxicity in rainbow trout. Arch Environ Contam Toxicol 19:380–385.

Organization for Economic Cooperation and Development, 1999. Final Report From the Oecd Expert Consultation Meeting, London, UK, 28–29 October 1998. Report 9906. Environmental Health and Safety Division, Paris, France.

Park YS, Chung NI, Choi KH, Cha EY, Lee SK, Chon TS (2005) Computational characterization of behavioral response of medaka (*Oryzias latipes*) treated with diazinon. Aquat Toxicol 71:215–228.

Shigeoka T, Yamagata T, Minoda T, Yamauchi F (1988) Acute toxicity and hatching inhibition of chlorophenols to Japanese medaka, *Oryzias latipes* and structure-activity relationships. J Hyg Chem (Eisei Kagaku) 34:343–349 (*in Japanese with English abstract*).

Smith EH, Bailey HC (1988) Development of a system for continuous biomonitoring of a domestic water source for early warning of contaminants. In: Gruber DS and Diamond JM (ed) Automated Biomonitoring: Living Sensors as Environmental Monitors. Ellis Horwood, Chichester, U.K., pp. 182–205.

Soldán P, Pavonič M, Bouček J, Kokeš J (2001) Baia Mare accident—brief ecotoxicological report of Czech experts. Ecotoxicol Environ Saf 49:255–261.

Suzuki K, Takagi T, Hiraishi T (2003) Video analysis of fish schooling behavior in finite space using a mathematical model. Fish Res 60:3–10.

Tsuda T, Kojima M, Harada H, Nakajima A, Aoki S (1997) Acute toxicity, accumulation and excretion of organophosphorous insecticides and their oxidation products in killifish. Chemosphere 35:939–949.

Yoshida F (2003) IT pollution problems in Asia. Econ J Hakkaido Univ 32:1–23.

The Effects of Earthworm Maturity on Arsenic Accumulation and Growth After Exposure to OECD Soils Containing Mine Tailings

Byung-Tae Lee and Kyoung-Woong Kim

Abstract Earthworm, *Eisenia fetida*, was exposed for 14 days to OECD soils mixed with As-contaminated mine tailings. Prior to exposure, earthworms were divided into three growth states (mature, formative, and immature) with both the development of clittellum and body weight. Toxic effects on growth of immature earthworms were caused by 5 and 10% proportions of mine tailings in OECD soils, while mature or formative ones did not show any adverse effects on growth. Moreover, only immature worms exposed to OECD soils with 10% mine tailings resulted in the loss of body weight, which was contrast to the result that body weights of immature worms increased mostly when exposed to the control. Arsenic concentrations in earthworms increased for 14-day exposure. Arsenic uptake patterns did not show any difference with the initial earthworm maturity. Arsenic accumulations were dependent on not the maturity but the proportion of mine tailings in OECD soils. The results implied that earthworm has an ability to resist As toxicity, and the degree of tolerance to As may differ with maturity. Young earthworms were more easily affected by As than mature worms.

Keywords Earthworm (*Eisenia fetida*) · Growth state · Mine tailings · Arsenic uptake · Resistance

1 Introduction

Earthworms play an important role in soil maintenance and fertilization. Earthworms stir soil surface, aerate soils and stimulate drainage by moving and growing and half of soil aggregates are formed by earthworm activities (Edwards and Bohlen 1996; Edwards 2004). Earthworms can accumulate contaminants in the body when exposed to contaminated soils (Spurgeon and Hopkin 1996, 1999; Cotter-Howells et al. 2005; Nahmani et al. 2007). The toxicant accumulations are assessed to characterize uptake or elimination and the results are co-evaluated with observed adverse effects over exposure time. One compartment model is widely used to characterize uptake or elimination of earthworms (Neuhauser et al. 1999; Peijnenburg et al. 1999a, b). The contaminants accumulated in earthworms can be transferred to birds, small mammals, and other soil biota through the terrestrial food web (Romijn et al. 1994; Spurgeon and Hopkin 1996; Stephenson et al. 1997). Because of these features earthworms have been used to evaluate toxicity of contaminations on terrestrial environment as a representative biomarker.

Toxicity is usually assessed through exposure to contaminants. Survival, growth, and reproductions are measured to evaluate toxic values like LC_{50} (median lethal concentration), EC_{50} (median effect concentration), and LOEC (lowest observed effect concentration). Recently, interest in the detoxification or resistance of earthworms to metals or As is being increased (Langdon et al. 2003; Vijver et al. 2004). Spurgeon and Hopkin (2000) showed that the pre-exposed generations developed the resistance to zinc resulted in higher toxicity values. Also, earthworms collected from the

K.-W. Kim (✉)
Department of Environmental Science & Engineering,
Gwangju Institute of Science and Technology (GIST),
Gwangju 500-712, South Korea
e-mail: kwkim@gist.ac.kr

Y.J. Kim et al. (eds.), *Atmospheric and Biological Environmental Monitoring*,
DOI 10.1007/978-1-4020-9674-7_21, © Springer Science+Business Media B.V. 2009

heavily polluted forest exhibited the higher cadmium tolerance for survival and the resistance to cadmium was heritable (Rozen 2006). Reinecke et al. (1999) demonstrated the development of the resistance to cadmium using pre-exposed or unexposed earthworms, suggesting that the pre-exposed worms survived at the concentration of cadmium which caused significant mortality on the unexposed worms. These are the obvious evidences that earthworms have an ability to resist toxicants and can develop the tolerance.

Most of these studies have used mature (adult) earthworms to evaluate toxicity or uptake kinetics. Unfortunately, little is known about the development of tolerance with earthworm maturity. Understanding the tolerance of earthworms with different maturity can improve the life cycle assessment of earthworms exposed to contaminated soils. The study emphasized the uptake of As and the adverse effects on growth of earthworms with different maturity exposed to OECD soils mixed with As-contaminated mine tailings. Initial body weight and the development of clittellum were measured to classify earthworm maturity. Body weight changes over time were measured to evaluate adverse effects of As from soils. Arsenic concentrations in earthworms were analyzed and uptake/eliminations were characterized by one compartment model. Earthworm body weight changes and As accumulations were compared with respect to the earthworm maturity as well as the proportion of mine tailings in OECD soils.

2 Material and Methods

2.1 Test Organisms and Classification of Three Growth States

Earthworms, *Eisenia fetida* (purchased from Carolina Biological Supply Co., USA), were exposed to test soils mixed with As-contaminated mine tailings. *E. fetida* has been widely used for toxicity tests and recommended as a representative terrestrial organism by OECD and International Standard Organization (OECD 1984, 2004; ISO 1993, 1998, 2003). Earthworms were raised for 6 months prior to experiment in the culture of peat and tree bark moisturized at 70% water holding capacity (WHC). The culture was

Table 1 Initial earthworm body weights of three growth states

Classification	Initial body weight (g)		
	M	F	IM
Max.	0.513	0.376	0.309
Min.	0.379	0.313	0.219
Average	0.432	0.347	0.272
(±SD)	(±0.037)	(±0.020)	(±0.027)
Number of worms	48	48	48

M = mature; F = formative; IM = immature; SD = standard deviation

maintained at 20°C, 80% ambient humidity with a 12 hr light/12 hr dark cycle. Earthworms were fed with arsenic free grain powder.

Earthworms were sorted and classified with maturity that was determined by body weight and the degree of clittellum development. Individual worm was weighed and divided into three growth states (M; mature, F; formative, IM; immature). M state worms had distinct developed clittellum. In F and IM state worms, however, clittellum was not present or, in part of F state worms, vestigial ones were found. The average initial worm weight for each state is presented in Table 1. Before exposure to test soils, earthworms were left on wet filter paper to void their gut for 48 hr. After depuration, body weight was re-measured to confirm the classification of growth states.

2.2 Test Soils and Experimental Setup

Mine tailings were collected at Myungbong, Korea using a hand auger at depth of 15 cm. Tailing sample was air-dried, disaggregated, and sieved to < 180 um. Previous study determined the concentrations of arsenic (As) and heavy metals (Cu, Pb, and Zn) and showed that As is highly elevated (3,021 mg kg^{-1}) in Myungbong tailings and is the major contaminant among other heavy metals (Lee et al., 2006).

Soil culture for control test was composed of 70% quartz sand, 20% kaolinite, and 10% sphagnum peat suggested by OECD as a standard soil for earthworm toxicity tests. For earthworm exposure to arsenic contaminated soils, Myungbong tailings were mixed at 10, 5, and 2% of control soils, therefore, three treatments and one control test soils were prepared. Soils were weighed to 30 g, mixed with distilled water to 70% WHC, and put into 50 ml plastic vial. Three replicates of each test (total 12 tests; 3 treatments and a control with 3 growth states) were prepared and left to stabilize for 4 days at 20°C, 80% ambient humidity prior to the addition of one earthworm to each vial. One earthworm was randomly selected from each growth state and introduced to each vials to satisfy the uniform distribution of initial body weight for all treatments. All vials were maintained as same conditions as earthworm cultures were kept before experiment. At times 5, 10, and 14-day exposure to test soils, body weight was measured. Three worms were collected from each test and allowed to depurate their gut contents for 48 hr prior to As analysis.

2.3 Earthworm Arsenic Concentration

After depuration, earthworms were killed by dipping in hot water (Piearce et al. 2002). Hot acid digestion method was used for earthworm samples (Davies et al. 2002). Earthworms were dried in the oven (50°C) to constant weight, and worms were weighed individually in 50 ml test tubes. Ten milliliters of nitric acid (>65%, Merck) were added to each test tubes and digested at 80 ± 10°C for 8 h. After cooling, solution was filtered with 0.45 um filter and diluted to 50 ml with distilled water. The solution was stored at 4°C before analysis. Arsenic concentration was analyzed by HG-AAS (5100, Perkin-Elmer) and duplicated matrix spiking samples were prepared at each batch of analysis for analytical quality control.

2.4 Data Analyses

The earthworm growth and arsenic accumulation were analyzed statistically using analysis of variance (ANOVA) among treatments and growth states. When significant differences were found, post hoc comparisons were applied to determine any differences in earthworm growth or arsenic accumulation. Tukeys' post hoc comparison was used to determine differences in arsenic accumulation and body weight changes with growth state for each treatment. Adverse effect of arsenic exposure on earthworm growth was determined by Dunnett t-tests with the control. Statistical analysis was performed by SPSS software.

To analyze the data from As accumulation in earthworms over time, one-compartment model was applied which is expressed by the formula

$$\frac{dC_{worm}}{dt} = k_u C_s - k_e C_{worm} \qquad (1)$$

where C_{worm} = As concentration in earthworms (mg kg^{-1}); t = time (days); C_s = the total As concentration in soils; k_u = uptake rate constant (kg$_{-soil}$ kg$^{-1}_{-worm}$ day^{-1}); k_e = elimination rate constant (day^{-1}). The curve fittings were performed using SigmaPlot 8.0. In the model, exposure route was not considered and uptake rate and elimination rate constant were estimated.

3 Results

3.1 Effects on Earthworm Growth

Figure 1 presents the body weight changes of earthworms for three growth states and four treatments including the control. Both weight gain and loss were observed after 14-day exposure. In the control, maximum weight gain ($112.2 \pm 2.3\%$) was observed for IM state earthworms while maximum weight loss was found to be $96.3 \pm 2.3\%$ in IM state earthworms of

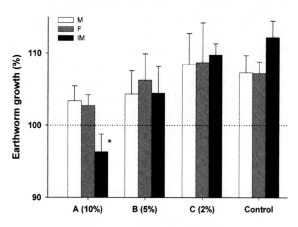

Fig. 1 Percentile body weight increase of earthworms exposed to A = 10% tailings; B = 5% tailings; C = 2% tailings; and control for 14 days. Results were averaged for each growth state (M = mature; F = formative; IM = immature) and error bar is standard deviation (*n*=3). Asterisk means the significant difference from the growth of control test (ANOVA, $P < 0.05$)

treatment A (10% tailings). Two-way ANOVA was performed for the weight changes by both treatments and growth states, which showed the significant differences by treatment ($P < 0.05$) but not by growth state ($P = 0.913$). Because interactions by both treatments and growth states had also significant effects on the growth ($P < 0.05$), the data was analyzed using one-way ANOVA for the weight changes by the combination of treatments with growth states. First, weight changes at each treatment were compared by three growth states. At the control, body weights increased over time for all growth states and significant difference was found by three growth states (ANOVA, $P < 0.05$). The weight increases of IM state worms were the highest among three growth states (Tukey, $P < 0.05$) and weight changes of M state worms were not different with those of F state worms (Tukey, $P = 0.952$). Secondly, weight changes of each growth state were compared by three treatments. No significant difference in body weight changes was found for all growth states at treatment B (5% tailings) and C (2% tailings). There were significantly different weight changes for three growth states at treatment A, the highest proportion (10%) of tailings (ANOVA, $P < 0.05$). Only IM state earthworms at treatment A brought decrease in body weight for 14-day exposure, which showed a significant difference for three growth states at treatment A (Tukey, $P < 0.05$).

Adverse effects of tailings on earthworm growth were observed through one-way ANOVA (comparison of body weight changes by three treatments) followed by Dunnett comparison with the control. Control test showed that body weights increased for 14 days for all growth states and IM state worms gained its weights more than other two state worms did (ANOVA, $P < 0.05$). Treatment A and B showed significant adverse effects on growth of IM state earthworms. The results showed that treatment B (5% of tailings) was the LOEC to IM state earthworms.

3.2 As Accumulation in Earthworms

At the start of earthworm exposure, arsenic was not detected in earthworms. No detectable level of As was found in the control during 14-day exposure. Arsenic concentrations in the earthworms increased with exposure time at all treatments. Figure 2 shows the increase

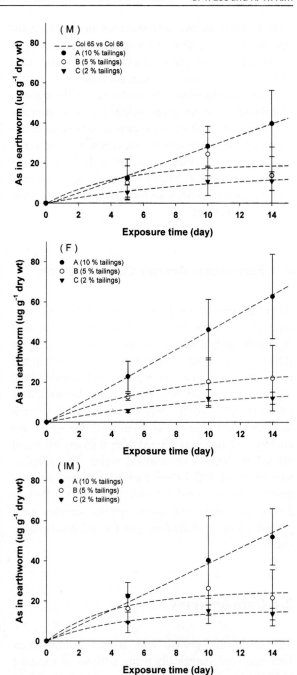

Fig. 2 Change over 14-day exposure in the mean As concentration (ug g^{-1}) $\pm$ SD in earthworms from A = 10% tailings; B = 5% tailings, C = 2% tailings. M = mature; F = formative; IM = immature. Error bar is standard deviation ($n=3$)

in As concentration of earthworms during 14 days for all treatments by three growth states. For all growth states, the highest increase in arsenic concentrations was found for treatment A. F state of treatment A

Table 2 Main characteristics of the As uptake in the 9 combinations of growth state with treatments

Growth-state	Treatment	As accumulation $C_{worm,t=14}$ (mg kg^{-1})	Modeled Uptake pattern	$k_u \cdot C_S$	k_e	R^2
M	A (10%)	39.78 ± 16.48	Linear [a]	2.816 ± 0.053	–	1.00
	B (5%)	13.87 ± 14.16	OCM	4.472 ± 4.857	0.233 ± 0.336	0.74
	C (2%)	11.04 ± 4.64	OCM [a]	1.539 ± 0.421	0.100 ± 0.059	0.98
F	A (10%)	62.76 ± 21.07	Linear [a]	4.534 ± 0.038	–	1.00
	B (5%)	22.02 ± 16.38	OCM [a]	3.597 ± 0.300	0.137 ± 0.020	1.00
	C (2%)	12.03 ± 3.06	OCM [a]	1.625 ± 0.490	0.093 ± 0.063	0.97
IM	A (10%)	51.79 ± 14.04	Linear [a]	3.871 ± 0.133	–	0.99
	B (5%)	21.58 ± 14.00	OCM [a]	6.121 ± 2.779	0.247 ± 0.145	0.95
	C (2%)	13.41 ± 2.88	OCM [a]	3.232 ± 1.046	0.212 ± 0.095	0.97
Average [b]				3.534 ± 1.124	0.170 ± 0.120	

a: $P<0.05$; b: Average value derived from the one-compartment model; OCM: One-compartment model

showed the highest As accumulation (62.76 mg kg^{-1}), and the lowest As concentration (11.04 mg kg^{-1}) was found at M state of treatment C (Table 2). The effects of treatment and growth state on As accumulation were determined by two-way ANOVA. Treatment, the proportion of tailings in OECD soils, had significant effects on As accumulation for 14 days (ANOVA, $P < 0.05$), but three growth states of earthworms did not affect the As concentrations. Any significant effect of interactions by both treatment and growth state was not found. At treatment A, arsenic concentrations linearly increased during 14 days for all growth states. In contrast to treatment A, steady-state As levels in earthworms were achieved for treatment B and C. One compartment model was applied to the data. Table 2 shows the characteristics of the models. For treatment A, elimination rate constant was assumed to be zero so that the linear regression was applied, which suggested that arsenic elimination rate was too low to decrease the As levels in earthworms. All models satisfied the significance ($P < 0.05$) except the treatment B with M state ($P = 0.142$). The model estimated average uptake rate and elimination rate constant, 3.534 kg$_{-soil}$ kg$_{-worm}^{-1}$ day^{-1} and 0.171 day^{-1}, respectively. The calculated elimination rate constant is similar to that of Peijnenburg et al. (1999a) even though earthworm species and the properties of soils were different. The distinct differences were found in the uptake rate ($k_u{}^*C_s$) among the 9 combinations of growth state with treatments. Uptake rate is dependent of organisms and arsenic bioavialability in soils which are the critical factor for uptake rate. Two possible reasons may explain the bias. One rea-

son is the different As bioavailability in soils. Even though the same mine tailings were mixed with OECD soils, the As bioavailability in soils to which earthworms were exposed may not be linearly proportioned to the amount of mine tailings. Another is the disturbance of the biological mechanisms that may regulate the accumulation in earthworms, caused by toxic effects of As. The different uptake patterns of treatment A as well as the high bias of the kinetic rate may result from the different As bioavailability. Unfortunately, these postulations have little scientific basis, therefore, these are remained for the future study.

4 Discussion

The study was focused on the changes of earthworm body weight and As concentration over 14-day exposure to OECD soils mixed with As-contaminated mine tailings. Earthworms were divided into three growth states (mature, formative, and immature) before exposure so as to characterize maturity effects on As accumulation and toxicity. Toxicokinetic model characterized the increases in As concentration of earthworms with earthworm growth states and the proportions of mine tailings in OECD soils.

Kim and Shin (2006) showed that 20% of Myungbong mine tailings in OECD soils was the LOEC on growth of clitellate *E. fetida* for 14-day exposure. In this study, three treatments had lower proportions of mine tailings than LOEC value and no adverse effects on earthworm growth would be expected. For M (mature) and F (formative) growth state earthworms, treatment A (10% tailings) did not cause any adverse effects on growth of mature or formative earthworms. Immature earthworms, however, were significantly affected by both 10% and 5% mine tailings in OECD soils that were below the LOEC values. Moreover, immature earthworms lost their weight on 10% mine tailings, which was the only decrease in this study. Weight loss of immature earthworms in treatment A contrasts with the result of the control at which immature earthworms showed the highest weight gain among three growth states (Tukey, $P < 0.05$). Contrast to the weight changes with growth state, no significant difference was found in As accumulation. The experimental setup in this study simplified the possible factors that may cause the toxic effect on growth and the major factor is As. Generally, it is accepted that the toxicity results from the toxicant accumulation (Weeks et al. 1996; Hankard et al. 2004; Ma 2005.). However, the study suggested that the toxicity on earthworm growth varied with the growth state of earthworms even in the similar As accumulation in body.

Earthworm has an ability to resist abiotic toxicants (Morgan and Morgan 1998; Langdon et al. 1999, 2001; Stürzenbaum et al. 2004; Vijver et al. 2006) and the As toxicity decreased by the methylation of inorganic arsenic or the bindings with methallothionein (Geiszinger et al. 1998; Lin et al. 1998; Langdon et al. 2002, 2003; Dombrowski et al. 2005; Spurgeon et al. 2005). The study implies that the toxicity of mine tailings on earthworm growth can be altered by the earthworm growth state and the young earthworms can be more easily affected even though the uptake and elimination of arsenic in earthworms is not dependent on the maturity. Toxic effects in immature earthworms with similar As accumulation to mature ones can be explained by the fact that young earthworms less develop its detoxifying mechanisms than adults do so as to be easily affected by As. Earthworm ecotoxicology is being spread from individual to population or generation levels (Fayolle et al. 1997; Edwards 2002; Spurgeon et al. 2005). Young earthworms early exposed to As are affected easily and the toxicity may be inherited to next generation by decreased growth or propagation power. Therefore, life cycle study for earthworm should consider the detoxifying ability with different growth state.

5 Conclusions

Earthworms, *E. fetida*, were exposed to OECD soils serially mixed with As contaminated mine tailings for 14 days. Prior to exposure, earthworms were divided into three growth states to observe the effects of maturity on growth and As accumulation. Immature earthworms showed the loss of weight, which was the significant difference in growth at the highest proportion of mine tailings (10%). As concentration increased with time at the highest proportion of mine tailings (10%) while in 5% and 2% of mine tailings, As levels reached the steady-state during 14-day exposure to be analyzed using one compartment model. Any difference was not found in As accumulation and uptake patterns with earthworm growth state. Eventually, toxicity was dependent on the initial growth state of earthworms and As levels in exposed soils. Young earthworms were affected by As on its growth even though there was no significant difference in As accumulations compared with mature worms. Considering the resistant mechanisms of earthworms to As, low levels of resistance in young earthworms brought toxic effects on growth.

Acknowledgments This work was supported by the Korea Science and Engineering Foundation (KOSEF) through the National Research Lab. Program funded by the Ministry of Science and Technology (No. M10300000298-06J0000-29810).

Abbrevations

OECD:	Organization for Economic Co-operation and Development
LC50:	Median Lethal Concentration
EC50:	Median Effect Concentration
LOEC:	Lowest Observed Effect Concentration
WHC:	Water Holding Capacity
HG-AAS:	Hydride Generation-Atomic Absorption Spectrometry
ANOVA:	Analysis of Variance

References

Cotter-Howells, J., Charnock, J.M., Winters, C., Kille, P., Fray, J.C., Morgan, A.J., 2005. Metal compartmentation and speciation in a soil sentinel: The earthworm, *Dendrodrilus rubidus*. Environ. Sci. Technol. 39, 7731–7740.

Davies, N.A., Hodson, M.E., Black, S., 2002. Changes in toxicity and bioavailability of lead in contaminated soils to the earthworm *Eisenia fetida* (Savigny 1826) after bone meal amendments to the soil. Environ. Toxicol. Chem. 21, 2685–2691.

Dombrowski, P.M., Long, W., Farley, K.J., Mahony, J.D., Capitani, J.F., Di Toro, D.M., 2005. Thermodynamic analysis of arsenic methylation. Environ. Sci. Technol. 39, 2169–2176.

Edwards, C.A. and Bohlen, P.J., 1996. Earthworm Ecology and Biology. Champman & Hall, London, UK.

Edwards, C.A., 2002. Assessing the effects of environmental pollutants on sol organisms, communities, processes and ecosystems. Eur. J. Soil. Biol. 38, 225–231.

Edwards, C.A., 2004. Earthworm Ecology. 2nd Ed., CRC Press, Boca Raton, USA.

Fayolle, L., Michaud, H., Cluzeau, D., Stawiecki, J., 1997. Influence of temperature and food source on the life cycle of the earthworm *Dendrobaena veneta* (oligochaeta). Soil Biol. Biochem. 29, 747–750.

Geiszinger, A. Goessler, W., Kuehnelt, D., Francesconi, K., Kosmus, W., 1998. Determination of arsenic compounds in earthworms. Environ. Sci. Technol. 32, 2238–2243.

Hankard, P.K., Svendsen, C., Wright, J., Wienberg, C., Fishwick, S.K., Spurgeon, D.J., Weeks, J.M., 2004. Biological assessment of contaminated land using earthworm biomarkers in support of chemical analysis. Sci. Tot. Environ. 330, 9–20.

ISO, 1993. Soil Quality – Effects of pollutants on earthworms (*Eisenia fetida*). Part 1: Determination of acute toxicity using artificial soil substrate. ISO 11268-1. Geneva, Switzerland.

ISO, 1998. Soil Quality – Effects of pollutants on earthworms (*Eisenia fetida*). Part 2: Determination of effect on reproduction. ISO 11268-2. Geneva, Switzerland.

ISO/CD, 2003. Soil Quality – Avoidance test for testing the quality of soils and the toxicity of chemicals – Test with earthworms (*Eisenia fetida*), ISO 17512. Geneva, Switzerland.

Kim, K.W. and Shin, K.H., 2006. Earthworm toxicity test for the assessment of arsenic and heavy metal-containing mine tailings, Korea. The 6th International symposium on advanced environmental monitoring, Heidelberg, Germany, (2006. 12. 07–09), p. II-A03.

Langdon, C.J., Meharg, A.A., Feldmann, J., Balgar, T., Charnock, J., Farquhar, M., Piearce, T.G. Semple, K.T., Cotter-Howells, J., 2002. Arsenic-speciation in arsenate-resistant and non-resistant populations of the earthworm, *Lumbricus rubellus*. J. Environ. Monit. 4, 24–29.

Langdon, C.J., Piearce, T.G., Black, S., Semple, K.T., 1999. Resistance to arsenic-toxicity in a population of the earthworm *Lumbricus rubellus*. Soil Biol. Biochem. 31, 1963–1967.

Langdon, C.J., Piearce, T.G., Meharg, A.A., Semple, K.T., 2001. Resistance to copper toxicity in populations of the earthworms *Lumbricus rubellus* and *Dendrodrilus rubidus* from contaminated mine wastes. Environ. Toxicol. Chem. 20, 2336–2341.

Langdon, C.J., Piearce, T.G., Meharg, A.A., Semple, K.T., 2003. Interactions between earthworms and arsenic in the soil environment: a review. Environ. Pollut. 124, 361–373.

Lee, S.W., Lee, B.T., Kim, J.Y., Kim, K.W., 2006. Human risk assessment for heavy metals and As contamination in the abandoned metal mine areas, Korea. Environ. Monit. Assess. 119, 233–244.

Lin, K.W., Behl,[1] S., Furst, A., Chien, P., Toia, R.F., 1998. Formation of dimethylarsinic acid from methylation of sodium arsenite in *Lumbricus terrestris*. Toxicol. Vitro. 12, 197–199.

Ma, W-C., 2005. Critical body residues (CBRs) for ecotoxicological soil quality assessment: Copper in earthworms. Soil Biol. Biochem. 37, 561–568.

Morgan, J.E., Morgan, A.J., 1998. The distribution and intracellular compartmentation of metals in the edogeic earthworm *Aporrectodea caliginosa* sampled from an unpolluted and a metal-contaminated site. Environ. Pollut. 99, 167–175.

Nahmani, J., Hodson, M.E., Black, S., 2007. A review of studies performed to assess metal uptake by earthworms. Environ. Pollut. 145, 402–424.

Neuhauser, E.F., Cukic, Z.V., Malecki, M.R., Loehr, R.C., Durkin, P.R., 1995. Bioconcentration and biokinetics of heavy metals in the earthworm. Environ. Pollut. 89, 293–301.

OECD, 1984. OECD guidelines for testing of chemicals: Earthworm acute toxicity test. OECD Guideline No. 207. Paris, France.

OECD, 2004. OECD guidelines for testing of chemicals: earthworm reproduction test. OECD Guideline No. 222. Paris, France.

Peijnenburg, W.J.G..M., Baerselman, R., de Groot, A.C., Jager, T., Posthuma, L., Van Veen, R.P.M., 1999a. Relating environmental availability to bioavailability: Soil-type-dependent metal accumulation in the oligochaete *Eisenia andrei*. Ecotox. Environ. Saf. 44, 294–310.

Peijnenburg, W.J.G.M., Posthuma, L., Zweers, P.G.P.C., Baerselman, R., de Groot, A.C., Van Veen, R.P.M., Jager, T., 1999b. Prediction of metal bioavailability in dutch field soils for the oligochaete *Enchytraeus crypticus*. Ecotox. Environ. Saf. 43, 170–186.

Piearce, T.G., Langdon, C.J., Meharg, A.A., Semple, K.T., 2002. Yellow earthworms: distinctive pigmentation associated with arsenic- and copper-tolerance in *Lumbricus rubellus*. Soil Biol. Biochem. 34, 1833–1838.

Reinecke, S.A., Prinsloo, M.W., Reinecke, A.J., 1999. Resistance of *Eisenia fetida* (Oligochaeta) to cadmium after long-term exposure. Ecotox. Environ. Saf. 42, 75–80.

Romijn, C.A.F.M., Luttik, R., Canton, J.H., 1994. Presentation of a general algorithm to include effect assessment on secondary poisoning in the derivation of environmental quality criteria: 2. Terrestrial food chains. Ecotox. Environ. Saf. 27, 107–127.

Rozen, A., 2006. Effect of cadmium of life-history parameters in *Dendrobaena octaedra* (Lumbricidae: Oligochaeta) populations originating from forests differently polluted with heavy metals. Soil Biol. Biochem. 38, 489–503.

Spurgeon, D.J., Hopkin, S.P., 1996. Risk assessment of the threat of secondary poisoning by metals to predators of earthworms in the vicinity of a primary smelting works. Sci. Tot. Environ. 187, 167–183.

Spurgeon, D.J., Hopkin, S.P., 1999. Comparsions of metal accumulation and excretion kinetics in earthworms (*Eisenia fetida*) exposed to contaminated field and laboratory soils. Appl. Soil Ecol. 11, 227–243.

Spurgeon, D.J., Hopkin, S.P., 2000. The development of genetically inherited resistance to zinc in laboratory-selected generations of the earthworm *Eisenia fetida*. Environ. Pollut. 109, 193–201.

Spurgeon, D.J., Svendsen, C., Lister, L.J., Hankard, P.K., Kille, P., 2005. Earthworm responses to Cd and Cu under fluctuating environmental conditions: A comparison with results from laboratory exposures. Environ. Pollut. 136, 443–452.

Stephenson, G.L., Wren, C.D., Middelraad, I.C.J., Warner, J.E., 1997. Exposure of the earthworm, *Lumbricus terrestris*, to diazinon, and the relative risk to passerine birds. Soil Biol. Biochem. 29, 717–720.

Stürzenbaum, S.R., Georgiev, O., Morgan, A.J., Kille, P., 2004. Cadmium detoxification in earthworms: From genes to cells. Environ. Sci. Technol. 38, 6283–6289.

Vijver, M.G., Van Gestel, C.A.M., Lanno, R.P., Van Straalen, N.M., Peijnenburg, W.J.G.M., 2004. Internal metal sequestration and its ecotoxicological relevance: A review. Environ. Sci. Technol. 38, 4705–4712.

Vijver, M.G., Van Gestel, C.A.M., Van Straalen, N.M., Lanno, R.P., Peijnenburg, W.J.G.M. 2006. Biological significance of metals partitioned to subcellular fractions within earthworms (*Aporrectodea caliginosa*), Environ. Toxicol. Chem. 25, 807–814.

Weeks, J.M., Svendsen, C., 1996. Neutral red retention by lysosomes from earthworm (*Lumbricus rubellus*) coelomocytes: A simple biomarker of exposure to soil copper. Environ. Toxicol. Chem. 15, 1801–1805.

Abbreviations

AATSR	Advance Along Track Scanning Radiometer		CDMA	Code Division Multiple Access
ACE	Aerosol Characterization Experiments		CHRIS	Compact High Resolution Imaging Spectrometer
ADP	Acoustic Doppler Profiler		CNES	le Centre National d'Etudes Spatiales
AERONET	Aerosol Robotic Network			
AGL	above ground level		CPR	Cloud Profiling Radar
AI	Aerosol Index		CrIS	Crosstrack Infrared Sounder
AMF	Air Mass Factor		CV	Cyclic Voltammetry
AMSL	above mean sea level		DDV	Dense Dark Vegetation
ANOVA	Analysis of Variance		DLR	German Aerospace Centre
AOD	Aerosol Optical Depth		DMEM	Dulbecco's Modified Essential Medium
AOE	average opacity error			
AOT	Aerosol Optical Thickness		DNA	Deoxiribo Nucleic Acid
APS	Aerosol Polarimetry Sensor		DO	Dissolved Oxygen
AQI	Air Quality Index		DOCS	Digital Opacity Compliance System
ARL	Air Resources Laboratory (a part of the USA National Oceanic and Atmospheric Administration – NOAA)			
			DOM	Digital Optical Method
			EarthCARE	Earth Clouds, Aerosols, and Radiation Explorer
ATBD	Algorithm Theoretical Basis Document		EC50	Median Effect Concentration
ATMS	Advanced Technology Microwave Sounder		ECM	ExtraCellular Matrix
			EF	enrichment factor
Au	Gold		EFDC	Environmental Fluid Dynamics Code
AVHRR	Advanced Very High Resolution Radiometer			
			ELISA	Enzyme Linked ImmunoSorbent Assay
BAER	Bremen Aerosol Retrieval			
BBR	Broadband Radiometer		ENVISAT	Environment Satellite
BOD	Biochemical Oxygen Demand		ENVISAT	Environment Satellite (http://envisat.esa.int)
BRF	Bidirectional Reflectance Factor			
CALIPSO	Cloud-Aerosol Lidar and Infrared Pathfinder Satellite Observations		EOS	Earth Observation System
			ERS	European Remote Sensing satellite
CCD	Charge Coupled Device		ERTS	Earth Resoureces Technology Satellite
CCD	Charge-Coupled Device			

ESA	European Space Agency (http://www.esa.it/export/esaCP/index.html)	LER	Lambert Equivalent Reflectivity
		LIDAR	Light Detection and Ranging
FET	Field Effect Transistor	LIDORT	Linearized Discrete Ordinate Radiative Transfer
FMF	Fine Mode Fraction	LITE	Lidar In-space Technology Experiment
FRESCO	Fast Retrieval Scheme for Clouds from the Oxygen A-band	LOD	limit of detection
FT	free troposphere	LOEC	Lowest Observed Effect Concentration
GACP	Global Aerosol Climatology Product	LUT	Look Up Table
		MBL	marine boundary layer
GLAS	Geoscience Laser Altimeter System	MERIS	Medium Resolution Imaging Spectrometer Instrument
GMES	Global Monitoring for Environment and Security	MISR	Multiangle Imaging SpectroRadiometer
GMS	Geostationary Meteorology Satellite	MODIS	Moderate Resolution Imaging Spectroradiometer
GOME	Global Ozone Monitoring Experiment	MSI	Multi-Spectral Imager
		N	number of samples
HG-AAS	Hydride Generation-Atomic Absorption Spectrometry	NASA	National Aeronautics and Space Administration
HYSPLIT (model)	hybrid single-particle lagrangian integrated trajectory (model)	NIR	Near InfraRed
IAEA	International Atomic Energy Agency	NIST	National Institute of Standards and Technology (a non-regulatory federal agency within the USA Department of Commerce)
IEPA	Illinois Environmental Protection Agency	NN	Neural Network
IGAC	International Global Atmospheric Chemistry Observation	NOAA	the National Ocean and Atmosphere Administration
ILAS	Improved Limb Atmospheric Spectrometer	NOAA	National Oceanic and Atmospheric Administration (a federal agency within the USA Department of Commerce)
IOE	individual opacity error		
ISCCP	International Satellite Cloud Climatology Project	NPP	National Polar-orbiting Operational Environmental Satellite System Preparatory Project
ITN	*Instituto Tecnológico e Nuclear* (official Portuguese designation of the Technological and Nuclear Institute)	NRL	Naval Research Laboratory
		OECD	Organization for Economic Co-operation and Development
JAXA	Japan Aerospace Exploration Agency	OMI	Ozone Monitoring Instrument
JPL	Jet Propulsion Laboratory	ORP	Oxygen Reduction Potential
k_0-INAA	k_0-standardised, instrumental neutron activation analysis	PATMOS	Pathfinder Atmosphere
		PCA	principal components analysis
KNMI	Royal Netherlands Meteorological Institute	PM	particulate matter
L1, L2	Level 1, Level 2	PM_{10}	airborne particulate matter with an equivalent, mass-median aerodynamic diameter of 10 μm or less
LC50	Median Lethal Concentration		
LEO	Low Earth Orbit		

PM$_{2.5}$	airborne particulate matter with an equivalent, mass-median aerodynamic diameter less than 2.5 μm	SPR	Surface Plasmon Resonance
		SRM®	standard reference material (from NIST)
PMF	positive matrix factorization	SSA	Single Scattering Albedo
PMT	Photo-Multiplier Tube	SSD	Space Shuttle Discovery
POAM	Polar Ozone and Aerosol Measurement	SST	Sea Surface Temperature
		SWIR	Short Wave Infra Red
POLDER	POLarization and Directionality of the Earth's Reflectances	SYNAER	Synergetic Aerosol Retrieval
		SYNTAM	Synergy a combination of MODIS/Terra and MODIS/Aqua
PROBA	Project for On-Board Autonomy	TARFOX	the Tropospheric Aerosol Radiation Forcing Experiment
READY	real-time environmental applications and display system (a web-based platform for accessing and displaying meteorological data, and running trajectory and dispersion models on the NOAA ARL web server)		
		TIR	Thermal InfraRed
		TMA	three-mirror anastigmat
		TMDL	Total Maximum Daily Loads
		TN	Total Nitrogen
		TOA	Top of Atmosphere
REF	reference element	TOC	Total Organic Carbon
RNA	Ribo Nucleic Acid	TOMS	Total Ozone Mapping Spectrometer
ROD	Rayleigh optical depth	TP	Total Phosphorus
RPI	*Reactor Português de Investigação* (official Portuguese designation of the Portuguese Research Reactor)	TRAQ	Tropospheric composition and Air Quality
		USA	United States of America
RSM	Reference Sector Method	USEPA	United States Environmental Protection Agency
SAM	Stratospheric Aerosol Measurement		
SC	Slant Column	USEPA	United States Environmental Protection Agency
SCIAMACHY	SCanning Imaging Absorption 5 spectroMeter for Atmospheric CHartographY	UV	Ultra Violet
		VC	Vertical Column
SCIAMACHY	Scanning Imaging Absorption Spectrometer for Atmsopheric Chartography	VeMAS	Vertically moving automatic water monitoring system
		VI	Vegetation Index
SD	standard deviation	VIIRS	Visual/Infrared Imager Radiometer Suite
SeaWiFS	Sea-viewing Wide Field-of-view Sensor		
		VIRS	Visualisation and analysis tool
SLSTR	Sea and Land Surface Temperature Radiometer	VNIR	Visible Near Infrared
		VOC	Volatile Organic Carbon
SNR	Signal to Noise Ratio	WASP	Water Quality Analysis and Simulation Program
SPOT	Satellite Pour l'Observation de la Terre		
		WHC	Water Holding Capacity

Index

A

Acid dissociation constant (pKa), 166, 176
Active DOAS tomography, 4
Active remote sensing, 17
Advance Along Track Scanning Radiometer
    (AATSR), 15, 16, 18, 23, 26, 31, 32, 303
Advanced Technology Microwave Sounder (ATMS),
    29, 32, 303
Advanced Very High Resolution Radiometer
    (AVHRR), 14, 15, 18, 19, 23, 26, 27, 28, 29,
    32, 303
Aerodynamic particle sizer, 86, 95
Aerosol Characterization Experiments (ACE), 13,
    32, 303
Aerosol climatology, 19, 27
Aerosol concentrations, 22, 88, 90, 93, 96, 97, 125,
    127, 137–154
Aerosol concentrator, 83, 84, 85, 86, 98–99
Aerosol index (AI), 14, 15, 19, 23, 26, 28, 29, 32, 303
Aerosol monitoring, 13–31
Aerosol optical depth (AOD), 14, 32, 303
Aerosol optical thickness (AOT), 14, 15, 18, 19, 20,
    21, 23, 26, 27, 28, 29, 30, 31, 32, 303
Aerosol Polarimetry Sensor (APS), 29, 32, 86, 87, 88,
    95, 97, 98, 99, 100, 102, 303
Aerosol Robotic Network (AERONET), 19, 20, 21,
    27, 32, 303
Aerosol sampling efficiency, 83–102
Airborne elements, 137–154
Airborne Imaging DOAS, 6, 8
Air mass factor (AMF), 65, 66, 68
Air-mass trajectories, 148
Air Quality Index (AQI), 32, 303
Algorithm Theoretical Basis Document (ATBD),
    32, 303
Arsenic uptake, 296, 300

Artemia salina, 244, 245
Aspiration efficiency, 84
Average opacity error (AOE), 46, 47, 48, 49, 303
Azores archipelago, 137

B

Bidirectional Reflectance Factor (BRF), 21, 32, 303
Bioaerosol tests, 97–98, 102
Bioassays, 207–208, 241, 242, 243, 246, 248,
    252, 254
Biobadge samplers, 98
Biomonitoring, 182, 183, 286
Bisphenol A (BPA), 162, 163, 164, 165, 166, 167,
    168, 169, 170, 171, 172, 173, 174, 175, 176, 229,
    230, 232, 233, 237, 275, 276, 279, 282, 283
Brachionus plicatilis, 244, 245
Bremen Aerosol Retrieval (BAER), 21, 26, 32, 303
Broadband Radiometer (BBR), 30, 32, 303

C

Cadmium, 208, 209, 210–216, 217, 253, 256, 296
Cell chip, 261, 265–270
CE-MS (Capillary electrophoresis-mass
    spectrometry), 208, 209–210, 214, 215, 216, 217
Chambers, 89–90, 106–118, 121, 122, 130, 131
Charge-Coupled Device (CCD), 9, 21, 40, 49, 64, 70,
    286, 303
Cloud-Aerosol Lidar and Infrared Pathfinder Satellite
    Observations (CALIPSO), 15, 16, 22, 23, 32, 303
Cloud Profiling Radar (CPR), 30, 32, 303
Cluster analysis, 146, 147, 152
Collection efficiency, 84, 85, 86, 88, 101, 102
Compact discs (CDs), 166, 176
Compact High Resolution Imaging Spectrometer
    (CHRIS), 15, 32, 303
Comprehensive impact value, 233, 235, 237

Concentration of adsorbents (M), 169, 176

Contrast model, 40, 41–42, 43–44, 45, 46, 47, 48, 49

Coulter multisizer, 86, 89, 94, 96, 97, 98, 101, 102

Crosstrack Infrared Sounder (CrIS), 29, 32, 303

Culturable fraction, 85

Cytarabine (CytR), 166, 167, 168, 169, 170, 171, 172, 173, 174, 175, 176

D

Daphnia pulex, 244, 245, 246

Dense Dark Vegetation (DDV), 21, 32, 303

Deoxyribonucleic acid (DNA), 166, 167, 176, 193, 194, 197, 199, 200, 201, 203, 207, 208–209, 210, 211, 212, 217, 219, 220, 221, 223, 226, 230, 231, 235, 238, 253, 254, 255, 257, 263, 264, 266, 270, 303

Detection, 14, 16, 17, 19, 23, 31, 40, 43, 51–62, 65, 83, 97, 100, 102, 119, 145, 146, 147, 162, 167, 173, 188, 216, 219, 221, 222, 224, 226, 235, 241, 242, 248, 252, 253, 254, 255, 256, 261, 263, 264, 265–270, 275–283

Dichlorodiphenyldichloroethylene (DDE), 162, 176

Dichlorodiphenyltrichloroethane (DDT), 162, 176

Differential absorption, 65, 69

Differential optical absorption spectroscopy (DOAS), 3

Digital Opacity Compliance System (DOSC), 40, 49, 303

Digital Optical Method (DOM), 40–44, 45–46, 47, 48, 49, 303

Digital versatile discs (DVDs), 166, 176

Dissolved organic carbon (DOC), 167, 168, 169, 170, 171, 173, 174, 175, 176

DNA microarray, 208–209, 210, 217, 219, 220, 221, 226, 230, 238

Dual-channel algorithm, 19

Dust emission factors, 74

Dust monitors (DM), 74, 75, 76, 80

E

EarthCARE (The Earth Clouds, Aerosols, and Radiation Explorer), 29–30, 32, 303

Earth Observation System (EOS), 29, 32, 303

Earth Resoureces Technology Satellite (ERTS), 32, 303

Earthworm, 295–300

Effective absorption crosssection, 66

Effluent organic matter (EfOM), 168, 170, 176

Eisenia fetida, see Earthworm

Electrochemical detection, 268–270, 275

Endocrine disrupting chemical, 162, 165, 167, 175, 176, 229, 275

Enrichment factors (EF), 143–145, 154, 303

Environmental protection agency (EPA), 176

EnviSAT (Environment Satellite), 15, 16, 22, 23, 31, 32, 64, 69, 70, 303

Equivalent background compound (EBC), 169, 176

17α-ethinylestradiol (EE2), 162–176, 230

EUPHORE chamber, 109, 111, 131, 132

European Remote Sensing satellite (ERS), 32, 303

European Space Agency (http://www.esa.it/export/esaCP/index.html) (ESA), 14, 16, 26, 29, 30, 31, 32, 203, 304

Exposure profiling, 74

F

Fine mode fraction (FMF), 20, 29, 32, 304

Fluorometer, 87, 88, 93–94, 96, 97, 101, 102

5-fluorouracil (5-Fu), 164, 166, 167, 168, 169, 170, 171, 172, 173, 174, 175, 176

Fractional aerosol coefficient, 125

Free troposphere (FT), 138, 139, 148, 154, 304

Freundlich coefficient (K), 169, 170, 176

Freundlich exponent (n^{-1}), 169, 170

FTIR, 31, 51, 52–55, 56, 57, 58, 61, 62, 109

Fugitive dust emissions, 73–80

G

Genomics, 207–217, 252, 256

Gentoxicity tests, 253–254

Geoscience Laser Altimeter System, (GLAS), 15, 16, 22, 32, 304

Geostationary Meteorology Satellite (GMS), 15, 18, 32, 304

Geostationary Satellite Algorithm, 18

German Aerospace Centre (DLR), 14, 32, 303

Global Aerosol Climatology Product, (GACP), 19, 32, 304

Global Monitoring for Environment and Security, (GMES), 30, 31, 32, 304

Growth inhibition tests, 252–253, 257

Growth state, 296, 297, 298, 299, 300

H

HepG2, 268, 269, 270

1,3,4,6,7,8-hexahydro-4,6,7,8-hexamethylcyclopenta-gamma-2-benzopyran (HHCB), 164, 176

High performance liquid chromatography (HPLC), 162, 167, 176

Horizontal satellite observation concept, 16

Hydrophilic interaction liquid chromatography (HILIC), 176

HYSPLIT model, 153, 148, 152

I

Imaging DOAS, 3, 6–10

Impaction, 85, 86

Impingers, 84, 86, 88

Improved Limb Atmospheric Spectrometer (ILAS), 15, 22–23, 32, 304

Individual opacity error (IOE), 46, 47, 49, 304

Indoor environmental camber, 109

Initial concentration of adsorbate (C_0), 169, 172, 176

Ink jet aerosol generator (IJAG), 91–92

International Global Atmospheric Chemistry (IGAC) Observation, 13, 32, 304

International Satellite Cloud Climatology Project (ISCCP), 19, 29, 32, 304

Intratracheal instillation, 220, 221, 222, 223

Isokinetic dust samplers, 73

J

Japan Aerospace Exploration Agency (JAXA), 14, 31, 32, 304

Jet Propulsion Laboratory (JPL), 32, 304

K

k_0-INAA, 139, 150, 152, 154, 304

L

Lambert Equivalent Reflectivity (LER), 19, 32, 304

le Centre National d'Etudes Spatiales (CNES), 14, 31, 32, 303

Level 1, Level 2 (L1, L2), 32, 304

Lidar In-space Technology Experiment (LITE), 15, 16, 22, 32, 304

Light Detection and Ranging (LIDAR), 3, 16, 23, 32, 40, 80, 304

Lignite coke dust (LCD), 166, 170, 171, 172, 173, 174, 175, 176

Limb sounding, 22–23

Linearized Discrete Ordinate Radiative Transfer (LIDORT), 66, 70, 304

Look Up Table (LUT), 18, 19, 23, 32, 304

Low Earth Orbit (LEO), 29, 32, 63, 304

Lung, 152, 219–226

M

Measured environmental concentration (MEC), 176

Medaka, 229–238, 285–293

Medium Resolution Imaging Spectrometer Instrument (MERIS), 15, 16, 18, 20–21, 26, 27, 31, 32, 304

Membrane damage, 194, 197, 199, 200, 201

Metabolic biomarker, 255–256, 257

Metabolomics, 207–217, 256

Methyl tertiary butyl ether (MTBE), 164, 176

Micellar electrokinetic chromatography, 275

Microchip, 275, 283

Millimolar (mM), 176

Mine tailings, 295–300

Mixing ratio, 6, 10, 66

Moderate Resolution Imaging Spectroradiometer (MODIS), 14, 15, 16, 18, 19, 20–21, 23, 26, 28, 31, 32, 33, 304, 305

Molecular weight cut-off (MWCO), 166, 167, 175, 176

Molecular weight (MW), 166, 167, 173, 175, 176, 208, 215, 266

Multiangle Imaging SpectroRadiometer (MISR), 16, 32, 304

Multi-angle-Multi channel Instruments, 21

Multidrug resistance, 254

Multi sensor satellites, 18, 23

Multi-Spectral Imager (MSI), 30–31, 32, 304

Municipal wastewater treatment plant (WWTP), 163, 165, 167, 168, 169, 170, 171, 172, 173, 174, 175, 176

N

Nano, 201–202, 219, 220, 221, 222, 262, 263

Nanofiltration, 161–176

Nanoparticles, 132, 193–203, 219, 220

Nanotoxicity, 194

National Aeronautics and Space Administration (NASA), 14, 22, 26, 29, 31, 32, 304

National Ocean and Atmosphere Administration (NOAA), 14, 15, 31, 32, 142, 152, 153, 154, 303, 304, 305

National Polar-orbiting Operational Environmental Satellite System Preparatory Project (NPP), 29, 32, 304

Naval Research Laboratory (NRL), 22, 32, 304

Near InfraRed (NIR), 20, 21, 22, 32, 304

Nebulizers, 86, 88, 90–91, 96, 97, 98, 100

Nervous system related gene, 233–235
Neural Network (NN) Approach, 23, 32, 265, 304
NO_2, 4, 6, 7, 9, 10, 63–70, 113, 114, 115, 116, 117,
   118, 119, 123, 128, 129, 134
n-octanol-water partition coefficient (K_{ow}), 166, 176

O

Ocean color algorithm, 20, 21
Open path laser transmissometer (OP-LT), 74, 80
Optical remote sensing (ORS), 73–80
Outdoor chamber, 107, 108, 109, 111, 128, 129–131
Oxidative toxicity, 197, 198, 199–201, 202, 203
Ozone, 14, 19, 22, 57, 63, 64, 69, 106, 108, 109, 111,
   118, 123, 125, 128, 129, 130, 131, 132, 133,
   134, 165
Ozone Monitoring Instrument (OMI), 15, 31, 64,
   70, 304

P

Particulate matter (PM), 39, 49, 73, 80, 115, 131, 147,
   154, 242, 304, 305
Passive infrared spectrometry, 51
Pathfinder Atmosphere (PATMOS), 18, 32, 304
Persistent organic pollutants (POPs), 162, 176
Pharmaceutically active compounds (PhACs),
   163, 176
Photochemical smog, 105, 131, 132
Pico mountain, 137–154
PICO-NARE observatory, 139, 150, 151
Pleiotropic drug resistance, 254, 255
PM_{10}, 73, 74, 75, 76, 78, 80, 139, 142, 143, 145, 150,
   154, 304
Polarization, 16, 17, 20, 67, 69
POLarization and Directionality of the Earth's
   Reflectances (POLDER), 15, 16, 20, 26, 31,
   32, 305
Polar Ozone and Aerosol Measurement (POAM), 15,
   22–23, 32, 305
Polychlorinated biphenyls (PCB), 162, 176
Polydisperse solid aerosol tests, 96–97
Positive matrix factorization (PMF), 147, 148, 153,
   154, 305
Powdered activated carbon, 161–176
Predicted environmental concentration (PEC),
   162, 176
Predicted non-effective environmental concentration
   (PNEC), 162, 176
Principal components analysis (PCA), 145, 147, 148,
   152, 153, 154, 304

Project for On-Board Autonomy (PROBA), 15,
   32, 305
Puffers, 86, 91, 96, 97

Q

Quantification, 40, 54–55, 74, 86, 87, 88, 167,
   246, 270

R

RAPID, 53, 55, 56, 58, 59, 60, 61
Rat, 219–226, 237, 243, 247
Rayleigh optical depth (ROD), 17, 32, 305
Reactive oxygen species (ROS), 196, 199–201, 253
Reference Sector Method (RSM), 66, 68, 69, 70, 305
Registration, Evaluation, Authorisation and
   Restriction of Chemical substances (REACH),
   251, 252
Remote sensing, 14, 16, 17–23, 31, 51–62, 63–70,
   73–80
Remote sources, 137–154
Resistance, 163, 201, 202, 211, 212, 254, 255, 264,
   277, 281, 282, 295, 296, 300
River pollution, 242
Royal Netherlands Meteorological Institute (KNMI),
   14, 31, 32, 304

S

Saccharomyces cerevisiae, 208, 213, 251–257
Sampling efficiency test, 85, 88, 89, 96, 98, 101, 102
Satellite, 3, 10, 13–31, 63, 64, 65, 66, 68, 69, 70
Satellite observation concept, 16
Satellite Pour l'Observation de la Terre (SPOT), 15,
   22, 33, 305
SCanning Imaging Absorption 5 spectroMeter for
   Atmospheric CHartographY (SCIAMACHY), 15,
   21, 22–23, 27, 32, 63–70, 305
Sea and Land Surface Temperature Radiometer
   (SLSTR), 31, 33, 305
Sea Surface Temperature, (SST), 14, 29, 33, 305
Sea-viewing Wide Field-of-view Sensor (SeaWiFS),
   14, 15, 18, 20, 21, 33, 305
Secondary organic aerosol, 108, 109, 119, 120,
   121, 129
Seven-wavelength aethalometer, 139
Short Wave Infra Red (SWIR), 31, 33, 305
Simulation of Atmospheric Photochemistry In a large
   Reaction Chamber (SPHIRE), 109
Single channel algorithm, 15, 18, 19
Single scattering albedo (SSA), 19, 21, 31, 33, 44, 305

Slant-column density, 9
Smog chamber, 105–134
Smog, *see* Smog chamber
SO_2, 4, 6, 7, 9, 57, 63–70, 111, 118, 130
Solar irradiance, 66, 69
Solar occulation satellite observation concept, 16
Solid phase extraction (SPE), 162, 167, 175, 176
Sonic nozzles, 86, 91, 96, 97, 98, 99
Source-attribution, 147
South Atlantic Anomaly (SAA), 69, 70
Space Shuttle Discovery (SSD), 15, 22, 33, 305
Spatially resolved DOAS measurements, 3–10
Spectral fitting, 64–65
SPR, 265–268, 270, 305
Stratospheric Aerosol Measurement (SAM), 14, 15, 32, 213, 305
Survival fraction, 85
Swimming behavior, 285–293
Synergetic Aerosol Retrieval (SYNAER), 23, 33, 305
Synergy a combination of MODIS/Terra and MODIS/Aqua (SYNTAM), 33, 305

T
The temperature-dependent absorption cross section, 65, 66
1-(5,6,7,8-tetrahydro-3,5,6,8-hexamethyl-2-naphthalenyl)ethanone (AHTN), 164, 176
Tetrahymena thermophila, 244, 245
Thermal InfraRed (TIR), 33, 305
Three-mirror anastigmat (TMA), 31, 33, 305
Tomographic techniques, 3, 10
TOMS algorithm, 19
Top of atmosphere (TOA), 16, 19, 23, 33, 305
Topographic Target Light Scattering, 4–6
Total organic carbon (TOC), 167, 176, 182, 183, 189, 305
Total Ozone Mapping Spectrometer (TOMS), 14, 15, 18, 19, 23, 26, 28, 31, 33, 63, 305
Toxic cloud imaging, 51, 52, 54, 55, 57, 58, 59, 60, 61
Toxicity bioassays, 241, 242, 243, 246, 248
Trace gas, 3–10, 63–70
Traditional impactors, 86, 87

Transmission efficiency, 84, 85
Transmission model, 40, 41, 42–43, 44, 45, 46, 47, 48, 49
Tributyltin (TBT), 162, 176
Tropospheric Aerosol Radiation Forcing Observation Experiment (TARFOX), 13, 33, 305
Tropospheric chemistry, 3, 63, 66, 70, 109
Tropospheric composition and Air Quality (TRAQ), 31, 33, 305

U
UCR-EPA chamber, 107, 108, 109, 115, 116, 117
Ultrafine nickel oxid, 220, 221
Ultrasonic atomizing nozzles, 93, 101
Ultra Violet (UV), 33, 165, 305

V
Vegetation Index (VI), 33, 305
VeMAS (Vertically moving automatic water monitoring system), 179–189, 305
Vertical satellite observation concept, 16
Vertical shape factor, 66
Vibrating orifice aerosol generator, 86, 87, 92–93
Vibrio fischeri, 197, 243, 244, 245, 246
Virtual impactors, 83, 85, 87, 99
Visibility, 13, 45, 73, 105, 106, 109, 132
Visible Near Infrared (VNIR), 33, 305
Visual/Infrared Imager Radiometer Suite (VIIRS), 14, 29, 33, 305
Visualisation and analysis tool (VIRS), 15, 33, 305
Visualization, 51–62, 232

W
Wall loss, 101, 115–117, 123, 124, 127, 134
Water-organic-carbon partition coefficient (K_{oc}), 166, 176
Water quality, 179–189, 242, 248, 285–293
Water quality management, 179–189
Water quality modeling, 181
Water quality monitoring, 179, 181, 182, 183, 185, 188

Y
Yeast indicator test, 254–255